D1088928

Nature's Oracle

For Dean Harvey Kahalas
with best wishes!

Chicago 2/6/2013

Ullica Segerstrale

Nature's Oracle

The life and work of W. D. HAMILTON

ULLICA SEGERSTRALE

OXFORD
UNIVERSITY PRESS

OXFORD
UNIVERSITY PRESS

Great Clarendon Street, Oxford, OX2 6DP,
United Kingdom

Oxford University Press is a department of the University of Oxford.
It furthers the University's objective of excellence in research, scholarship,
and education by publishing worldwide. Oxford is a registered trade mark of
Oxford University Press in the UK and in certain other countries

First Edition published in 2013
Impression: 1

British Library Cataloguing in Publication Data
Data available

Library of Congress Cataloging in Publication Data
Data available

ISBN 978–0–19–860727–4

Printed in Great Britain by
Clays Ltd, St Ives plc

PREFACE

Writing a biography is an interesting experience. The author has to be in two places at once, both "inside" and "outside" the mind of her subject. She needs to see things from the subject's perspective, feel his feelings and reconstruct his reasoning. At the same time she has to consider the views of others who intimately knew her subject—family members, colleagues, students—as well as people who took a completely external perspective. Processing all this she has to come to her own conclusions. So, too, have the readers of this book. The question to those close to my protagonist will be: does the person that emerges represent the Bill Hamilton they felt they knew? I hope so, though I also hope there will be at least some surprises. My basic aim with this book has been to keep Bill Hamilton alive for all who knew him and loved him, and to introduce him to those who didn't.

In this endeavor I have been assisted by a great number of people. They have provided materials, told me about Bill, explained some piece of science, read my manuscript, corrected errors, or just encouraged me to carry on with a difficult task. How, indeed, does one write an intellectual biography of somebody who constantly gets new ideas as he simultaneously works on half a dozen projects, interrupted by conference travel and by moving house between continents? I found that I was enormously helped by immersing myself not only in Bill's own writings and personal letters, but actually physically experiencing what it was like to live at Oaklea, his childhood home in Kent. For making this possible and for all our deep conversations I want to cordially thank Mary Bliss, Bill Hamilton's older sister. I also want to thank his younger sister Janet for her invaluable help with important information and documents, and his wife Christine and daughters Helen, Ruth, and Rowena for their hospitality and willingness to talk to me. Indeed, I believe that through living in the houses of the Hamilton siblings (including Robert and Margaret in New Zealand, where I visited close cousins and Ian Prior, an enormously helpful family friend), I familiarized myself with a certain "Hamiltonian" outlook on the world. I want to especially thank the New Zealand relatives of Bill Hamilton for their warm hospitality.

Among Bill Hamilton's many scientific colleagues, my thanks goes foremost to Alan Grafen for his wonderfully positive and helpful attitude. His most important contribution to the book was actually his very positive reaction to a first draft of my manuscript at a point when I badly needed validation. (Needless to say, he also addressed some scientific misunderstandings and typos as he read my manuscript—twice!) I also want to thank Richard Dawkins, Robert Trivers, Sarah Hrdy, Robert Axelrod, and Bernie Crespi for valuable comments.

But the manuscript was only the last step of a long process. My research for this book has involved a great number of scientists in various countries, colleagues and students, and even an old teacher of Bill's. Especially important for the book—and not only due to their valuable insights about Bill and his science—have been Bernie Crespi, who started me off with a list of possible people to contact; Warwick Kerr, who told me about Bill in Brazil; Steve Frank, who presented me with a bunch of important materials; Peter and Rosemary Grant, Bill's close neighbors and colleagues in Michigan; and Paul Schmid-Hempel, who kept me up to date with developments.

Chapter 3 in the book—the account of Bill Hamilton's school years at Tonbridge—could not have been written without Martin Jacoby's remarkable detective work which brought together Bill's fellow Tonbridgians' impressions of him as a schoolboy at the time; and chapter 4 owes much to the valuable information I got from RA Fisher's student, Anthony Edwards. In addition, I want to cordially thank Richard Alexander, Francisco Ayala, Patrick Bateson, Laura Beani, Walter Bodmer, Jacobus Boomsma, Dieter Ebert, Marcus Feldman, John Hajnal, Peter Henderson, David Hughes, Laurence Hurst, Pierre Jaisson, Tim Lenton, Stephen Levinson, James Lovelock, Marian Luke, Robert May, Nancy Moran, Pekka Pamilo, Naomi Pierce, Fred Ratnieks, Andrew Read, Mark Ridley, Carl Simon, Albert Somit, Steven Stearns, Staffan Ulfstrand, DS Wilson, EO Wilson, Michael Worobey, and Marlene Zuk for providing important insights. Sarah Bunney (Bill's copy editor) made many astute observations about Bill, and Jeremy John and Anne Summers graciously helped me with archival information at the British Library at an early stage of my work. There are many more who have discussed Bill with me or assisted me in some other way, too many to mention here. (A longer list of acknowledgements can be found at the end of the Preface.)

In conclusion, I would like to say how impressed I am with the clarity of mind, generosity of spirit, and sense of excitement that I have felt exists in the huge group of scientists who knew Bill Hamilton closely. The scientist in me is truly happy to have become acquainted with this group and experienced that special sense of camaraderie that seems to prevail among Bill Hamilton's students and colleagues at large.

Sadly, some persons whose input has been crucial for this book are no longer with us. I want first to mention Richard Southwood (Bill's sponsor at Imperial College and Oxford), Colin Hudson (Bill's close friend and correspondent from his Cambridge years), and Luisa Bozzi (Bill's companion of his last six years). My long and frank discussions with them about Bill were truly illuminating. I am also grateful for my conversations with John Maynard Smith, Rainer Rosengren, George Williams, and Ernst Mayr. Last but not least: Yura Ulehla, Bill's long-time friend and close correspondent ever since their time together at Silwood Park (Imperial College), generously provided me with a large number of letters, which have proved indispensable for this book.

I also wish to thank my first editor at Oxford University Press, Michael Rodgers, who outlined my writer's task with great inspiration. My second editor, Latha Menon, knows better than anyone the fundamental role she has played in making this book a reality.

I am gratefully acknowledging the support I have received for this project from the following sources: the John Simon Guggenheim Foundation, the American Philosophical Society, the Rockefeller Foundation, the Sloan Foundation, and the Finnish Society for Science and Letters. These grants enabled me among other things to travel and meet colleagues and relatives of Bill Hamilton and gather letters and photographs. And this brings me finally to the recognition of my permanent hero VJ Martin, for—incredibly—driving me all over England and Scotland, New Zealand top to bottom, both coasts of the United States, parts of Switzerland, France and Italy, and even a bit of Brazil during my research for this book.

Additional acknowledgements

I would like to additionally thank the following persons for their valuable help:

Rauno Alatalo, Erik Arnquist, Christopher Badcock, Anders Björklund, Sebastian Bonhoeffer, Jacobius Boomsma, Andrew Bourke, Arthur Burks,

Robert Boyd, Gary Breitenbach, Geoffrey Boys, Michael Cohen, Fred Cooke, Marian Dawkins, Francisco Dessi, Paul Ehrlich, Steven Emlen, Ville Fortelius, Wayne Getz, Richard Grossberg, Torsten Gustafsson, Clive Hamlin, Ilkka Hanski, David Harper, Paul Harvey, Claire Henderson, Edward Hooper, John Holland, Dan Hrdy, Hugh Ingram, Ian Jamieson, Jeff Joy, Jeyaraney Kathirithamby, Hanna Kokko, Curt Lively, Bobbi Low, Michael Lynch, Veronika Medana, Richard Michod, Darlyne Murawski, Randy Nesse, George Oster, Gwen Owen, Robert Page, David Penney, Ghillean Prance, Will Provine, Kern Reeve, Vernon Reynolds, Tom Seeley, Paul Sherman, Montgomery Slatkin, Robert Smith, Nils Stenseth, Lotta Sundström, David Sutton, Eörs Szathmary, Victoria Taylor, Michael Treisman, Michael Wade, Mary Jane West Eberhard, Gereon Wolters.

CONTENTS

INTRODUCTION

THE MAN WITH X-RAY EYES

Selfish genes, altruism, kin selection—these are household names by now. At the beginning of the third millennium it has become generally accepted that what matters in evolution is not the survival of the individual organism, but its genes. From a gene's eye point of view it may pay for an animal to behave altruistically—that is, show behaviour that benefits others—if, in this way, it can help others that are likely to carry copies of its genes; in other words, its relatives. Here, then, we have the solution to the long-standing problem of animal altruism. Why is it that a bird lets out a warning call and thereby makes itself more easily identifiable by the predator? Why does a bee sting invaders of the hive and by this action lose its own life? Darwin had wondered about such behaviours, but did not come up with the answer.

The man who worked out the solution in precise, mathematical form, was William Donald (Bill) Hamilton. His 1964 paper 'The Genetical Evolution of Social Behaviour', which he wrote as a graduate student, started a paradigm shift in science. And he went on to solve other important problems in biology and open up ever new research fields. Among these were the evolution of sex, host-parasite co-evolution, mate choice, cooperation between non-relatives, sex ratios, dispersal, senescence, the evolution of insect sociality, and conflicts within the genome. For his contributions he received a great number of international prizes and honours, including the most coveted of them all, the Crafoord prize (the would-be Nobel for biology) and the Kyoto prize. Hamilton has been called 'the primary theoretical innovator in modern Darwinian biology, responsible for the shape of the subject today', and 'Darwin's heir'. Many have observed his uncanny similarity to Darwin in general reasoning style.[1]

But this is all in hindsight. Bill Hamilton's recognition came late. His career represents a classic case of misunderstood genius, or what the sociologist

Bernard Barber would call 'resistance by scientists to scientific discovery'. Every step of the way was a struggle with editors and journal referees who could not follow his way of thinking. This lack of timely acknowledgement led to bouts of loneliness and self-doubt.[2]

It was not only the content of his science, but also the form that was often difficult. Hamilton's mathematical language remained obscure to many of his more traditionally trained naturalist teachers and colleagues. But he wanted to present his ideas in the most modern and rigorous biological language he knew: Neo-Darwinian population genetics (which expresses evolution as a change of gene frequencies in populations). Also, when writing, he was oriented strictly toward his scientific colleagues, not the general public. As a scientist he knew that when it came to novel ideas, the ones that that first needed convincing were his scientific peers. It was left to others—younger colleagues like Richard Dawkins and Matt Ridley—to give his ideas catchy names ('The Selfish Gene', 'The Red Queen') and present them lucidly to a broader audience.[3]

There was a film in the 1960s called 'The Man with X-ray Eyes'. The special gift of this man was to see things that were hidden from others. He could also see into the future. The man didn't necessarily like this special gift and actually it scared him a little to have it. Bill Hamilton is very similar to this man. He also saw things that were hidden from others, because he formulated questions of a type that people usually did not ask, questions that had evolutionary answers. For Bill, an explanation was not satisfactory if it lacked the evolutionary dimension.

Bill Hamilton had evolution on the brain. Actually, he described himself as an 'evolution sufferer'—he could not stop constantly thinking about the world around him, endlessly theorizing about it.[4]

When he was in pursuit of an idea he lived in a different universe—he was indeed the typical absent-minded professor. Visitors to his home in Wytham village near Oxford would see piles of papers everywhere, Bill scribbling away somewhere in the middle of it all. His daughters would see him sitting late into the night at his computer, trying out some new simulation—and wisely realize that this was definitely *not* the time to disturb Dad. In keeping with the absent-minded professor style, he also had a tendency to mumble when he talked. This often got worse when he lectured.

But anyone who thought that this was what Bill Hamilton was all about would have been badly mistaken! Bill was at the same time a brilliant naturalist.

His expeditions into the Wytham Woods were legendary: he could name every single animal that crawled up under the bark that he kicked off rotting tree trunks, every plant that his students saw on the way. There are reports of him spotting the smallest insect in, say, a hole in a work of art, or identifying a wasp from five metres away in order to help out a student, struggling to do this with a microscope.[5]

Bill's naturalist zest often took extreme forms—he was notorious, for instance, for sticking his hands into every hole in the ground, and he is said to have proudly reported that he had been stung by over 1,000 wasp species. He liked to sleep on the ground, lulling himself to sleep with evolutionary fantasies, and he didn't mind that his sleeping bag was slowly becoming saturated with water from the moist sphagnum moss (one of his favourite organisms).

Many entomological colleagues would claim Bill Hamilton as a wasp or insect man, but the truth is that he loved all of life. He felt a profound empathy with all living things—be they birds, insects, plants, or later in his career, even parasites and pests. He felt the bravery of the yellow ragwort flowering in the London asphalt. He laughed with trees. He thought of what strategy he would use if he was an Ebola virus. (He may, though, have had a particularly soft spot for insects.)

Most of all he loved the Amazon, the nature there and the native people. The Amazon is an inspiration for naturalists because of its abundance of species, its majestic proportions, and the sense it gives of nature being totally in control. No doubt Hamilton's affinity with the natural world helped him in his research. The other side of the coin was a corresponding awkwardness with the human world, which he repeatedly declared he did not understand, or even much care for.

Bill's nature observations, though, were never just that, they also represented data points for further reference. Bill's unusual mind was capable of an interesting sort of data analysis: it could simultaneously make comparisons on several variables across a number of different species in order to quickly assess whether a new theoretical conjecture of his was likely to hold water. Throughout his life he developed a number of creative strategies and tools for thinking, some of them baffling to even close colleagues.[6]

Hamilton was a '24-hour' scientist, always working on projects, always curious about new phenomena, always interested in unusual ideas. One of his most interesting features is his absolute disregard for implicit rules of the academe—for instance the apparent rule among some professors not to pay too much attention to mere students. He was totally idea-directed—unusually free of what in social psychology is called 'the fundamental attribution error', that is, our tendency to use various external indicators, such as style of dress, as guides when it comes to assessing others. He didn't care much about academic rank either. Bill was interested in what a person actually knew or thought, not who the person supposedly was. This also meant that he welcomed others' interesting-sounding ideas, a feature that made him remarkably 'un-resistant' to other scientists' scientific discoveries. Among his friends and close associates were some eccentric types, like George Price, the self-taught genius with whom Bill collaborated in the 1970s.[7]

Hamilton's vision of science was clearly more idealistic than that of many other academics. For him, science was the pursuit of truth, and truth should be told, no matter what. It was quite inconceivable for him, for example, that certain scientific facts ought not to be publicized for political reasons. His outspokenness often got him in trouble with the authorities: the scientific and medical establishments, and even the Church. And let it be told: Bill Hamilton was a rebel and enjoyed being one. He liked going against the grain and had a special affinity for underdogs and underdog theories. He felt the need to challenge the System—but was sometimes unprepared for his own emotional reaction when the System struck back.

How may we explain a person such as Bill Hamilton? Obviously there is an interaction between the idiosyncracies of a particular individual's personality and the environment in which he or she grew up, 'environment' here seen as a complex combination of family, school, physical environment, intellectual climate, and larger historical situation. For Bill the scientist, it is necessary first to trace the path to his chosen speciality as formed by his own preferences and choices and the stimuli and obstacles provided by family, friends, and teachers. Following his further trajectory, we need to add such factors as advisers, scientific journal editors and referees, colleagues, collaborators and competitors, and students. We will especially look at Bill Hamilton's bittersweet interaction with the scientific environment over half a century, and the various creative strategies he developed over time to make his mark.

***

People with unusual gifts may perceive the world differently and so make special contributions to knowledge. So it was with Bill Hamilton, the man with evolution on the brain. For Bill, though, the problem was not being able to turn his extraordinary power off! His mind kept taking in observations, analysing them, and developing theories. Bill never stopped exploring and theorizing. He kept on opening up new fields—it was as if his mental machete was made for cutting a path for others to follow.

One drawback of Hamilton's X-ray vision was that he got to see things he didn't like, developments that worried him. As a scientist, he felt it was his responsibility to alert mankind to the downside of certain popular beliefs. At the same time he was more than aware that people do not want to hear certain things, and that the messenger is often killed. Bill, to his credit, spoke up about dangers that he saw on the horizon, and asked for a general discussion about uncomfortable issues having to do with the future of humankind. He saw himself as Nature's messenger, or rather, as the privileged interpreter of Nature's mind. He did believe that we as humans could learn from studying Nature, although Nature was speaking to us in a language that needed deciphering:

> It does seem to me that amazing clues are made gifts to us in the book of nature around us. Clues to almost everything we can want to know seem to be there. Hence I believe that enough effort will give the answer to almost any question. Thanks to a path that seems to have been paved (for us in particular?) with fragments of a giant Rosetta Stone that when reconstructed may eventually unlock for us the text of our nature and the whole universe, there seems to be nothing that is at all interesting that we cannot hope to find out.[8]

Bill Hamilton was Nature's oracle. This was a privilege and a heavy responsibility.

This book is the story of the life and work of a scientific pioneer, an intellectual risk-taker and warrior for truth, who followed his scientific conviction wherever it took him, whatever the cost. It is also the story of a scientist who was acutely aware of the implications of his own theories, and who saw his own work as part of a larger struggle for Good over Evil. In fact, drawing out the implications of his own discoveries spurred him on to find ever new solutions to what he was truly interested in: proving that it was possible for humankind, using science and reason, to find a basis for human solidarity and chart a path to a future that might just avoid disaster.

I

Growing up at Oaklea

It is certainly not without consequence that Bill Hamilton grew up in Kent in the south east of England, on Badgers Mount, close to the North Downs. The landscape is undulating, with a particular type of flora and fauna stemming from its chalk foundation, the chalk lying bare in some places. A wonderful place to go butterfly collecting and a treasure trove of interesting fossils. A short distance from home Bill could stand high up on Well Hill and look around, orienting himself in relation to neighbouring towns, the North Downs, and the Thames Estuary. In fact, somewhere over those fields, not too far away, was Darwin's Down House.[1]

As a small child, Bill was deeply affected by the fact that the chalky ground around the house had once been a sea floor. So that was the reason for the funny fossils that he had been collecting in a nearby plowed field! His mother Bettina was to come up with even more wonderful revelations: her garden cabbages were actually cousins to turnips and wild cardamine (look at their similar cross-like flowers!) and the potato originating from far away in the Andes was a relative both to the tomato and the native bittersweet, a common hedge weed (look at their fruits!). Bill was electrified. He now realized that it was possible to compare plants and animals across time and space, that they were all united by a grand evolutionary plan. There was a system to all of life which also included the human species.[2]

Another of Bill's early childhood impressions was his strongly emotional response to colours, as he encountered them in beautiful flowers and the insect life buzzing around them. The feeling was so intense, he himself later recalled, that it was akin to love. He felt an overwhelming desire to be around all this beauty and to understand it. One of his favourite plants was the yellow ragwort, crawling with brightly patterned caterpillars. He soon started learning the Latin names of flowers and the insects that visited them, finding that he had a facility for this. Childhood friends remember that he was fascinated by orchids in particular.[3]

What to do with such a boy and his strong reaction to nature's beauty? His mother prepared a butterfly net and a killing bottle for her son and encouraged his naturalist interests. This meant that even as a small child, Bill found himself straying away from the house and garden and into the unknown territory beyond. (As he admitted later, going off on butterfly hunting expeditions was also a way to avoid household chores, such as feeding his mother's chickens).[4]

The Hamilton family lived in the Kent countryside, not too far from Sevenoaks and within walking distance from the train station at Knockholt. Later this general area would become part of the green belt of London. From Badgers Mount, Archibald (Archie) Hamilton, Bill Hamilton's father, a civil engineer and inventor, commuted to his London office, while his mother Bettina stayed at home, taking care of her growing number of children.[5]

Both of Bill Hamilton's parents were from New Zealand. They had met on a boat returning from England, Bettina from her training as a medical doctor and Archie for a visit after completing his latest project. They were married a few months later. Bettina, who had already bought a set of medical instruments, now decided to devote herself fully to her husband and her family instead, and so she did, raising six children, Bill being the second oldest after Mary. There were only 14 months between Bill and Mary, which made for a very close lifelong bond between the two siblings. They also developed a strong sense of being the firstborn, especially since they had often to assume responsibility for the group of younger siblings, Robert, Janet, and Margaret. Little Alex (or Leco) was the real baby, born more than a decade after Bill. (The story goes that for a birthday present Bettina asked Archie for one more baby.)[6]

Bill Hamilton was born in Cairo, Egypt, in 1936, where his father was stationed with the Corps of Royal Engineers before taking up a position in England. Archie Hamilton was famous for the road to Kurdistan whose building he had singlehandedly supervised, in charge of a great number of men. His published book *Road to Kurdistan* told the dramatic story of how he learnt the language of the Kurds and Arabs, overcame various adversities, and formed a strong friendship with the people. Throughout his life, Archie continued working on similar large projects, especially bridges, and travelled extensively.[7]

It could be said that travelling and writing about it was part of the Hamilton family tradition. Bill's grandfather, a lawyer, had made a grand tour of travel to Europe together with Archie's younger sister Rosemary, and also by himself to visit Herbert, Archie's older brother who was stationed with the military in Iraq. Reports of both these trips were published privately and distributed to family members in green covers, as was a history of the Hamilton family compiled by Archie's only male cousin, David Pullon. (Pullon, interestingly, lived on Hamilton road in Hamilton, New Zealand.) There was a strong sense of family and family tradition, which was also kept alive by Bill. He later used to tell his own daughters: 'Remember that you are a Hamilton!'[8]

Archie was a family man, keeping contact with his siblings in New Zealand, and taking particular responsibility for the four daughters of his sister Jane who had died quite young. Uncle Herbert, also a world traveller, but without the sunny disposition of his brother Archie, had all kinds of dramatic stories to tell, no doubt lapped up by the younger members of his audience. One of his claims to fame was to have spent some time in a Japanese prison during the Second World War. Herbert was married but had no children himself.[9]

New Zealand was very much a presence in the Hamilton house in Kent, through memories, lively correspondence, and visiting cousins and friends; so much so that two of Bill's siblings, Robert and Margaret, later returned to New Zealand to live, and their childhood neighbour friend Gwen emigrated there for a short time. I asked Gwen what it was that led her to make this decision and learnt that she had been inspired by Bettina's descriptions of the beauty of the landscape. Everything was so green, the sky was so blue. Bill's sister Margaret also clearly remembers how her mother told the children when they were playing a game of sending stick 'boats' down a river that it was only in New Zealand that one could find the right kind of weed for such boats. And Bill was actually travelling on a New Zealand passport until he was about 20 years old.[10]

The family house on Badgers Mount was called 'Oaklea' after Archie's own childhood home in Waimate on the east coast of the South Island. That house, which still stands, is rather modest, but what is extraordinary are the oaks around it and the open views. Archie was probably looking for something similar, and he did find it. Oaklea in Kent, a two-storey brick and mortar house of four small bedrooms, sits on top of a broad slope. When Bill was growing up, the view from the top of the garden was unobstructed, creating a great sense of space and freedom. From her kitchen sink (a good stainless steel one, which she had asked her husband for as a wedding present) his mother could supervise the children playing on the lawn and in the garden.[11]

With six children, there was an emphasis on frugality and good habits (bedtime was at 10 pm). Anything 'posh' was usually frowned on (among those things considered to be posh were, for example, tablecloths), making Bill later dismissive of such luxuries in everyday life. (His mother did use tablecloths for important occasions, though.) Bettina managed the house herself, with the help of a neighbour, the wonderful Mrs Jess, who became something of a second mother to the children, and also an invaluable companion to Bettina. (They did, however, continue to call themselves Mrs Hamilton and Mrs Jess.) Bettina never said so, but it seems likely that she missed New Zealand greatly. Perhaps that is why she tried to recreate a romantic image of her homeland in Kent. Oaklea was turned into something of a would-be pioneer vision of a New Zealand house; 'intentionally modest' is perhaps the expression that best captures the place. Bettina was shy when it came to people she didn't know. She did not become an active part of British society but rather kept herself at home, although she gladly received visitors and often helped out neighbours in medical matters.[12]

One of the most striking features about her, apart from her love for children and her great role as the family communicator (she kept in close contact by letter with kinfolk and friends everywhere and an open house for visitors), was her extreme modesty when it came to herself. Bettina was seemingly oblivious to fashion, typically wearing the same dress continuously. It was as if she wanted actively to indicate that these things were not important. (Bill picked up on some of this himself—people remember, for instance, his sometimes too-short trousers.) The family had the usual interaction with neighbours, school, friends, and business associates and led a more or less typical English life. Archie and Bettina often attended church on Sunday, sometimes taking the children with them. This was more a nod to convention and social expectations than a reflection of genuine religious feeling. Archie and Bettina considered knowledge of the Bible to be part of one's general education.[13]

In the Hamilton childhood home there was a great emphasis placed on creativity and self-sufficiency. Schoolmates and visiting friends remarked on the diverse activities of the children: 'Everybody always seemed to be doing something.' The children were encouraged to take up various hobbies, and to undertake various construction projects of their own invention.[14]

There is a famous New Zealand expression: 'Anything can be fixed with a number 8 wire.' This refers to a particular type of fencing wire, and is a metaphor for the general 'can do' spirit that is part of the New Zealand general culture. It is easy to imagine that in this sparsely populated land, people would often have to rely on their own ingenuity in their homes, which is why resourcefulness would have been encouraged. Many early settlers, too, were engineers and other professional men from Europe and for them this kind of attitude would have been natural or at least congenial. This kind of spirit is best picked up by visiting the country, which is something that I did in order to get a better understanding of Bill Hamilton's background. I believe the New Zealand connection explains at least part of his personality. This heritage may also have meant that the taken-for-granted background values that Archie and Bettina transmitted to their children may in some respects have been more Victorian than those of the English society in which they lived, as these values made their way from England to New Zealand and back.[15]

'Anything can be fixed' seems to have been the attitude, too, at Oaklea, both practically and metaphorically. What could be repaired was repaired and this applied to broken china, too. Archie used to repair china with glue that eventually turned brown, but his wife bravely kept using the repaired dishes over and over. This spirit was picked up by their offspring, too. A visiting friend reported that she once found a number of the Hamilton children huddled over the kitchen table, all holding together a broken drinking glass that they had just triumphantly glued back together.[16]

The Hamilton family was one where individuality was respected, but where family solidarity and cooperation were taken for granted. Visitors noted the family's unusually high level of education, with discussion of books, art, and ideas. It was a given that the children should become familiar with the best of world literature and art. Bettina used to order books that were delivered to the home, and these were then read and discussed. Music does not however seem to have played a major role, at least in the sense of family members practising music themselves. Archie was musical and once had a piano brought in, but it was eventually taken back. Bill was tone deaf and Mary did not show any special aptitude or interest. There were a few home concerts, though, with Archie playing flute and piano together with Marian, the daughter of his close New Zealand engineer friend. Marian was an only child, who often stayed with the Hamiltons and was treated as a daughter in the house.[17]

A factor that affected everyday life was the relative lack of indoor space. As more Hamilton children came into the world, the house was quickly filling up. Crowded out by younger siblings, Bill and Mary soon found themselves in new sleeping quarters in Archie's enormous garden shed. This shed was Archie's inventor's workshop, in which he used to experiment with various ideas in his spare time. (This shed had in fact been one of the features that originally attracted him to buying the place, the other being its five hectares of land.) 'Hardy' is probably the right word to describe life at Oaklea and the children who grew up there. In the winter little Bill and Mary would run barefoot in the snow to their sleeping compartments in the shed. Obviously, the shed was unheated.[18]

Another phrase that depicts Oaklea life might be 'no-nonsense'. As a medical doctor, Bettina believed in minimal intervention, except in obvious emergency situations. There is of course potentially a lot of wisdom in such an approach, but in Bettina's case, it was probably due to the early teachings of medicine. In any case, at Oaklea there was an emphasis on the 'natural' as better than the artificial or medically assisted. This attitude seems to have carried over to Bill, too, who avoided doctors, and, interestingly, to some extent at least, also to his two sisters Mary and Janet, who both became medical doctors themselves. Bill and Mary later developed a serious suspicion of modern medicine and its dependence on the pharmaceutical industry.[19]

Archie and Bettina encouraged their children to express themselves, to explore, and to build houses of their own in the garden. Archie taught the children to use his various tools and they were welcome to borrow them. Bill soon became quite proficient in using tools correctly. Mary, inspired by stories about brick-making out of clay and hay built her own 'mud house'—which is exactly what it turned into in the English climate. Her father and mother discreetly rebuilt it of better materials after it collapsed. There was, of course, also a tree house. A metal sheet-covered, round-roofed Nissen hut was used for storage of bigger items. Later there was Robert's shed. Add to this a rather deep mine shaft (Robert's exploratory creation with a little help from his friends and a tractor), a huge water tank (built by Archie to do experiments for the navy), and a workable tennis field (created by steamrolling), and one wonders how there could have been space for anything else. But there was more, because what dominated was Bettina's wonderful flower garden, and behind it the kitchen garden and orchard, sloping gently downhill, eventually meeting a farmer's field at the bottom. Bettina also kept chickens and indeed, Oaklea was to a large extent self-sufficient in regard to food.[20]

Bettina loved nature, and early on told her children that nature was there to be enjoyed. She took them with her on all kinds of excursions and picnics in the surrounding countryside. A favourite destination was the 'Shooting Hill', where Archie used to practise with the Home Guard. During these excursions Bettina used to arrange games and competitions for the children. One of her favourites was: 'Who can collect the best flower bouquet?' Off the children went and back they came, each one clenching their own unique bouquet. It was their mother's task to judge and declare the winner. Bill always tried his best—but was time and again exasperated when his flower collection was not declared the winner. What was wrong? He knew his was the best! But his mother kept thwarting him—and incidentally anyone else who badly wanted to win, because Bettina's principle was to find something valuable in each and every bouquet. The children's results were all wonderful in their own way and there was no clear winner.[21]

Bettina tried very hard not to show favouritism to any of her children. She tried to encourage each of them to develop their individual talents and to help them with their particular needs. There was, arguably, a sort of indirect favouritism, since Bill through his butterfly excursions was often able to avoid household chores. But when he was around he might be asked to do tasks such as washing the dinner dishes with Mary, a job which they sometimes turned into an opportunity for poetry recitation.[22]

Bettina taught the children the names of flowers and insects, as she took them all over the neighbouring countryside, 'combombulating' (her word) with anyone they met in order to detract attention from the fact that they were actually trespassing! (Much of the English countryside cannot be roamed freely; it consists of fenced-off areas separated by narrow roads. This is certainly true for the hills around Oaklea.) Well, if mother combombulated, so too would her son later on. Or he would just run. It is amusing to think that many of the butterfly catching expeditions in fact technically involved trespassing.[23]

Bill and his brothers and sisters were in principle allowed to go where they liked (the girls somewhat less than the boys), and encouraged to take risks. This is an interesting feature of the Hamilton children's upbringing, and one wonders about the parents' exact thinking behind it. It clearly worked—Bill and Robert certainly turned into systematic risk-takers for life. This started early: Bill nearly lost his life in an explosion when he was 12 years old (more about this later), and later Bill and Robert used to travel together, egging each other on, placing themselves in difficult situations and getting themselves out of them. A tragic outcome of this risk-taking behaviour was to be the death of Bill's youngest brother Alex (Leco) at age 18 in a mountain-climbing accident in Scotland. If we assume that this particular way of proving Hamiltonian family valor continued into the next generation, we may also put the head injury of Mary's younger son James as a child falling from a tree and the death of Bill's favourite nephew Richard, Janet's son, in a canoe accident, in the same category. (This is, in fact, what I have heard from some family members.)

But is this really the correct description of what was going on? Bill's parents did not seem to encourage reckless behaviour—in fact, they appeared to discourage it, or at least encourage the children to take responsibility for their own actions. One example is what happened to Bill when he was quite young. Bill had been happily climbing a tree in the garden, getting close to the top. But looking down he got scared and he couldn't see how he would get down. He shouted to his mother to come and help him. His mother heard him and came, but she did something unexpected. She just sat down quietly at the foot of the tree, knitting, waiting for Bill to come down. After some time (hours!) he overcame his fear and was able to bring himself down. And what about Bill's father? One good example of his approach is how when Bill was in Brazil and planning a major road trip across the country with an old car, Archie urged him to accept money from home in order to buy a safe, used car. My impression is one of tempered encouragement of initiative and risk. Perhaps what happened was that a more radical subculture developed among some of the Hamilton children themselves, quite outside parental control, one in which there may have been a different interpretation of 'family valor'. In any case, when it comes to Bill, there is no doubt that he often engaged in what seemed at least to others to be reckless behaviour.[24]

Bill and his sister were home schooled for a brief time after which they attended ordinary schools. Bettina would read aloud books from world literature suitable for their age, often adventure stories of various kinds. The children loved it. This is how they developed an early taste for Rudyard Kipling, Walter Scott, and Henry Walter Bates (the explorer of the Amazon). Their mother also read to them *Pilgrim's Progress,* the allegorical, rhymed 17th century adventure story, in which the main character, Christian, on his way from the City of Destruction to the Celestial City has to cope with various challenges. But when their father came home in the evening, the fun abruptly ended, since Bettina's rule was then to devote herself totally to her husband, spending time with him in front of an open fire in his study. Bill used to complain about this, saying that their father was mother's 'First Boy'.[25]

The impact of the Second World War was very much felt at Oaklea. When it began Bill was almost three years old and only four years old during the Battle of Britain. There is no doubt, though, that Bill was affected by what he saw and felt, and that this would stay with him for the rest of his life. Imagine having to take shelter each night inside a large wooden contraption in the living room (this was Archie's design for the children—the parents slept under the sturdy formal dining table), or, when things got rougher, in a dug out shelter in the garden. It was touch and go as to whether the family should be in the garden shelter or at home, this depended on the alarms. The Hamiltons were lucky. At the entrance of their garden shelter there is a bomb hole. That night they decided to stay in the living room.[26]

At night one could watch the planes. Bill remembered seeing at least one plane shot down. Also, near Oaklea was one of the many balloon stations established to guide aircrafts during the war. Archie was very much involved, he was in charge of training the Home Guard and making explosives. At some point even his much beloved large garden shed was taken into use—for a fund-raising event involving a whist tournament. Even card games could serve the war effort.[27]

The irony was that Archie and Bettina had in fact planned to go back to New Zealand with the whole family to be safe from the war, but it was already too late and no more ships were leaving. Mary and Bill were for a short time evacuated to wealthy relatives in Edinburgh during the war. Bettina, who took them there by train read aloud to them from PG Wodehouse during the journey. Mary remembers how they laughed and laughed. Wodehouse was later to become one of Bill's favourite authors, and indeed his own more essay-style writing occasionally has Wodehouse-like aspects. (One of Wodehouse's trademarks is to leave his chief character slowly sinking in the mud while the author indulges in lengthy philosophical musings.) The war also brought with it food shortages and rationing. Here again the solidarity of the larger family showed itself: one of Bill's New Zealand cousins described how they used to pack steel canisters with preserved eggs and other food wares to send to Oaklea.[28]

But the Second World War and its aftermath affected Bill Hamilton on a much more general intellectual and emotional level. It made him an avid individualist, distrustful of any 'system'. It was also responsible for his self-proclaimed later 'allergy' to anything that could be even faintly associated with totalitarian-sounding ideas. (This for instance later affected his attitude to group selection.)

For the Hamilton children, Oaklea was the centre of their world. For Bill especially, Oaklea and his early experiences seem to have left an indelible stamp on his mind. I believe that this was his true home for life. Also, it seems to me that the values and attitudes that he absorbed in this house in his childhood were those that he continued to stick with throughout his life, unless he had serious reasons to rethink them, which he usually didn't, being absorbed by his research. Oaklea and the life there can partly account for his general modesty, his strong sense of kin solidarity, informal style, creative and exploratory spirit, self-reliance and endurance, and appreciation of art and literature. Incidentally, it was not only Bill who was so influenced by Oaklea. Janet's house in the Ascot area seems largely modelled on Oaklea, situated in a similar setting and with the same geology map on the kitchen wall, and Robert's house in New Zealand also had a number of sheds in the garden reminiscent of his Oaklea childhood. Mary, arguably, beat them all as she was later able to move back to actually live in their childhood home. Oaklea does have a particular atmosphere that impresses upon you, even as a visitor. For the Hamilton children it was an environment that was at the same time safe and yet full of adventure, and it helped to mould Bill and his siblings as they grew up.

2

Finding Life's Pattern

At home at Oaklea one of Bill's favourite occupations was constructing outdoor miniature landscapes and imagining himself inside them as a miniature person. The landscapes were quite complicated, and not only because of their natural features—there were bridges connecting abysses, and cranes and cable cars transporting stick figures around. The scenario was dynamic. The stick figures took part in fantastic adventures—climbing mountains or braving bamboo forests—and it was Bill and his playmate Gwen who invented the stories.[1]

From an early age, Bill was regularly using his father's tools for his creations. He was also absorbing basic principles of engineering. He understood how an engineering model was constructed and what it took for a bridge to stand firm. Bill was in fact getting seriously involved in engineering practice, including the criteria for design, so much so that his own miniature constructions and their viability started to preoccupy his mind. He learnt to think and feel—and worry—like an engineer. Archie quietly encouraged Bill's mathematical and puzzle-solving interests, hoping that his eldest son would follow in his footsteps.[2]

But was engineering the right profession, Bill wondered? No, it would not leave enough time to do what he really wished, which was to explore nature and think about evolution. Better to have a less demanding profession, a practical one, such as a carpenter (Bill had become rather good at carpentry, using his skills mostly to build cages for butterfly breeding) or a school teacher. Engineering was just too absorbing. He knew that from his own experience:

> I knew well how even at night a design would never leave me until it was completed—and then immediately, if I was in the profession, there would be another... If an engineer is to be creative he must as far as possible concentrate an inward vision on what he means to make and must anticipate in his mind all the difficulties that his design is likely to encounter. In short, he must test his construction mentally so far as possible before he even starts to put it together.[3]

Fair enough, but was it only such rational considerations that led Bill away from his father's career? As he admitted himself, other issues may have been involved: 'Perhaps my father, being so good at it was also a factor, and perhaps there was a bit of Freudian perversity and competitiveness towards him combined'. Probably so.[4]

Meanwhile he didn't know that the profession of 'biologist' even existed, a job where he could expend his energies exactly on what he wished to do. As it later happened, it was not Bill but his sister Mary who would follow directly in her father's footsteps. She invented a hospital bed that could be used to avoid bed sores. It was an ingenious construction which carefully turned the patient so that different parts of the body came under pressure. The bed, however, was unexpectedly expensive to produce and unfortunately did not find widespread use.[5]

Likewise, his brother Robert, who later trained as a mining engineer, would turn out to be something of a general engineering wizard, solving the most unexpected problems in ingenious (although typically not elegant) ways. Bill considered Robert's engineering to be slapdash, and he often criticized him for it. Usually, though, his solutions worked.[6]

But did Bill really escape from engineering? We noted his complaining that all those construction models were taking up too much of his thoughts. They had to be worked out very carefully and they kept you awake at night. But was this not exactly the kind of thing that would indeed preoccupy Bill later on? It seems to me that Bill ended up doing the same type of calculations with his abstract models that engineers did with their physical ones. Bill Hamilton continued living in a world of engineering models—with cranes, cable cars, and all.

As a scientist, Hamilton typically made his models as visual as possible. Sometimes he was able to understand potential mechanisms even better if he could imagine a physical model or build one out of tangible materials and play with it. In his calculations he plugged in various values and saw where these took his curves. He made reasonable assumptions about the various factors affecting the system, aiming for a solution that was approximately right. Later, computer simulation was his way to extend his experimentation. Bill knew exactly what he was doing—just like engineering models, his evolutionary models had to be constructed so that they would hold up in the real world. Possible errors had to be found early on and corrected. The idea was to come as close to Nature's true intent as possible. After all, his mother had once said that Nature was a precision engineer.[7]

Later during Bill's career those among his colleagues who had a similar engineering training background—such as George Price, John Maynard Smith—would perfectly understand what Bill Hamilton was trying to do, and would themselves try to modify or extend his ideas. But other colleagues, those trained in a purer sort of mathematical modelling rather than engineering calculations, did not relate at all well to Hamilton's approximations. It was to take a long time for this group to be convinced.

Bill had been thinking about his future career since he was a young boy. Deciding not to become an engineer was certainly an important step, but how did he find his true calling? Geology was an obvious profession, ever since Bill realized that one could actually predict the flora and fauna of a landscape from its underlying geology. The kitchen at Oaklea sported a large geology map of England just above the dining table, and this was regularly consulted by the family. After all, as a civil engineer, Archie Hamilton needed to know the nature of the underlying ground material in order to factor it in to his overall calculations. Starting with Kent's familiar chalk landscape, Bill learnt to 'read' different landscapes. And Bill probably better than anyone else understood the true meaning of Darwin's beautiful little piece about the 'tangled bank' at the end of *The Origin of Species*. He had been walking past the chalky banks that Darwin wrote about on a daily basis.

Bill's early reflections on a career in geology were soon superseded by his next 'knight in armour' phase. Where did Bill get such an idea? A guess is that it came from the books he himself was reading, or that his mother read to him and Mary, books about knights undertaking dangerous missions and heroic deeds. A third idea, one of becoming an astronaut, derived from his early reading about the stars and skies and his own study of them from the garden at night with his father's little telescope. Bill read voraciously about all kinds of science. Astronomy probably resonated particularly well with his developing romantic imagination and story telling skills. Bill was a great story teller. His younger sister Margaret remembers how once when babysitting her, he convinced her that it was indeed possible to go to Mars—through a hole in the wall! Bill had pointed to a small hole that actually existed and told her that he would be able to go to Mars through it. Here we had an imaginative astronaut with rare persuasive powers.[8]

What Bill Hamilton didn't realize as he was going through these various scenarios as a child was that he was already preparing himself for his future profession. Bill had started as a general butterfly collector, catching what he could, but soon he became more serious. The turning point was reading the book *Butterflies* by EB Ford, which his parents gave him as a birthday present in 1946 when he turned nine. He had coveted the book in the shop for a long time and was overjoyed finally to possess it. Now he learnt not only how to properly collect and classify butterflies and moths, but more than that, the book opened up for him a whole new way of thinking. In *Butterflies* Ford also provided a popular introduction to Neo-Darwinism (which uses population genetics to express evolution as a change in gene frequencies in a population). Bill was later to understand the mathematics better, but even at this point, he realized that although both collected butterflies, there was a clear difference between amateur naturalists and professional biologists. He wanted to belong to the latter.[9]

And so he ran around in the countryside, chasing these beautiful creatures, the one more splendid than the other, always elusive, a joy, and a challenge for the young collector. But Bill didn't always have to go far to find his prey. He cleverly found that the prey would even come to him at home. Next to a terrace at Oaklea was a huge 'butterfly tree', whose flowers were popular with a number of colourful species. Bill had only to stand there and be ready with his net. (On the terrace he sometimes surprised his little sister Margaret who had hidden away to read a book. The point at Oaklea was to be invisible, because someone, like big sister Mary, might just put you to work). At night, again, windows could be opened to lure in various species of moths.[10]

How and when did Darwin enter the picture? The year that Bill was nine Bettina took him and the other children with her on the five-mile walk to Darwin's Down House. It was quite a distance for the younger ones and surely says something about the sturdiness of the Hamilton children. One might assume that this early visit to the famous man's house would have given Bill the idea that one could have a whole career as a naturalist. Not so. On the contrary, seeing Darwin's house and study, and especially his garden—tended to by a gardener—Hamilton says he got the vivid impression that naturalists had to be gentlemen and follow a certain genteel lifestyle. And this he was not interested in, especially since he gathered that one would have to be independently wealthy to afford this kind of life. Bill later commented that this trip did not serve to bring him closer to Darwin, rather the opposite. It was only a few years later, as he actually read the *Origin of Species* and became familiar with Darwin's reasoning and observations that he overcame his original reaction. (At least in part, this may have been an early example of rebellious reaction to established authority, many more were to follow.)[11]

***

But Bill did not spend all his time learning how to become a serious collector. The garden at Oaklea was lovely and this was the obvious place for the neighbours' children to come and play. Bill enjoyed participating in all kinds of usual childhood games and in making up little theatre plays and acting them out. The children used to have sleepovers in one of the small garden sheds, telling fantastic stories to each other in the dark. Oaklea was clearly a childhood paradise. Paradoxically, this very feature of the place may have had a drawback. Bill's sister Mary later reflected on the fact that the Hamilton children were always on their own home ground and together with each other. This may have left them socially underexposed. According to her, it would have been good for them to have had to negotiate early more challenging social situations away from home. Bill at least, although self-confident as a youngster (in pictures, he always looks straight into the camera), later on had problems with polite social small talk. How does one talk to a girl, for example? A New Zealand girl visiting Oaklea remembered Bill, after a very long silence, suddenly opening a conversation by showing her a beetle. (Later Bill was to profusely thank his Czech colleague and room mate Yura Ulehla for helping him improve his social skills.)

Bill was shown early on a model for how to combine work with interesting travel. Sometimes the whole family went camping, travelling in the family car with a trailer behind, in which sat Bill and Mary. There is a picture of these two at about 10 and 11 years old. The aim of these family excursions was typically to inspect some of Archie's bridges and usually also some relative was visited. (The Hamilton children picked up on this camping style of life and would later transmit the tradition to their own families.)[12]

But one aspect of family life cast a shadow over the otherwise generally happy life of the Hamiltons and made a deep impression on the young Bill. This had to do with the core of the family's income: the patent that Archibald Hamilton had obtained for the so-called Callender-Hamilton bridge. (Callender was the firm which had employed him.) This bridge is composed of portable parts, which makes it easy to transport and assemble even in terrain without proper roads. It largely looks like a blown up version of a Lego-type toy bridge or a railway bridge. Archie held the patent for this type of bridge, and it was a popular one in many parts of the world, especially less industrialized countries. It was also popular with the army.[13]

But soon a dark cloud appeared on the horizon. An army engineer—General Bailey, who had earlier been involved with the army's choice of the Callender-Hamilton bridge—unexpectedly came up with his own patent for an alternative bridge. And Bailey had obvious insider status. The Bailey bridge consisted of bigger ready-made sections. Because more parts were ready-made, the bridge construction was presumably faster, but on the other hand, one would assume that the size and weight of the pieces made for poorer portability. It was a bridge with different practical requirements and applications (eg, it required better roads); the choice between the two may have depended on what was wanted in a particular terrain. In other words, the two bridges could have co-existed, and they could each have had their own patents.[14]

Archie, however, viewed Bailey's bridge not as an alternative but as a direct appropriation of his original idea, without due recognition. 'Recognition' in this case does not constitute the reward for a discovery, as in science, but means getting concrete royalties. If Bailey had presented his design as an improvement on Archibald Hamilton's design, Archie would automatically have been eligible for a percentage of the royalties. But Bailey had acted as if the whole bridge idea was his from the beginning. That was very upsetting to Archie. He decided to take a strong course of action: to challenge Bailey's patent claim. Archie won, and was awarded a substantial economic settlement. But this whole event deeply affected the entire Hamilton family.[15]

The patent challenge soon became fixed in the Hamilton family history as a story about an evil competitor and his dirty schemes, and was packed away in this form. Although a reasonable man, for the rest of his life Archie was never willing to admit that the Bailey bridge could have had any possible advantage over his own. He could not find strong enough words to denounce General Bailey and his bridge. So, even though Archie did successfully fight the patent and win, the Bailey bridge incident remained a traumatic memory for the Hamilton family. It became a dark model of unfair professional behaviour, of a colleague stealing an idea from another, and trying to present it as his own. Archie could perhaps have reasoned that through his patent challenge he had in fact demonstrated that engineering was a proud and fair profession with a clear system for litigation. The rules had worked, and he had won. But what true inventor would be expected to show such calm? Archie was emotionally involved with his bridge. This was his design, his life. This was the bridge that had been transported in bits and pieces to remote areas by mules in Kurdistan, the only type of construction that could have made that road building possible. Bailey's bridge somehow diminished his own pioneering achievement. In the end, for Archibald Hamilton, it was a matter of professional recognition.[16]

***

Patterns, patterns everywhere. This was what little Bill Hamilton saw all around him. Even better, he could generate patterns himself, beautiful colours spreading out from oil drops from the household can, if he could only surreptitiously lay his hand on it. And what were those intricate braided patterns that were formed in the silt by the rainwater? How did those come about? But the patterns that really captivated Hamilton were the ones on butterfly wings. How splendid they were, and how intricate, sometimes patterns within patterns, seemingly repeating themselves. Hamilton did not of course yet know about fractals, but when he later learnt about Benoit Mandelbrot he felt as if he was coming home.[17]

Butterfly collection turned out to be an ideal occupation for this growing boy, combining as it did two sides of Hamilton: his endless curiosity and his quest for beauty. In fact, in awe of all the wonderful colour patterns on butterfly wings, Bill's first impulse was to try to somehow emulate them, or better, outdo Nature with the help of science, by using even more vibrant pigments, produced by himself. He learnt about the pigments of butterfly wings from his great new information source, EB Ford: melanin, carotenoids, and flavones. By combining such knowledge with knowledge of genetics and the production and dispersion of these pigments, he hoped to 'synthesize new wonderful wings for butterflies, new designs Rhopalocera had never seen'. This was also why chemistry was yet another of Bill's early career considerations.[18]

The butterfly wing was the foundation on which young Bill thought he would build a general theory of pattern:

> I was deciding to work upwards, as I saw it, out of my primary inspirations, like the butterflies were flying, toward a theory of pure pattern. My aim would be to

> understand patterns generally, but I would start with the patterns laid out on my butterflies' wings...[19]

But what about the butterfly's flight—was not there, too, a pattern to be found? It seemed that the butterflies sometimes were just too clever to be caught, just as you reached them they changed their expected trajectory. What was going on? Could their flight be predicted at all? Bill felt a strong desire to form a theory of the pattern of the butterfly's avoidance style. He had studied this in great detail already:

> How easy...I found the speckled wood (*Pararge aegeria*) to catch; how different and more difficult the clouded yellow (*Colias*), which I used to chase in the fields of red clover or lucerne. Especially it was difficult when a missed first swipe had 'turned on' your prey...It seemed as if literally some 'worse-than-random' generator of dodges had switched on the working of the butterfly's wings—a pre-programmed switchable option in how it was built to fly.[20]

Though he did not know it, this was Bill's first encounter with game theory in action. Later, game theory would become one of the corner stones of sociobiology.

In the meantime, there were patterns everywhere—patterns of relationships, patterns of behaviour, all of these beckoning to be sorted out and explained. And these were all part of Nature's big plan, an unfolding pattern that it was Bill Hamilton's personal challenge to unravel. There were cues all around him, if he could only succeed in deciphering them—every plant, every animal was a potential puzzle for a curious naturalist, and each was a piece of a bigger pattern. 'All science is finding patterns', he was later to conclude. For Hamilton, science was a total engagement, and patterns were at its core.[21]

Patterns also lay at the intersection of two sides of Bill Hamilton's personality, the emotional and intellectual, the artistic and the scientific. Patterns could be enjoyed for their beauty *and* they could be studied. Pattern recognition was to become a central part of his life. We later find Bill on board airplanes enjoying and analysing the landscape below him (think, for instance, of the American Southwest), or during a dinner party reacting to the unusual patterning on a polished wooden table: 'Aha! A tree parasite!'[22]

He was fascinated by maps, especially old maps, which he used deliberately as guides to modern cities. His explanation was that he was looking for older, more authentic, underlying patterns. And if there was no pattern, he could imagine one. For instance, as a mental exercise and entertainment he once decided to juxtapose the countryside of his childhood Kent on to the streets of London as he walked along. In this way, instead of treading the asphalt, he could imagine himself walking in the landscape around Oaklea that he knew so well.[23]

Pattern thinking was also a driving force in his theorizing. Over the years we typically find Bill Hamilton looking at some phenomenon that reminds him of some other thing, on a different continent, or having to do with a totally different species; there is something they have in common, they are both part of a larger pattern, and he is confident that he will be able to discover what it is. Nature's plan can be revealed. Sometimes these connections require a long period of mental 'incubation'. In such cases the workings of Hamilton's mind is reminiscent of that of a detective in a mystery story: there is something that reminds him (or her, think of Miss Marple) of something else, but it is not immediately clear what it is. Suddenly the connection is illuminated, a pattern emerges, and the mystery is solved.

With his butterfly collecting having become more professional, a new world opened up for young Bill Hamilton. There was the thrill of the hunt and the delightful surprise of finding a rare specimen in an unexpected place. Bill's trips (on foot or by bicycle) away from home became longer and longer, and sometimes he stayed overnight and came back the next day. In this way he set up a system of venturing into new unknown areas, prospecting for ever new, ever more beautiful and original specimens, ever farther from the safety of home. This was to become a parable for his scientific life as a whole:

> What tortuous paths my life has taken through dark and frightening forests of reason, and paths of what different kinds! Woodland lanes I used to wander barefoot as a boy on Badgers Mount, Kent, in shorts, with an insect net in my hand, a killing bottle carried like a false pregnancy in my shirt, seem in retrospect almost like a model for all I have done later.[24]

This metaphorical statement does seem to capture the life of Bill Hamilton. By following scientific truth wherever it took him, he often seemed to reach conclusions that defied moral and political conventions. As a scientific pioneer in a number of fields he certainly wandered many different paths, but barefoot and in shorts he was vulnerable to unknown difficulties of the road ahead. One of the greatest obstacles on his journey would be the referees of his journal articles. He had set himself the goal of catching and bringing back to science new and beautiful theories, but there was suspicion about what exactly he had under his shirt. And just in the same way that his killing bottle was once laughed at and smashed by a group of boys, so journal referees would later shatter his precious findings. But for Bill, there were always new butterflies to catch.

3

Schoolboy at Tonbridge

Tonbridge is among the famous British public schools, with a proud history that goes back to 1553. Its stately grey stone buildings are surrounded by a multitude of green sport fields. It is exactly the kind of place where the expression, 'we were at school together' has deep meaning, especially since it is one of those schools that has traditionally prepared its students for entrance to Oxford and Cambridge, where such friendships can grow stronger still. Of course Tonbridge has changed with the times, but one encounters its long history everywhere one steps. A good example is the mysteriously named 'Upper Fifty' and 'Upper Hundred' areas. These refer to particular playing fields where rugby at Tonbridge was originally played with a much larger number of participants than is currently the case.[1]

The term 'public school' sounds like a misnomer, since these schools are in fact private, that is, outside the usual state controlled educational system. The public schools got their name from the alternative to home tutoring which they offered at the time. During Bill's time at Tonbridge (1949–54), entry to grammar school was by the notorious 11+ exam, which sorted the wheat from the chaff as far as an academic future was concerned. For those planning on going to 'public school' the entrance exam was at age 13+, a common exam for the whole public school system. Students were not necessarily accepted by the school of their first choice. Bill was a well-prepared student, but had been set back because of an accident when he was 12 involving an explosion (more about this later). But his exam results were good enough for Tonbridge, and his father had written an explanatory letter to the headmaster, indicating that Bill was catching up on his schoolwork and would also soon be able to participate in sports. Tonbridge had the added advantage of being located in Kent, a bus ride away from Badger's Mount.[2]

During Bill's time the boys at Tonbridge entered the school at about age 13 or 14 and finished around age 18. Most were boarders, but some were 'day boys', commuting from their homes. Bill had chosen to be a day boy (he had read about life in public schools and was not thrilled by it), which meant that he had to get up very early in order to catch the only bus connection. He used to rush up to the upper front seat in the double decker, sitting there hunched over homework until the bus arrived at Tonbridge at five to eight. Over time, he became friends with a number of boys taking the same bus. The some 500 students at Tonbridge were organized into nine houses, of which Smythe House was Bill's house. Smythe was one of the two houses for the one-fifth or so of the students who were day boys. This was a place where commuters could keep their things and use as a base between classes. Being a day boy did not entail special privileges: if there was some gathering or event at school during the weekend, the day boys had to attend.[3]

In other respects, too, Bill was participating fully in everyday life at Tonbridge. Just like the other boys he was subject to the many rules and regulations of the place. Bill could now verify for himself that some of those seemingly strange and petty regulations that one read about in books about boarding school life actually existed. Moreover, at least at Tonbridge, they were ferociously enforced. Let's take a look at the punishments for disobeying the rules. Small misdemeanours earned such things as potato peeling or gardening, or a punishment run; bigger rule breakings resulted in more runs, and really horrendous deviations from the rules were punished by caning. What kinds of offences were those? During Bill's first year he was caned for leaving his gym shorts on the changing room floor. He took it quietly, but it certainly contributed to a growing sense of rebellion and created solidarity with other boys who had become victims of the 'system'.[4]

An initiation to membership in a house was the 'social test' in which one had to demonstrate knowledge of everybody's name and nickname, including teachers and school leaders, the colours of each house, and the colours of the sports. Among the housemasters were men with nicknames such as Toad, The White Rabbit, The Crab (the name of the school Chaplain), Puddle (the name of the Commanding Officer of the Combined Cadet Force), Deadly Bones, and The Arch (the headmaster, who was a clergyman). Tinfin was a physicist who had lost an arm in the war and replaced it with a metal one. Three of the masters were ordained in the Church of England.[5]

The official school uniform was a red jacket (called a Lovat jacket) and a straw hat with a ribbon in the house colours. The first day at school Bill made a faux pas by bringing his books in a knapsack, not a briefcase, a mistake that was quickly corrected. But it was not enough to be properly dressed and equipped. There were specific rules on how to wear the school jacket. This, in turn, had to do with the number of terms a student had been at Tonbridge—the basis of the school's social order. The first-term students were required to keep all three buttons buttoned, and were not allowed to keep their hands in their pockets, no matter how cold it was outside. The second-year students were allowed to use one button and have one hand in their trouser pocket. *Praes* or *praepostors* (prefects or student leaders) could have their jackets open and keep both hands in their pockets. Moreover, there were different places to walk or play for boys from different terms and also different common rooms and areas for personal storage. Great care seemed to be taken to avoid the mixing of students from different terms or years.[6]

One reason for this might have been the fact that Tonbridge was also training students for the military, a place where strict hierarchy of rank is enforced. Belonging to the Combined Cadet Force at the school was mandatory, starting the very first year. Bill too went through this training. There is a picture of him in uniform as a not too happy looking cadet. It seems that the cadets were chronically bored, which is not surprising, considering that they mostly learnt to polish brass and parade around in khaki uniforms holding 1905 bayonet rifles. As we shall see, Bill later found a way to cheer up the cadets, at least in the short term.[7]

But separating the students by term or year was not the only principle at work. There existed at the same time a countervailing tradition of signalling with bright colours in bands, badges, scarves, ties, blazers, caps, etc which particular house, or sport, or club pupils belonged to. Many of these items indicated special prominence rather than mere belonging—for instance, achievement in sports was advertised by 'first team colours' on ties and blazers—but in other cases the paraphernalia indicated membership in something that actually cut across the term hierarchy. For instance, 'house colours' were displayed on silk or woolen scarves, as well as in hat ribbons and ties. A count of all the Tonbridge items that around that time could be acquired yielded the following astonishing list: 17 different barge-bands (the straw hats were called barges), 23 ties, 13 squares (silk scarves), 10 scarves, 11 blazers, and 16 caps. No wonder this system sometimes received sarcastic commentary in the *Tonbridgian*, the school newspaper.[8]

Tonbridge prided itself on a sort of self-government system in regard to school discipline, in which older boys were put in charge of younger ones.

Every house had a head of house (a senior student) assisted by a handful of *praes* to keep order and enforce rules both in the house and in general. The house *praes* gathered in the gym to discuss more serious cases, and these might well result in 'two to six lashes with a cane, administered to the buttocks of the offender by the head of the house'. Bill himself had to go undergo such a punishment. Eventually he went on to become head of Smythe House as a senior during his last term at Tonbridge. But Bill was a benevolent head and the spirit of the house improved under his time. His fellow students remember the little book in which he 'with his spindly handwriting' was keeping track of crimes and punishments. He also reformed the punishment runs, usually considering runs around the cricket field sufficient, and even arranged it so that a number of smaller punishment runs could be 'traded in' for a longer (and more enjoyable) cross-country run.[9]

A personal challenge for Bill were the daily morning gatherings in the Chapel. The problem was not that Bill did not care much for Church rituals (although he did not), but there was all that singing! Bill really was tone deaf, at least in the sense that he could not hold a tune himself. Meanwhile around him everyone was supposed to loudly praise the Lord. Of course, Bill could have just continued doing what he usually did during these sessions: study the lyrics of the different hymns to see if any of these could pass as poetry (very few did) but that was not enough to satisfy a persistent music teacher, who insisted that he at least 'mouth' hymns. As Bill later put it: 'Having been always totally unable to sing any song of correct note or harmony, I would when pressed by a prefect open and shut my mouth silently like a fish, mouthing hymns as required...'[10]

How did Bill cope with all these constraints and controls? The answer is that he was largely 'pre-formed' already before entering Tonbridge—he knew who he was and what he wanted. He had his own inner world and was not too dependent on other people or their opinions of him. He concentrated on his studies; that was, after all, why he was at school. He realized that it helped to be good academically in order to instill respect in others. It also helped to be good at sports. Bill applied himself to both of these fields of endeavour. Although he was quite contemptuous of all the petty school rules, he did not let them upset his inner composure. (Later, though, he confessed to one of his daughters that he once put holly on the music teacher's chair—his revenge for all the 'mouthing' that this teacher had put him through.)[11]

How did Bill fare as a student? At Tonbridge incoming students, dependent on their ability, were usually placed in streams within each subject. More advanced classes were typically taught by better teachers. Classes at Tonbridge were small, between 10 and 25 students in each. Bill usually found himself in the upper streams. In their first year, all students took English, history, Latin, French, mathematics, and geography. Then followed two sets of two-year courses in a limited number of subjects. These might have been classics, languages, mathematics, or a number of other things. Bill concentrated on physics, chemistry, and biology. He wanted to do things himself. He was no longer (if indeed he ever had been) an obedient schoolboy listening to teachers—he was a curious investigator and all round experimentalist. (As a scientific experiment he once fired a blank bullet into his boot during a cadet exercise!) Indeed, subjects other than science were almost entirely lecture-oriented and one can only guess how boring some lectures probably were to Bill. In addition to his science subjects, Bill was still obliged to take English, mathematics, Latin, and French.[12]

One of his schoolmates commented that 'Bill doodled incessantly'. This might be taken as an indicator of his boredom. It could also, of course, have been an expression of Bill's pattern fixation. The doodles were intricate abstract black and white patterns, the schoolmate noted, so fine that he actually tried to copy Bill's doodles as models for his own.[13]

Each of the two-year sequences finished with a major exam. The first one, the General Certificate of Education Ordinary Level Examination ('O' Level) was not graded, students being marked only pass/fail, but the second, the General Certificate of Education Advanced Level Examination ('A' Level) was graded, but unofficially (however, students could find out their grades). At Tonbridge the instruction in science was especially good, and students early on became familiar with the kinds of problems that were solved in typical Oxford and Cambridge entrance exams. Bill took his study of science very seriously, as evidenced by neatly written and meticulous-looking science lab reports, with more important points underlined in red ink. His fellow students would not have known the circumstances under which those careful reports were sometimes produced: in winter Bill would do his homework sitting in a sleeping bag in the cold living room at Oaklea (whose fireplaces are not of the heat-retaining kind). During his time at Tonbridge Bill did not have a room of his own, he used to sleep on a sofa in the living room.[14]

Bill's chemistry was good, but it was clear early on that his best science subject was biology. The type of biology taught at Tonbridge was mostly of the physiological and anatomical kind, rather than modern evolutionary biology. But even anatomy could be made interesting. Bill once suggested to a schoolmate visiting him at Oaklea that dissecting a rabbit might be a good preparation for their A Level exam—whereupon they actually went out into the forest and shot one. (Bill was handy with a shotgun, having practised rabbit shooting around Oaklea as a 'deputy' for his father, who had a permit.)[15]

Bill's school reports from different terms and years (or rather 'forms', numbered from IIIrd to VIth with a mysteriously labelled 'Upper Vth' form between Vth and VIth) show his grades for each subject and comments from his housemaster. (Bill had the same housemaster, T (Tom) Staveley for almost all his time at Tonbridge.) Bill certainly did not show an overall record of top grades in every subject. His performance was typically good, but not dazzling. Biology stood out as his top subject, and other sciences and mathematics were also good. But it was clear, for instance, that his French was poor (in fact, he later failed his French A Level). Indeed, perhaps because Bill did not stand out (or, especially, act) as some kind of model student, some of Bill's schoolmates were genuinely surprised at his later eminence. Here is one comment: 'It is extraordinary that this unkempt apparently oafish boy should have slipped past all his contemporaries without too much of a hint of his future brilliance, never responding to any of the ribbing which he got, and he left Tonbridge to carve such an extraordinary career for himself...' Another classmate noted that Bill's name had not been on the list of distinctions in science in 1954, the year when he graduated.[16]

But there were those who used the word 'genius' in their commentaries, and others who recalled aspects of Bill's personality and behaviour that they saw as associated with his later fame: 'He certainly was "different" because of his extra beta [brain] cells and could be in a world of his own—but this didn't prevent warm friendship' or 'I...recall his physical presence sitting at the back of the benches in the laboratory and my feeling that he was something special'.[17]

Bill's sister Mary had a simple explanation for why Bill did not stand out academically:

> I don't think it is at all surprising that his genius did not manifest itself at school. I would have thought that must be the rule rather than the exception. By and large, people, including school children, who shine early on are those who swot [study intensely] or who are naturally good at assimilating and regurgitating information. People with original ideas need much longer to think about them and may even be particularly unsure of themselves. Bill knew that he was intelligent, but it was not the intelligence to get all-round high marks and could be quite compatible with failing in French.[18]

As a rule, she suggested, Bill did not do well in subjects which 'involved learning a lot of stuff that he was not interested in'. Indeed, this may well be the case, because some of the same pattern was later to be repeated at Cambridge. Bill simply would not spend time on things in which he had no interest, focusing instead on what he himself thought that he should be learning. Bill was a very self-directed learner. He did have a phenomenal memory, but did not put it to use in memorizing things that he perceived to be of little value. In other words, even then he acted more like a determined scientist than a dutiful schoolboy.

But there is an additional characteristic of Bill's which became evident as early as Tonbridge: his particular exam writing style. This may have been partly related to this same drive of Bill's to focus on his own interests instead of mastering the required material. Bill typically never answered more than a few questions of any exam. Those that he chose, he answered well, and in depth (typically not more than three out of five). It is not clear if he was not prepared to give quality answers to the questions he left out because he had not studied that material (perhaps finding it uninteresting), or if he simply ran out of time, because he could not inhibit his habit of going into great and time-consuming detail in his answers. Probably both. Whatever the reason, this selective behaviour of Bill's must clearly have affected his grades in the case of those teachers who could not—or would not—appreciate such an idiosyncratic approach. On the face of it, choosing to answer only a limited number of exam questions would seem like a risky strategy. But it was Bill's way of coping with known constraints.[19]

Perhaps it was a calculated risk. It seems that Tonbridge was the kind of school that prided itself on paying attention to the particularities of its students, trying to provide for each individual an intellectually nurturing climate. There was no attempt at streamlining. After all, one of the aims was for the students to be so capable that they would be able to pass the Oxford and Cambridge entrance exams. Probably without realizing it, at Tonbridge Bill encountered the kind of climate that he would continue to need throughout his life in order for his special talents to develop.[20]

One unusual thing about Bill that his schoolmates clearly noticed had to do with his love of literature and poetry. As already mentioned, Oaklea was well stocked with books, many of them from the 1930s. (One visitor suggested that Bill's particular taste in poetry may largely have been derived from the books that he came across on his parents' bookshelves. This may well have been the case.) Schoolmates also noted that Bill enjoyed language and playing with words. Bill wrote poetry himself, and a couple of his essays won school prizes. One of these, 'Red Death', was published in the *Tonbridgian*. It was for that essay, written at the age of 14, that he received the school prize that provided him with a copy of Darwin's *Origin of Species*. He also made an impression by reading aloud in class another essay he had written, a dramatic story having to do with a sheep's skull on a deserted moor, time lapses, and déjà vu.[21]

It is strange to read Bill's prizewinning essay, written as a 14-year-old. It conveys the same apocalyptic sense as some of his essays written in the 1990s. The difference is that his later essays are reasoned arguments, this one is pure fiction. Still, both appear to come from the same inner source of Bill's personality. It seems that even as a teenager he could appreciate the existential angst of being a scientist potentially in charge of human life and the future of the world, as depicted in more fanciful books he had read. He could project himself into this role, and he could visualize what could happen. Scientific and ethical matters converged and had to be resolved together. He had grasped the Frankenstein-like glory and horror of science—its seduction—and the underlying tension between science and values that is a driving force for many. At 14 he was not yet grappling with practical issues about what exactly the concrete threat was and how this might be avoided—that was to come later.

'The Red Death' is the direct antithesis to the lyrical naturalist writing that Bill also enjoyed and later tried to emulate. Here emerges an early, darker side of Bill:

> The red death was creeping over the land. From bush to bush, from tree to tree, it moved. All was dying under its merciless jaws. The corn was bitten from head to root, and now it stood brown as the earth it grew from, dead... On wood and crop and grass and garden the red death crawled, and they died.[22]

The essay goes on to describe the catastrophes caused by scientists who tried to make life:

> Now they wandered among the stricken crops and stamped it underfoot hopelessly. They sprayed it with fire, dropped dust from the air, did everything conceivable to kill it, but they could not. A thing that has seven generations in a day, and that will eat anything green, is not easy to kill. Only two trees it refused as food: the Yew and the Cypress. Oh, horror of horrors, why those trees of all the thousands?[23]

This has an apocalyptic, almost Biblical sound to it, while it is juxtaposed with a real scientific observation: trees like the yew and the cypress do in fact have very few predators. We see here the germs of some themes that were to follow Bill throughout his life: life as an arms race, with mankind using science and technology to expand but not being able to control the result which strikes back, impending dangers for mankind which science may be helpless in fighting, and the need to find sources of natural resistance to such threats. At the time of course, he did not know about AIDS, but he had the 'template' ready for worrying about it. These themes come up again in his chapter prefaces in *Narrow Roads of Gene Land*, his two volumes of collected works written in the 1990s. Maybe we in the 'Red Death' theme can also see a proto model of his 1980s parasite theory of sex (sexual reproduction evolved because of the need to escape pathogens).[24]

It is clear from 'Red Death' and also a poem that he later wrote (see the beginning of next chapter) that for at least part of Bill's scientific thinking, the inspiration is not coming from science itself, but from a deeper source. Some of this may have to do with the looming threat of the potential invasion by the enemy he experienced as a child (the Germans may be coming and taking over England; invisible danger is everywhere). After all, Bill was very young during the Battle of Britain and probably internalized a general feeling which he could not clearly articulate. There is a strong sense of coming apocalypse in both these early and his much later writings. It is interesting that he does not see rationality and science as the source with which mankind will conquer these invaders and threats, whatever they may be, but regards science and reason as mere tools used by an organic human body whose life is being threatened, at least in the long run. His perspective is long range: he is viewing humankind as a biological species.

As a literary genre, of course, this type of writing was available to Bill at the time, especially in the form of science fiction books by HG Wells and others, which he surely read. Bill's sister Mary noted that some of the elements in 'Red Death' may have been inspired by books that she knew Bill had read at home; she suggested among others the work of Edgar Allan Poe, collections of stories by Dorothy Sayers, and a nasty short story involving ants eating up everything alive.[25]

At the same time 'Red Death' reflected an intellectual stance that Bill was deliberately assuming at this time. He had adopted a purely rational and scientific, evolutionary approach to reality.

One can imagine that this early determination may easily have made him scoff at things that he considered to be false, while he himself stoically preferred the nasty truth to pretty falsehood. This kind of attitude may conceivably have been reflected in his face, too. Bill's housemaster once made a comment about Bill in a term report that he was 'unfathomable'.[26]

Bill's schoolmates also observed this quiet and reticent aspect of his character. One schoolmate stated that Bill remained 'rather an enigma. We were good friends, as much as it was possible to be counted as a friend, he was a reserved character, self-sufficient and to some extent keeping himself to himself'. Bill was said to have 'a quiet and peaceful air about him'—even in the front row of the rugby scrum. But, importantly, Bill was described as a 'decent chap, utterly devoid of malice', and as having an 'affable and essentially friendly nature', and 'a lovely smile'.[27]

At Tonbridge, as in other public schools, sports were mandatory. Students had to practise one major and two minor sports. Bill chose rugby as his main sport. Rugby was the roughest sport around, and since this is New Zealand's national sport, it probably represented a special honour issue for him. His favourite minor sport was cross-country running, which he enjoyed since it took him out into the open and entitled him to go 'out of bounds', according to his sister Mary.[28]

Bill was part of a team of 15 players, more particularly part of the second row of the so-called 'scrum' (a formation of three plus two plus three players whose aim was to prevent the players in a similar formation on the other side from getting the ball as it was dropped between the two teams at the beginning of a game). Rugby (or 'rugger') at Tonbridge was played without protective gear. The only protection allowed was a 'scrum cap' for the heads of the members in this massive front formation. There were traditional matches against other schools, such as Harrow. One photograph shows Tonbridge–Harrow 1–1, with Bill and others jumping high in the air. Bill played in the second, and then the first, 15 in 1953 and 1954, participating in matches away from home. He was often mentioned in the *Tonbridgian*, for instance 'Hamilton, W. D. was a great-hearted second row forward who tackled well and ran hard (though he seldom passed [the ball to others on his team]'). It was also reported that he had 'won his [first team] colours'.[29]

Bill's classmates describe him as a 'fierce competitor and good sport', 'formidable, fearless and ferocious' and 'not violent'. He was able to put on a fearsome 'caveman' face on the rugby field to good effect, and although he was relatively short, he was extremely strong and 'a difficult man to stop'.[30] (Incidentally, most people believe that Bill Hamilton as an adult was taller than he actually was. This is due to people's tendency to extrapolate to the whole body from the size of the head, and Bill did have an unusually large head. He had a hard time finding hats.)[31]

In general he was seen as calm, and not easily perturbed. He took others' nicknames for him, 'caveman' or 'ape man', in his stride, and used his somewhat unusual looks to his advantage in some theatre plays, for instance once playing a gangster who was also an ape man in a comedy play.[32]

The friendly aspect of Bill's personality manifested itself early on at school. He helped classmates in areas such as mathematics or science, and he was kind to 'underdogs', having a nice word to say, for instance, to a team member who did not make the team one year. Bullying was relatively common in that public school environment. Bill, if anything, was an 'anti-bully', and stayed that way for the rest of his life.[33]

Of course, in order not to be bullied yourself, you needed some points of strength and you needed to establish a power basis. What was it that impressed the other boys? One area of strength was clearly academic ability, and Bill certainly was doing well especially in science and impressing his fellow students with his writing. Another area was skill in sports. Bill excelled in rugby. He even did 'homework' of his own invention. Oaklea visitors report Bill being out in the garden practising various types of runs and feints and tackles. But finally, Bill was no sissy. He was not quick to anger—he had a quite high tolerance threshold—but once made angry, he was fierce. One of his school friends describes Bill as follows:

> [F]or the most part Bill was quiet-spoken and quite diffident in manner... But if he was seriously riled by something it was another matter. Then his customary quiet-spoken diffidence would be replaced by a look of calculated hostile determination which Tom Staveley [the housemaster of Smythe House] said 'made him nervous'. If disagreement turned to mobbing [fighting], I recall Bill would pitch you to the floor or bounce you off the walls before you even knew that hostilities had turned physical. His large head, somewhat simian features and robust build... enhanced both the display and attack phases of these rare incidents. It was in this state of outrage, one imagines, that in Brazil he later tackled a mugger armed with a knife—and got considerably injured.[34]

Bill could certainly hold his own but there was in fact a heroic subtext of Bill's engagement with rugby. He was playing this game with a protective metal plate sown in under his rugby shirt. His schoolmates were impressed with the large scar on Bill's back and the missing top portions on two fingers. 'He obviously was an inquisitive sort of guy', noted one of his contemporaries. There were various rumours about the explosion that almost took Bill's life. Some assumed that Bill belonged to the Exploders, a revered group of boys at Tonbridge. This group, working in the well-developed chemistry labs, was busy striving to perfect the mixtures for explosives. The exploders' main claim to fame was to 'have changed the Fourth Termers' Walk forever', an event which was marked with a bronze plaque on a tree. Of course Bill had already been his own 'exploder' before Tonbridge, and this gave him considerable clout. He was even consulted as an explosives expert by a fellow student. The problem was how to spice up a particular Cadet Corps exercise, giving it a sense of verisimilitude and battleground atmosphere. Bill gladly obliged. The result was an unforgettably realistic 'attack on Smythe House', involving coordinated explosions of the house's several metal trash cans. The lids of the cans are said to have flown 30 feet high into the air before clattering noisily down.[35]

What was the real story behind Bill's accident? The best and briefest version of events is probably that of his brother Robert:

> I was with him that day. He was using empty .303 cartridges we had collected from the rifle range where our father went once a week to shoot his rifle with other Home Guard neighbours from Badgers Mount. Bill was filling one of these empty brass cartridges with powder he had mixed himself from sodium chlorate, sulphur and aluminium powder and squeezing over the open end in a vice when the explosion occurred. (He had intended to throw the 'bomb' onto a bonfire to demonstrate it to a friend, and had successfully detonated a similar explosion the day before. Fortunately I was not in the shed at that moment and the friend had retreated to the far end.[36]

In other words, Bill was conducting an experiment with an explosive mixture, which backfired (literally). Through quick action of his mother and the Denmark Hill Hospital staff he was saved, although the explosion cost him the top parts on two fingers and located shrapnel in his body. Bill himself remembered how two young interns used a syringe the size of a bike pump to try to suck up blood from his lungs: 'It seems to me now that I must have been both witnessing and feeling what it is like to be killed by a rapier thrust several times during that morning. And yet I watched and assessed it all in a rather detached, accepting way, merely longing to see them succeed...'[37]

Bill was often mixed up with two other students with the same last name at Tonbridge, one who was also at Smythe House (JB Hamilton) and another who was on the same rugby team (PJH Hamilton). JB, 'Jock' or 'Shock' played pranks, and had once sprinkled explosive on the Chapel floor so that the Chaplain as he walked in the aisle created a series of miniature explosions, producing violet vapour. Because of its necessary solemnity the Chapel was an almost irresistible place for pranks. A typical boyish prank was to bring in a box with a May bug and let it loose at a strategic moment when its detection would cause maximum havoc. Bill was certainly involved in such pranks, but not the JB one, which some gave him credit for.[38]

There were a number of clubs at Tonbridge, but Bill did not join them. He was not a very gregarious person, and one can assume that the Natural History Society, for instance, had little to offer someone who was as advanced as Bill. Informally, however, Bill went on nature excursions with school friends, and also read and discussed the essays published in the semi-popular journal *New Biologist* by Haldane and others. For budding biologists, it was that journal, rather than the textbooks studied in class, that was their real source of knowledge. Bill also used to discuss recent journal issues with his sister Mary. Mary was an excellent student and interested in all kinds of intellectual matters.[39]

Bill's special talent for biology had been spotted by his biology teacher, AL Thomas, generally called 'Crump'. There is a true story that his mother arranged a meeting with him to discuss future plans for her son, all the time addressing him as 'Mr Crump'. 'Crump' did not alert her to her mistake. He did have a lot of faith in Bill's academic prowess and ended up recommending him personally to his own college at Cambridge, St John's.[40]

Still, there was the matter of the Cambridge entrance exam. Bill stayed at Tonbridge an extra term, the winter term of 1954, in order to take this exam and compete for a state scholarship. It turned out to be a good decision. During this term he advanced to become the head of Smythe House, and to play in the first 15 in rugby. He was also invited to join the newly formed Athena Society, a serious society for advanced students of special eminence in art and literature. Getting a state scholarship would mean that at least some of his tuition and other fees would be taken care of by the state for the period of his undergraduate studies. This was important to Bill, because he wanted to depend on his father as little as possible. By now Bill had his own capital in the form of his achievements and accumulated knowledge. This had to be put to use. The task ahead would be to find sponsors for his further studies.[41]

A lot of anxiety was created around this Cambridge entrance exam, because for Bill it was not only a matter of prestige and money, but also a reality test. Bill in his idiosyncratic way did not answer all the questions in the exam, and for some time he was not sure whether he had passed. In a letter to his godfather Charles Brasch in New Zealand, he wrote about of his difficulties and worries. Brasch, a man of literature and art and the editor of the cultural journal *Landfall*, was to become something of a sounding board for Bill during his university studies. For Brasch, Bill represented the son he himself would have had, had he married Bettina as planned. After Bettina met Archie, that was not to be, but after her marriage, Charles Brasch remained a loyal friend of the family, taking special interest in Bettina's children. He exchanged letters with Mary and Bill, encouraging Mary's artistic talents and Bill's writing interests. In their uncle Charles they had a person who believed in and supported them without formally being a relative.[42]

A poem that Bill wrote during the winter of 1954, his last term at Tonbridge, may be signalling a sense of claustrophobia, that it was time to move on. The poem, published in the *Tonbridgian*, is called 'The Bracket Fungi' (a particular fungus growing in the beech woods surrounding Oaklea). At the beginning of the poem everything appears calm and beautiful in the sunset. But as the sun sets in the final verse, we see the parasite fungi with their white hands already sneaking up on the stately beech trunks:

> Quietly encircle, quietly pull and kill.
> And when that cobweb-shattered sun
> Sinks huge to a yet more distant hill
> The pallid hands grow fleshy, pink and fill
> That beauty with their growth.[43]

Of course, such a poem may have been produced as a pure exercise within what might be called Bill's 'dark paradigm' of absolute evolutionary rationality. He knew that the visions that he produced were not pretty, but they certainly related to a respectable poetic genre. Bill loved the melancholy poems of *A Shropshire Lad* by AE Housman. A more targeted description of the universal struggle in nature was 'The Woodsman' by Robert Louis Stevenson, a poem that Bill liked for what he called its 'stark realism'. Indeed, in Stevenson's poem every organism in the forest seems to be preying on every other. It is a Hobbesian world out there among the innocent-looking plants.[44]

Bill took the intellectual position he had already assumed as a young man very seriously. He had reasoned himself into a view that substituted religion and other irrational-seeming beliefs with a purely rational scientific credo. This entailed refusing to accept hypocrisy and telling the truth about life, however harsh or ugly. This position of uncompromising rationality would become Bill's intellectual source of strength throughout his life. It was his dark citadel. It was regularly to clash, however, with the gentle side of his personality, with his idealism and romanticism, his observations of other people, and his own actual behaviour. His 'dark paradigm' was a place that he visited mostly in his theoretical work, it was part of his intellectual persona and his self-presentation in writing. In everyday life, the 'Bill' people knew was somebody quite different.[45]

What was, then, the plan after Tonbridge? One obvious way for Bill to get away and even travel for free would be by doing his National Service, the two-year mandatory military training for all young men at the time. He hoped to be sent by the military to far away countries and in this way to see some of the world. His father, after all, had served in the Royal Corps of Engineers, which had taken him to remote places, and his uncle Herbert had also travelled widely. So had his grandfather. So why not do his military service immediately, the sooner the better?[46]

The other option would be to continue directly to Cambridge, for which he had passed the exam which had given him a state scholarship and earned a place at St John's College. In that case, he would do his army service later with a Bachelor's degree in hand, perhaps while planning his future career. But that meant postponing his travel dream by three years, and having to make a break in his studies later. One can just imagine Bill going through the pros and cons of the different alternatives. The urge to travel won out. Bill decided to postpone his college studies and opt for the army.[47]

What he had not counted on was how the army would judge his overall record. Bill in his own calculations saw himself as very suitable for military service. He was trained as a cadet, had learnt how to be in command of others (as a Lance Corporal in charge of 10 boys, and as the head of Smythe House). He knew how to handle tools and cars, and was clearly strong and fit: after all he was a formidable rugby player, walker, and runner. But he had forgotten one thing: that old explosion accident. There were still pieces of shrapnel in his lungs, and there were those missing fingertips on his right hand. From the army's point of view Bill was not fit to be in active service, and certainly not to be sent overseas.[48]

This was an enormous blow to Bill. Even worse, instead of travelling himself, he was put in charge of planning and dispatching equipment and food to army units overseas. His father had managed to arrange for him to be in the Royal

Corps of Engineers, which meant that he ended up spending his time of service in Chatham and Gillingham, where the Corps had their headquarters.[49]

The biggest irony was that Bill could have avoided the National Service altogether. Born in Cairo to New Zealand parents, he was a New Zealand citizen, and therefore exempt. His younger sister Margaret never understood why Bill so much wanted to change his citizenship in order to go to serve for two years in the army. There may of course have existed other motives than the mere wish to travel, for instance a strong sense of fairness and duty. His home was, after all, in England and his birth abroad more of an accident. Also, the idea of every man doing his duty was certainly ingrained in him from the atmosphere of the Second World War.[50]

Bill was under no real pressure during his time in the army. He was, after all, with the Royal Corps of Engineers. It was just that the whole atmosphere of authority, hierarchy, and subordination in the military was so totally alien to Bill's personality. He hankered for freedom. The mere being in the army triggered wild bouts of rebellion in Bill and landed him in strange situations of his own making. He seems to have felt an urgent need for escape—even of a physical kind. At one point, it seems just to assert his freedom, he left the army barracks and stayed away all night. He gave himself up in the morning.[51]

On another occasion, his assertion of independence took a more aggressive form. One of his duties was to be in charge of the officers' dinner. He chose to serve them pheasant—no doubt a delicacy anywhere, and a welcome change from the regular army menu. When his superiors complimented him on the nice choice, he informed them that the pheasant they had just eaten was actually roadkill. He was simply telling them the truth.[52]

Incidentally, it may have been that for Bill, fresh roadkill was fair game. At least this appeared to be the attitude of his brother Robert, whom I visited in New Zealand during the research for this book. (Robert explained the good-looking bird on his kitchen counter with a trace of embarrassment.) Perhaps he felt that nobody needed to know where the fowl came from; it was good meat. In the case of the officers' dinner, though, Bill was playing a game of his own invention: serve and tell.[53]

Bill's letters home from that period are full of complaints about army life. This was also the first time he was away from home for an extended time. During this time he kept in close touch with his family, as was his habit. His mother's letters reported on her successes or disasters in raising Bill's butterflies in the garden shed—Bill was trying to control his breeding experiments remotely—and kept him informed about the goings-on at Oaklea. His little brother Leco wrote to Bill in large childish handwriting about his own butterfly experiments. The times when Bill could actually visit his family were precious, and he became accustomed to taking very late trains back to his army quarters from Waterloo station. From that time on, Waterloo station was to take on a special meaning for him.[54]

But after a while Bill decided to turn his army time to his advantage. One way in which he learnt to cope with his army existence was by using it for field research. He started taking extensive notes on the surrounding area and its flora and fauna, ending up filling several notebooks.

However, the main credit for keeping up Bill's spirits during his army time goes to Aunt Prue.[55] Aunt Prudence, his great aunt (his mother's mother's sister), lived at a reasonable visiting distance from his army barracks. For Bill, she represented to Bill all that was refined and elegant. The atmosphere in her house was much more 'posh' than Bill was used to at home where a simple lifestyle was the rule. 'Posh' was a term used in the Hamilton home for things that were somehow too cozy or fancy, and they all knew what that meant. (For a big family with limited living space, certain niceties were also simply impractical.) At the same time the Hamiltons took a certain pride in their own particular Oaklea lifestyle. They formed a little world around themselves and were proud of their level of cultural and scientific knowledge. Some thought they were snobbish, but all this was really a natural part of their family life.

Aunt Prue did not only have elegance and good taste, she was also well-travelled and had interesting collections of things, including an insect collection that she kept in boxes in a special cabinet. Seeing Bill's interest when he was growing up, Aunt Prue later kindly presented him with her whole beetle collection. Bill, who by then had internalized the difference between true collectors and amateurs, took a look: the species were clearly labelled, but there was no locality or date information. So, as Bill confessed:

> I wanted the space badly for my own collection and with the self-righteousness of a True Collector following principles stated in my Books, which held that a specimen without locality and date was Worthless, I threw all her collection out, keeping only a few spectaculars and a still smaller few that had been fully labelled.[56]

Later on he reflected on what he called his 'vandalism'. He regretted what he had done, realizing that he had lost important information. It may have been less of a moral regret than an intellectual one: it turns out he would have liked to compare her specimens, collected at an earlier time, to those that he had himself observed and collected in the same larger area (Berkshire in this case), despite the lack of specific details on locality and date.[57]

Bill soon discovered that his aunt also had a fascinating library, containing among other items popular translations of the books by Jean Henri Fabre, the famous French naturalist. First he used to borrow these books; later he got the whole set as a present from his aunt.[58]

Clearly Fabre had stimulated Bill Hamilton's genetic imagination:

> But what astounding discoveries Fabre had made! I was completely enthralled by my aunt's collection of the translated stories from the Souvenirs Entomologiques. The world of insect behaviour that Fabre revealed...and a reworked edition of the Fables of La Fontaine, was amazing. It was robotic, totally without sentiment, strange in its acts and achievements almost beyond belief. And above all, there was no difficulty, no heresy (as came to be very important to me later) in my trying to imagine genes for all the behaviours he reported![59]

As a person, too, Fabre seems to fulfill a number of criteria for being a favourite with Bill: marginal, of modest means, and with incorrect 'Victorian' style. Even better, Bill had encountered Fabre, this strange man, in his eccentric aunt's library. ('Eccentric' was Bill's own term for her and it was a compliment. Probably Aunt Prue was not especially eccentric, she was just independent, unmarried, and well to do, able to do what she liked.)

During his time in the army Bill used to visit his aunt quite frequently, always sending her postcards announcing when he would be coming. He felt great affection for her. At one point he even wrote to her: 'You are like a mother for me'. When she died in 1958, he expressed his sorrow by writing a poem—that was one way for Bill to handle emotions. Later, he kept Aunt Prue's memory alive by caring for her Velotta lily in his student apartment.[60]

4

Fisher Found and Lost

Bill Hamilton was an unusual Cambridge undergraduate. He partly arranged for his own education, which was quite divergent from the official one. He practised this self-education especially during his second academic year when he was in fact supposed to be studying zoology, botany, and physiology for Part I of his Tripos exam at the end of that year. But he was bored with the largely physiological approach in biology education and was craving evolutionary reasoning. Evolution was not much on the agenda at Cambridge. Bill's teachers worked on the general background assumption that animals acted 'for the good of the species', an assumption which was not analysed more closely.[1]

A good example of the general attitude at the time could be found in the textbook by Professor Sir Vincent Wigglesworth. He was a superb scientist in his own area, the functional anatomy of insects (and Bill later recognized that he had received very good training from him), but he limited himself to the purely mechanistic aspects of behaviour. Still, he felt free to include a general statement about the ultimate evolutionary aim of insect behaviour:

> Insects do not live for themselves alone. Their lives are devoted to the survival of the species whose representatives they are... Indeed, we have now reached the aim and purpose... of the life of insects.[2]

In other words, Professor Wigglesworth was an unreflective group selectionist, a 'good for the species' thinker. So were Bill's various teachers and supervisors (tutors) at Cambridge, as he discovered quite early. But there were books and libraries to explore—he didn't have to depend on what his teachers told him. Bill set out to find resources outside the formal educational arrangements.

It was during his first year at Cambridge that Bill made a discovery that would have repercussions for his whole intellectual life. Coming across Ronald Aylmer Fisher's book *The Genetical Theory of Natural Selection* in the library of St John's College he experienced almost an epiphany. Excitedly he wrote a postcard to his sister Mary:

> I have discovered a grand new source of arguments. We started a course on zoology lectures on evolution last Wednesday and I felt rather dissatisfied with Carter's limited approach. So today, I got out of the college library Fisher's 'Genetical theory of natural selection'. Fisher's approach is, on the contrary, broad, mathematical and rather beyond me. It occupies the first half of the book; the second half is only loosely connected with the subject and expressly concerned with our old topic, the decline of civilizations. I now know why they decline, and how to stop them, and that after having read only two chapters. Did you realize that the invention of currency was one of the greatest progressive steps in human evolution? Do you know the evolutionary importance of the Family Allowance? That the most nearly perfectly progressive society is an oligarchy? That modern social structure is associating Intelligence and Initiative with Infertility? That bees do not suffer declines of *their* civilizations because they have disassociated their social strucure from Mendelian inheritance? That bees have no judges or lawyers? You can probably not make head or tail of all this, but I haven't time to explain further. Though written in a very mild and erudite manner it is intensely provocative. There are many echoes of my old opinions (which I never really abandoned, merely stopped thumping because I couldn't face the unpopularity they caused). All this was written in 1930, and alas, I am just too late: Fisher retired this year, although I believe he does still lecture occasionally.[3]

Fisher had indeed retired at the end of September 1957, but he was still around. However, Bill was not prepared to approach Fisher before he had thoroughly familiarized himself with *The Genetical Theory*. And that was no easy matter.

It was the second part of Fisher's book that particularly captured his imagination. Here was someone who was able to explain a number of big puzzles having to do with mankind and its future largely based on reasoning from science. It seemed to Bill that Fisher's scientific training gave him a basis for analysis of the past, present, and future, and even for policy suggestions. To Bill, the field of population genetics appeared increasingly relevant as a way of identifying evolutionary trends.

But there was the first, tightly reasoned part of Fisher's book that Bill would have to try to understand. Later, Hamilton was to characterize Fisher's book as a 'minor grail'. But we also learn that Fisher's reasoning 'took days, or weeks, to decipher' and that it was because of his study of Fisher that Bill Hamilton's degree was 'knocked down a notch' at Cambridge.[4]

Fisher is notoriously difficult to read. According to Fisher's own last undergraduate student, Anthony Edwards, today books like this would contain various aids, such as diagrams, pictures, and boxed text. There would be metaphors lightening the text. But, continues Edwards, 'Fisher's book, by contrast, is all argument. Each sentence must be *considered*, from the very first one... The style belongs to Charles Darwin's age, not ours, and it is illuminating to think of *The Genetical Theory* as a kind of mathematical-Mendelian appendix to the *Origin of Species*. Fisher was probably the best-read Darwinian of his generation, and it is not surprising that it affected his style.'[5]

There was another problem with the book. Fisher's mathematics was not the usual kind. It was usually left unclear where the ideas came from and how they were derived. Fisher didn't derive his results in a step by step manner for everyone to see. This was a habit that he had already developed as a child. His eyesight was very bad, and therefore he had been advised to do mathematics in his head. Fisher's mathematics, then, may have looked misleadingly intuitive, while it was actually a product of a mental calculation by a superb mathematical mind.[6]

Hamilton's Cambridge instructors at the time had told him that Fisher was not relevant for biologists. This of course made Bill even more keen on reading Fisher's 1930 book and cracking its mathematical reasoning. Ironically, Bill's teachers didn't know how deeply Darwinian Fisher actually was. As a school prize Fisher had asked for and obtained all 13 volumes of Darwin's collected work, which he later took with him to Cambridge and reread as a student there.[7]

Ignoring his instructors, Bill continued educating himself in the most modern mathematical methods he could find:

> I have taught myself the rudiments of matrix algebra which seems to be essential in many modern fields, and I have found it quite enlightening, for example, it has given me insights into preliminary standardization of Fisher's reproductive value before I set out to calculate these for actual human populations.[8]

But was he doing the right thing? It seemed that none of his teachers had read Fisher. Was there something he should know?

> ...I have alarmed my zoology supervisor with a summary of Fisher's theory of the inversion of the birth rate: he has the book on his shelves but seems never to have heard of his theory. What worries me about the theory is the fact that absolutely no one has heard of it: I begin to wonder whether Fisher himself has publicly retracted it...perhaps...I'll insinuate myself in the Genetics Department...and try to find an opportunity to ask him about it.[9]

Hamilton seems to have succeeded in 'insinuating himself', because later he wrote to his sister:

> I see Professor Fisher very occasionally now: having committed myself to do genetics, I have to try to lay foundations for my arrival: if I make sure they are keen to have me, it won't matter if I crash in Tripos part I, as I am almost certain to do. Sordid, I know—anyway, I have this ecological problem arising from some work that Colin and I did in Semerwater: a question of analysing statistics...my own analysis has turned out to be phoney, but it is still undecided whether my *findings* are wrong (it is a question of whether an orchid population is randomly distributed). It seemed to be beyond the Botany Department to help me so I took it to the Genetics Department this afternoon. My tame statistician was out, but someone else offered to have a shot and before long about five people had given advice. We went to the library, looked up several papers: the answer was in a paper by Fisher, of course, but the mathematics abstruse, so with trepidation they decided to ask the Prof. (everyone stands in awe of Fisher up there—no wonder). It was teatime and he was chatting with some students, carpet slippered, white haired, white bearded, senile—God, he is the very image of Gorki's Tolstoy in his old age in every detail of his manner. He sits in that office and listens to the clatter of the calculating machines he has set in motion all over the world and smiles—I swear he does. He is urbane, almost supercilious, everything he says has profound weight—I begin to dislike him, he must be laughing at us. He took my problem a little testily—what is all this about?—and grasped it immediately. I expected him to answer my question at once, from his immense experience, and I suspect he could have done so; but no, he suggests that I go and throw some more quadrats, then it will be really obvious. Unfortunately, I say, the place is in Yorkshire, and he laughs and makes some light hearted remark about getting... [here the postcard stops or becomes illegible].[10]

This must have been some time in the spring of Bill's second year at Cambridge, before that first big exam, Tripos I. We see how Bill even as an undergraduate acts very independently and professionally—consulting experts, going to the library to look up things, and finally asking the ultimate authority, Fisher himself.

We see also a typical Bill strategy. He believes that he can continue at Cambridge in the same way as he did at Tonbridge, where he was appreciated by his science teachers and got good grades despite the fact that he did not answer all the questions in the exams. He wants the Genetics people to respect him as a fellow scientist, not as a mere undergraduate, well in advance of his becoming their student for Tripos II during his third and last year.[11]

It is clear that Bill was interested in more interaction with Fisher. Here he now had a real problem that he thought Fisher would appreciate. After all, it was Fisher who had invented the methods that were used to assess the distribution of plants in the field (one method was 'throwing quadrats', that is, delineating a particular area, dividing it in squares and counting the specimens within each square), and even more importantly, he was the world's statistics guru, having invented a number of highly useful approaches to data analysis (among others the Analysis of Variance). But Fisher was not responding enthusiastically at all. Rather, he was following his own strategy, which was to put students to work and figure things out for themselves. (For instance, he once answered a question about sex ratios from Anthony Edwards by telling him to go and read a certain book in Italian.)[12]

Bill's reference to Semerwater in the quote above involves a first-year students' field trip to Yorkshire in June 1958 arranged by the Botany Department. It was during this excursion that Bill met two of his lifelong friends, Colin Hudson and Hugh Ingram. They were housed in the same farmhouse for the duration of the stay. It may have helped, too, that all could claim Scottish ancestry.[13]

But what of the other aspects of Cambridge life—clubs and other social activities? It turns out that Bill was uninterested in clubs. In contrast, his friend Colin, Mr Congeniality himself, was involved with a number of clubs. Colin admitted that the primary function of the clubs seemed to be to get members to arrange parties for each other—he remembered an occasion when he had to produce an enormous number of sardine sandwiches for a reception in his room. He tried to get Bill involved in the Naturalist Club, which he thought Bill would enjoy but Bill was not keen on becoming a member. He visited once, but did not come again. Colin believed that Bill may have been too shy to give a presentation to the club members (everyone had to give a talk about their research before being considered for membership). The fact was probably that, just like at Tonbridge, Bill was again too advanced to gain any intellectual profit from belonging to this type of club.[14]

Bill arranged his social life himself, outside the official social arrangements at Cambridge. He seems to have assiduously avoided staying in dormitories. Whenever he craved company he could contact his friends Colin and Hugh. Bill was learning to live on his own. Colin remembered how when he visited Bill in his 'digs' how cleverly Bill had arranged his pots and pans on top of one another so that he was able to produce a three-course meal on a single burner. Bill also had close connections to a family visiting Cambridge from New Zealand, he even became the baby-sitter for their little daughter Sarah. In a letter to his godfather Charles Brasch he movingly reported how the little girl used to fall asleep on his shoulder. It is interesting to imagine Bill Hamilton as baby-sitter. But of course, he had had plenty of practice taking care of younger siblings and telling them bedtime stories.[15]

Bill did not have much time for a social life anyway. He was busy with Fisher. And when he wasn't struggling with Fisher, he was busy with his other hobby: reading Russian novelists and analysing their characters (especially Tolstoy, Gorky, and Turgenev). Later he tackled Kafka. In comparison with Fisher, Kafka was a piece of cake! As Hamilton later wrote: 'Not even Kafka, whom I was reading at the time, could depress me so much as *The Genetical Theory of Natural Selection*'.[16]

At the weekend, again, Bill could simply travel home, sometimes bringing his friends with him. There are brief letters from Bill telling his mother that he is coming, say, 'Thursday', but these letters are otherwise undated. His mother, too, appears sometimes to have the same 'Thursday' habit, and even Bill's little brother Alex adopted it in his letters to Bill. Indeed, one of Bill's characteristics throughout his life is not always to date letters, or not give the year (or even to give the wrong year). Bill may have lived partly in a timeless space. One of his students later remarked that Bill sometimes had no idea what day of the week it was.[17]

This is in interesting contrast to the enormous effort that Bill put into writing letters and postcards to family and friends. His postcards are typically filled to the very edge of the page—and often spilling over into letters. The card—often an art card, whose meaning the recipient is expected to ponder—is completely full of Bill's miniature, scribbled handwriting. A small boxed square is created somewhere—anywhere—for the stamp and the address of the recipient. Meanwhile the text may well be crossing itself at least twice diagonally, and often in different colours. (This kind of cross-writing was a Victorian invention, perhaps a paper-saving device?) As a result Bill's postcards were extremely difficult to read—and the address difficult for the postman to decipher. Considering how many of these cards actually seem to have arrived to their destination, English postmen of the time must have been an especially resilient—and sporting—breed.[18]

At Cambridge, then, Bill saw it as his real job to pursue questions of evolution that interested him and to learn to formulate these in the language of Neo-Darwinism. This was a lonely task, since, as he told his sister Mary in a postcard: 'all my supervisors and lecturers so far are strongly anti-mathematical biology'. Among the ideas that began to interest him as he advanced further into *The Genetical Theory* were Fisher's discussions of senescence and the sex ratio. In addition he had a project of his own that excited him more and more: finding the genetic basis of social behaviour.

Engaged in all these other interests, Bill was clearly doing the 'wrong' thing in regard to studying for his Tripos I exams in botany and zoology at the end of the year. No wonder that before that exam Bill wrote Mary:

> I am getting a bit alarmed about the exams: I have done no revision... I know nothing...[19]

His Tripos I was not a great success, but he passed it well enough, with an upper second class degree. This seems to have somewhat disappointed him because later he kept referring to 'not getting a First'. Obviously the teachers at Cambridge were more of sticklers for answers to their questions than his Tonbridge ones. Bill realized that he would have to sharpen up and study the official curriculum a little harder during his final year in order to have a chance for a research career.[20]

But now it was summer, and time for adventure. Let's follow Bill on one of his early expeditions. In the summer of 1959 Bill, together with his brother Robert and their friend David Harris, decided to drive across Europe all the way to Constantinople. Wise from an earlier trip on bikes to Eastern Europe (during which they never reached their goal), the brothers had now borrowed their father's 23-year-old Vauxhall.[21]

A letter from Bill to his godfather Charles Brasch dated 9 October 1959 includes the following report:

> I went no farther than Istanbul... but Robert went on to Ankara for a few days, hoping to go on to a mining job in W. Kurdestan, but K. is practically closed to

> strangers owing to political troubles and he came back. Meanwhile I had taken a job in a school of Forestry just outside Istanbul, which consisted mainly in teaching English to one of the officials, but I caught malaria and since life was very inconvenient out there came back to town where David and Robert looked after me. By the time the malaria was under control I was feeling pretty deflated, and we were all besides pretty fed up with Istanbul... so we set off back.
>
> ....
>
> Just short of Thessaloniki, we had an accident, I overturned the car in a ditch and Robert broke his wrist. Then followed a very confused period, myself in jail and being tried, Robert having had his hand mended trying to find us at the village and David trying to find Robert in Thessaloniki visiting me, trying to stir up the British Council and so on.... The British Consulate was most unhelpful but we were met with extreme kindness by the Greeks, first from a doctor who had worked in England and interested himself in Robert, and from a lawyer to whom he introduced us who advised us on the car and the insurance claim (all for nothing). Secondly we walked into a private house, mistaking it for the student-hostel next door and demanded beds. The family, a set of brothers and sisters about our own age, seemed surprised but gave us beds and we stayed there a week as guests. It was not for several hours that we realized our mistake.[22]

Bill in jail? What was that about? I consulted his brother Robert for more detailed reminiscences. After the accident Robert had been taken to a hospital in the nearest town, leaving Bill and David with the car. There it was, overturned in a ditch. What to do? To their surprise, a passing bus stopped, everyone got out and spontaneously pulled the car out of the ditch. Someone also called the local police to help tow the car. Things were looking up, or so it seemed. Here is Robert's account:

> The police towed them alright but straight into custody. After a night in the local cell they were escorted into Saloniki and before they had realized it they were marched through the jail gates up the sentry patrolled corridor and the cell gates clanged to behind them. David was later freed when it was ascertained that he was not the driver.[23]

Bill remained in jail, awaiting trial. Robert continues:

> At last the trial day arrived. David and I were given a seat in the court and picked out Bill among a motley collection of other criminals in the closely guarded dock. Late in the day Bill's turn came and in turn David and I with hand on the

> Greek Orthodox bible solemnly repeated the oath after the judge and gave our statements. The final verdict was that he was guilty of careless driving and hurting me but he was let off with a caution. Our interpreter told him to thank the judge but he was fed up well beyond thanking anyone.[24]

No wonder Bill was fed up. He felt unfairly accused, Robert tells us. The whole 'reckless driving' incident had been an accident. A bee had flown in through the car window, making Bill lose control. The car veered to the right, off the road, and when Bill tried to get it back on the road, he steered too sharply, causing the car to skid on two wheels, fall down a six-foot gully on the left side and land upside down. (It was amazing that there were no more serious injuries than Robert's broken wrist.) Did Bill lose control as he was trying to flick the bee off his trousers, as one story had it, or was it rather so that he got too interested in this insect?! I have different suggestions from different sisters. Robert later confirmed the trouser story.[25]

In any case, now that Bill was out of jail, the car had to be fetched from the village and brought to Thessaloniki for repair. It was in terrible shape but at least it still moved:

> Bill let in the clutch and the astonished crowd of policemen and villagers gave way as our car, with doors tied back with string, crooked radiator adorned with headlamps dangling from the end of their cords, crushed mudguards and wheels that were bent so that they were only just able to rotate, got underway. Soon the wind began to whistle through the shattered windscreen and splintered roof onto which we had tied the bonnet top. We set course for Saloniki two hundred and fifty miles away.[26]

With a provisionally repaired but progressively disintegrating car the trio headed for home, coping with challenges as they arose. Incredibly, Bill finally got the old Vauxhall onto the boat for Dover. The high point came when well back in England the car was stopped by the police who wanted to know why the car had no licence plate. 'We lost our windscreen in an accident in Greece' was their answer, and this was absolutely true and could even be proven. 'Oh, excuse us, gentlemen', said the police, 'We thought you just picked up this car from a junk yard'.[27]

In many ways, this trip in 1959 is a metaphor for Bill's scientific career. Bill was keen on meeting challenges, and did not easily give up. Ingenious solutions were found to problems which arose during the trip. He was misunderstood by the authorities, treated unfairly and suffered, but was finally vindicated and able to prove his point. Good and helpful people regularly made their appearance. Bill, meanwhile, showed helpfulness to others, in this case giving lifts to 'stranded' people (this happened quite a number of times; juggling the space given the car's already existing two extra passengers, who were being given a lift to Turkey). Just like his brother, Bill had a general 'fixer' attitude and enjoyed finding solutions to unexpected problems. It may also have been true, as suspected by later observers, that Bill and Robert enjoyed getting themselves into trouble for the sheer pleasure of finding clever ways out.

It was shortly after his return from this trip that Bill wrote his condensed travel report to his godfather. The letter also contained more serious matters. This was now his third and last year, and he had chosen Genetics for his Tripos Part II:

> I wanted to get on with my work as soon as I got home, but haven't done much yet. I didn't do well at all in Pt. I of the Tripos, but did not expect to, since I have spent much more time on Genetics and connected interests than the syllabus. I am hoping that this will pay off with a 1st in Pt. II; I hope so since I won't be able to do research if not. This is all something of a gamble, but I remain convinced, even obstinately, that I am capable of doing good research. Some of my supervisors, and even the Genetics Department seem to have some confidence in me but I am afraid that I may find that they have been disillusioned by the results of my last exam. The Genetics Dept., where I will be working next term, have [sic] just acquired a new professor which provides another unknown factor. I am fearful that Professor Fisher towards whom I have aimed the whole of my university career may disappear before I have time to learn anything from him. He has been retired for several years but up to now he has gone on working at the Dept. He is a proud and undiplomatic man and I am afraid that he and Professor Thoday may fight like queen bees—the personal battle which the new queen always wins. I think he is still in Australia at the moment: he makes himself conspicuous now and then by dogmatic public assertions about Eugenics, Lung Cancer, decline of intelligence and so on, much of which you may possibly have heard. He seldom bothers to explain how he arrives at his conclusions, but they are never without foundation. I reckon him the brilliant man in Cambridge at present; genetics is really only a side [of] him, properly he is a mathematician and modern statistician and modern statistics probably owes more to him than to anyone else.[28]

Fisher was indeed to 'disappear' relatively shortly thereafter to Australia (he was probably staying on there already during the visit Bill mentioned). After officially retiring in the autumn of 1957, Fisher had remained in Cambridge for almost two years, finally leaving the very academic year when Bill formally started studying Genetics II ending with Tripos II in June 1960. As part of his studies during his last year, Hamilton was now expected to work as an apprentice to a research student. He was assigned to Fisher's student Anthony Edwards, who was writing his PhD dissertation on the sex ratio in mice. Fisher's famous theory of the sex ratio predicted that it always eventually readjusted to become 50:50 male/female.[29]

This is Edwards' account:

> Here is where my involvement with Hamilton starts. All the questions raised by *The Genetical Theory*, not only the question of the sex-ratio, were going round in my head when Hamilton was allocated to me, and I had decided to use my experience of running a mouse experiment to try to modify the murine sex-ratio. The plan, which I describe in my Ph.D. thesis p. 82, was to kill off half the males at birth and wait for natural selection to 'bring back the sex ratio at the end of the period of parental expenditure to its original value'. (Some hope!)
>
> I explained Fisher's theory to Hamilton, and my plan to test it, and together we started the experiment. I taught him how to sex new-born mice... and my single image of those days is the two of us in a dark room attempting the task...
>
> Hamilton was a very thoughtful companion, and after a week or two of his involvement with my experiment he came to me and said that he thought it could not work anyway. He pointed out that sex-differential infant mortality is irrelevant...
>
> It would be interesting to know exactly when Hamilton first saw a copy of *The Genetical Theory*.... Whenever it was, we must have had many discussions of the theory of sex-ratio selection and of the role of parental expenditure and perhaps of other topics as well in which he was later to make his name. I have just one more memory which reinforces this. We must have discussed altruism and its evolutionary consequences (and the 'we' may have included others beside Hamilton and me, for departmental tea still continued the Fisherian tradition of intellectual debate) because I have a very distinct memory of realizing that it would be selectively advantageous to be able to recognize one's close relatives. In man, the similarity of facial appearance and expression might be important. So seized was I with the importance of this idea at the time that I can remember

> the precise point where I thought of it, on the walk between my college and the Department.[30]

By then Bill had been reading *The Genetical Theory* for about a year. He had come to better understand Fisher's ingenious sex ratio reasoning but also wondered about its limitations. At this point in time, however, he was more involved in the genetics of altruism. He had found some hints about this in Haldane and Fisher, but these did not help him along very much. The arguments were brief and verbal. Haldane in his famous 1932 book *The Causes of Evolution* had indeed included an Appendix in which he tried to derive altruism mathematically from group selectionist assumptions but, as Hamilton was to declare in his first published paper in 1963, Haldane's model did not work under its own stated conditions.[31]

It was Fisher who provided Bill with a more stimulating idea, reminding him about the brightly coloured bad tasting larvae that predators learnt to avoid. Fisher's important formulation (in his *Genetical Theory of Natural Selection*)

> The selective potency of the avoidance of brothers will of course only be half as great as if the individual itself was protected; against this is to be set the fact that it applies to the whole of a possibly numerous brood[32]

was to stay with Bill as his clue to the evolution of altruistic behaviour. Surprisingly, it seemed that Fisher did not want to expand his argument further. Bill on the other hand wanted to develop as general a model as possible.

But to do this, Bill felt that he needed more information. What was known in general about altruistic behaviour? He considered his own family, consulted literature he had read. He needed material to work with. One place that occurred to him was the Anthropology Department. Perhaps he could get help there? Surely they must have collected a lot of data on the social behaviour of various human groups. Perhaps he should try to take Anthropology as one of his subjects? Was that possible? (The genetics degree required the student to select some subsidiary subject and pass an exam arranged by the respective department.)[33]

Bill's reception by Anthropology was a disappointment. He was told by both his own department and the Department of Anthropology that such a combination was simply not possible. Of course this was the time of the Two Cultures (CP Snow's book with that name came out in 1959) and Bill may have been caught unwittingly in the middle of academic politics. It was clear, however, from what was to happen later in regard to sociobiology, that anthropologists were simply not keen on their field being 'taken over' by natural scientists. One can understand how they would have been careful to avoid setting dangerous precedents by granting student requests. Their interest may have been exactly to keep the Two Cultures apart. Bill, being stubbornly oblivious to such seemingly minor 'political' issues, looking instead for the big rational picture of how the academic community *should* work, was very angry with the decision:

> I think I will let things slide and attend what lectures I please... All I really want is a sensible critic and someone to guide me in the literature. I am beginning to find Cambridge intolerably oppressive: I don't care about the coming exams; I don't want to do research here. I think I will give up the hope of making headway against all this and take up school teaching and do my research on my own—after all, it involves hardly anything but reading. Unfortunately, the G. Dept. still thinks quite highly of me and I am on very good terms with the Professor. I like them more and more... but this makes it only the more difficult and I am gradually finding myself in a false position where I hardly ever dare mention the real nature of my objectives. Their quaint and homely organization can scarcely reconcile itself in my mind with my theories of ethics, and since I can't hate them, I must end by hating myself, it seems. Nevertheless, these theories are progressing faster than the genetics, I fear.[34]

This quote raises some puzzling questions. Why is it a problem that he is liked by the Genetics Department? Does he know the 'real nature of his objectives' at this point? And how dangerous is that, if by his 'theories of ethics' he means the genetics of altruism? At this point Bill appears to be doing a number of additional things (such as thinking about the sex ratio), following in Fisher's footsteps.

Bill seems to offer at the same time quite a number of different reasons for leaving Cambridge: they like him too much, he likes them too much, he cannot tell them his ideas, and finally: he cannot study anthropology, which angers him so much that he 'decided to immediately leave Cambridge'. Is he casting himself as a hypothetical rebel, potentially misunderstood? Or does he somehow intuitively feel that people are getting too interested in him? He is here on Fisherian territory. For Bill to get proper credit for any Fisher-inspired work he is planning he would *have* to get out in order to distance himself. He cannot for his own sake become a research student at Cambridge, although they would like to have him. He has to go some place where people do *not* do what he does.

When we follow Hamilton further in his career it will become apparent that he typically delights in the fact that nobody in his immediate workplace is doing what he is doing—be it altruism when people do mainstream genetics, or evolution when people do insect ecology, or modelling when nobody else does that. He likes working on his own puzzles with his own methods and at his own pace.[35]

In his autobiographical writings Hamilton gives the impression that he met Fisher very little ('only twice'). My impression is that he attended at least some departmental teas. At least one fellow student remembers Bill conversing with Fisher 'in a way that went over his [the fellow student's] head'. It is important for Bill that the world does not believe that Fisher actually told him anything important—which, judging from the quadrat story, may well be the case.[36]

5

The Struggle for Altruism

Bill had two strategies after Cambridge. One was to find a place for graduate studies in a genetics department. His backup plan was to study for a diploma of education in order to become a secondary school teacher. He thought that would be a suitable profession—there had been a number of important scientists who had worked as school teachers, including Ronald Fisher and David Lack. He also believed that studies in education would be easier than genetics and give him free time to continue developing his thinking about altruism.[1]

The puzzle that Bill had already been trying to solve at Cambridge concerned altruistic behaviour. Bill was sure that there existed a genetic basis for altruism, however counterintuitive this behaviour seemed in regard to individual fitness. Why would an animal ever behave so as to reduce its biological fitness by putting itself in danger (say, letting out an alarm call), or by forgoing reproduction (like the workers of many social insects)? Why does a bee sting invaders of the hive and lose its own life? These kinds of behaviours certainly did not give the individual animal any evolutionary advantage.[2]

This was a real challenge. But this counterintuitive altruistic trait could actually spread under certain conditions, Hamilton reasoned. What was required was simply that the benefits of an altruistic act did not fall on random members of a population but on individuals who were genetically related to the donor. Bill was sure that one could produce a general mathematical model of the workings of this universal principle. The bird giving out the alarm call would be sacrificing itself for its own relatives. This also meant that the concept of fitness would need to be completely re-thought. It was not the fitness of the individual organism that counted, but the fitness of the whole group of relatives who also carried the individual's genes.

Hamilton still smarted somewhat at having received an upper second class degree, not a first, in his natural sciences degree from Cambridge, but that should hardly matter when it came to secondary school teaching. Consequently, he was very surprised not to be accepted on his first choice of education programme, Moray House in Edinburgh. Bill was told that his genetics degree would qualify him only for teaching at the elementary school level, not the secondary one. There must be some mistake. Perhaps Moray House thought that he only knew genetics? Bill immediately set out to explain that the genetics part of his BA degree only occupied the last year of his studies and that he actually was well trained in fields such as botany, zoology, and physiology. But despite Bill's protest—in a number of letters—the decision stood. According to Hamilton, Moray House simply 'stonewalled' him. Worse, he went on to be rejected by London's Institute of Education, his second choice.[3]

So much for his backup plan. How did Hamilton's genetics prospects look at this time? An obvious choice was the Genetics Department at University College, London. However, Bill had already had an unfortunate experience there. Dr Lionel Penrose, whom he had approached, had been quite disapproving of the genetics of altruism as a problem for research. Penrose had said that he doubted that there was such a problem to be studied in the first place. Instead he invited Bill to come and work on one of the standard genetics topics that his lab was pursuing. Of course, that was not at all what Bill had in mind. But no genetics department seemed to want to have him and his project.[4]

All these rejections! Here was a young man with an interesting, potentially revolutionary idea, and people were putting all kinds of obstacles in his way. First Edmund Leach in Cambridge; then Penrose, the leading geneticist at the Galton Laboratory, as well as other genetics departments; then Moray House and the London Institute of Education. Hamilton felt dismissed by everybody. And all the while, all he wanted was to explore a scientific puzzle that so intrigued him: the genetics of altruism.

How, then, did Bill ever become a graduate student? The answer is, with a little help from his friends. Bill's Cambridge friend Colin Hudson's father, a professor of horticulture at Leicester, who knew about his research interests, was especially instrumental in suggesting that he approach Professor Glass at the London School of Economics (LSE). The person who interviewed and took a friendly interest in Bill at LSE was Norman Carrier, a human demographer trained in mathematics, generally interested in puzzles, and a good listener.

This is how Bill described Norman Carrier's impact to his uncle Charles just before Christmas 1960:

> Although I have been interested in demography for a long time and it is quite relevant to what I have been studying so far and to what I want to study, I must admit that I should not be at L.S.E. if several other more genetical sorts of establishments had not refused to have me except on terms of their own ideas of research. I was really very depressed by the time I got round to trying the L.S.E. and I was so dubious about it I doubt whether I would have persisted with my enquiries if I hadn't chanced on a man of such extraordinary enthusiasm and kindliness as I did. I don't think that even now he really understands what my research projects are about since he is one of these people who interprets all one's explanations in terms of what he has just been thinking about himself, and he has handed me over to another lecturer who knows more about genetics, a much colder, more level-headed person who knows fairly well what I am after and criticizes my approach to it unmercifully, although I am glad to say that he has become more favourably inclined recently.[5]

Bill describes John Hajnal, also a demographer and competent in genetics, as someone who

> was not particularly sanguine that my ideas would find application to anything in nature (could genes really affect altruism, he wondered, was there evidence?), let alone could apply to humans, but he liked the sheer puzzle of how a self-destructive trait *might*, in a pure world of genetics, evolve—perhaps on Mars or something like that—under the influence of natural selection.[6]

It was decided that Bill should take a course in demography to somehow 'anchor' his apparently interdisciplinary research topic and be jointly supervised by these two professors. Carrier was even able to find funding for his student—he alerted Bill to the existence of the Leverhulme scholarship. Bill was baffled about the lack of procedure:

> Anyway, greatly encouraged by Mr Carrier, I attended an interview with some board of the L.S.E., argued with them briefly about my proposals, and about a week later I learned that I had been awarded this scholarship. This was very astonishing since what I had proposed was quite [out] of the run of things normally covered by the L.S.E., & I hadn't submitted a written summary of it beforehand as the scholarship is supposed to require; in fact so far as I know they only

> had my word for it that I had a Cambridge degree at all! Anyway, the scholarship doesn't commit me to any particular higher degree & provides me with a grant just enough to make me independent of the Parents so that so far as I am concerned it is ideal.[7]

This scholarship was subsequently renewed, and later Bill was also granted a Medical Research Council fellowship. But every year it was touch and go as to where he would find funding for the next year, and this contributed to a very unsettled feeling during his three-year intense exploration of the topic of the genetics of altruism. (He probably could have been supported by his father if he had asked, but this was something that he did not even want to consider.)[8]

At the end of 1960, because of the genetic nature of Hamilton's project, it was arranged that he would be part-registered at University College and have a supervisor from the Galton Laboratory, the geneticist and statistical specialist Cedric Smith. Here he was now in the Galton Lab, which had been so unaccommodating a year or so earlier when he made inquiries.[9]

Life at the LSE was a new experience. As Bill told his uncle Charles at the time:

> It is difficult to imagine anything more different from Cambridge than the L.S.E.: it has, I have been told, about 4000 students, which is 1/2 the student population of Cambridge, & the place swarms from early morning till late at night with the most fantastic mixture of ages, races, nationalities and social classes it is possible to imagine, & in spite of the numbers everyone seems to be on speaking, or shouting, terms with everyone else, or at least with a clique of, say, like nationality. As usual I am under the impression that I am the only person who doesn't know anybody, who is politically indifferent, who isn't engaged in some research or study of an approved and pragmatic kind. But this impression follows me everywhere & no doubt follows everyone of like disposition including dozens of inconspicuous figures at L.S.E. Altogether there is a pleasant absence of the polite and quiet exclusiveness of all the social circles in Cambridge. It will be interesting to see how U.C. compares.[10]

So now Hamilton had a place for graduate study, but he was full of doubts about his own ability. He wondered about the best way for him to continue his research. Feeling pressured to produce was not his style. He oscillated between self-confidence and despair. He suffered from depression:

> Actually I am still very dubious myself, not so much about the importance of the ideas I am investigating, since I am convinced that they are very important, as

> about my own ability to investigate them, so I am not particularly sanguine about my prospects at the end of this year. I may very likely have to revert to my idea of school-teaching which I must say still attracts me considerably. I should then feel assured that I was doing something useful, & it wouldn't matter so much if nothing was to come of these other ideas, & under such circumstances I feel that I would be able to work at them more steadily & methodically (in my spare time) than I do now when I feel all the time I must hurry to produce something concrete & worthwhile to justify the confidence that has been placed in me. As it is my life is rather hectic & I seem to alternate between great joy & self-confidence and acute depression. Nevertheless I like London much better than Cambridge perhaps simply because one can just get depressed much more anonymously.[11]

He felt the urgent need to be creative, to produce. But he was looking for the right kind of work environment. It was hard to work at home, libraries were better. What about the university? Was it the right place for work or not? Would it constrain his freedom? He shared his concerns with Uncle Charles:

> My digs are very unsatisfactory for working in so I spend a lot of time wandering from library to library, but now my new supervisor at U.C. is trying to find me a desk to work at in the Galton Laboratory, so I may soon be absorbed in a more family sort of academic atmosphere once again: I don't know whether I am happy or apprehensive about this.[12]

Much later, when writing his autobiographical notes, Hamilton conjures up a sad picture of his life during these years:

> Although supervised at the Galton laboratory for 2 years I never had a desk there nor was ever invited to give any presentation to explain my work or my occasional presence to others. I think virtually no one I passed in the corridors or sat with in the library knew my name or what I was doing. Probably this was largely my own fault. I was a student who, if not offered or invited, wouldn't request, and I think I didn't ask for desk space either there or at LSE. In fact, I had no idea at the time what was normal for graduate students... I just wanted access to libraries plus some pittance of support; these together would give me freedom to follow my puzzles.[13]

Well, there does appear to have been at least an attempt to get Bill a desk. Cedric Smith seems to have been trying to help where he could. Probably he did not quite understand what Hamilton was after (few did), but sensed that he needed assistance. Another one of his initiatives was to introduce his new advisee to John Maynard Smith, his colleague in the Zoology Department at University College. That meeting yielded zero result. While Hamilton years later vividly recalled the meeting, Maynard Smith had no recollection of it. He guessed that Hamilton had not succeeded in clearly explaining his project. In any case, Hamilton's topic was quite outside his own area of interest at the time.[14]

The fact remains that Hamilton did not spend much time at University College. According to Sheila Maynard Smith, Maynard Smith's wife and herself a geneticist, he was never around. Indeed, Bill was mostly working in his rented London bedsit, feeling terribly lonely. This is abundantly clear from his autobiographical notes. Indeed, for anyone having struggled for days with a difficult problem, it is easy to sympathize with young Bill Hamilton. Bill describes the scary and alienating feeling of being all alone with a particular idea, without being able to share it with another human being:

> At times I was sure I saw something that others had not seen... At others I felt equally certain that I must be a crank. How could it be that respected academics around me, and many manifestly clever contemporary graduate students that I talked to, would not see the interest of studying altruism along my lines unless it were true that my enterprise were bogus in some way that was obvious to all of them but not to me? I could think of many bogus ideas that I had followed enthusiastically for quite long periods and then later had 'seen through'.[15]

Meanwhile Hamilton was trying to teach himself mathematical population genetics. He found himself constantly making mistakes when he tried to repeat what had been already been done in published papers. All this was very frustrating:

> Most of the time I was extremely lonely. Sometimes I came to dislike my bed-sitting room so much that, when even late libraries such as Senate House or Holborn Public closed and I was still in a mood to continue work, rather than return to my room I would go to Waterloo station, where I continued reading or trying to write out a model sitting on the benches among waiting passengers in the main hall... Although I virtually never spoke to anyone in such places or in the libraries I frequented, or in my departments, having people around me... seemed to soften loneliness a little.[16]

It seems that Hamilton had rather specific requirements for his optimal working environment. Being around people didn't necessarily mean that he wanted to interact with them. He needed to see people, but they had to leave him alone. In this way he got a verisimilitude of company and a welcome boost to his creativity. The hustle and bustle of Waterloo station with its constant movement may have been especially good for this purpose. Another of Bill Hamilton's favourite working environments was the top of buses. Riding around with regularly changing scenery and without any particular goal may have helped to stimulate his thought process, or distract him just enough—whatever he needed at that particular moment. Reading in buses was also something that Bill had got used to as a schoolboy. Perhaps the re-creation of this kind of familiar setting had a comforting effect. Sometimes he tried park benches, but there he usually found too much distraction for effective work, or his papers blew away.[17]

Back in his bedsit we can imagine Bill Hamilton crouched on what he was later to christen his 'kin selection chair', a brown patent leather car seat. He was becoming increasingly frustrated. The mathematics was not going well. What exactly was he trying to do?[18]

He was struggling with the paper that would eventually become his famous 1964 contribution 'The Genetical Evolution of Social Behaviour'. He was looking to formulate more generally how one could expect the genetic relatedness of individuals to affect their behaviour toward one another.[19] But this would be the ultimate goal. He started off by looking at different types of pairs of relatives, building trial models from which he later hoped to be able to generalize to a universal principle. (The genetic relationship between pairs of relatives can be calculated as the probability that they have inherited a copy of the same gene. For instance, on average half of the genes for full siblings are identical by common descent, the same goes for parent and child; the proportion is an eighth for first cousins, a thirty-second for second cousins, and so on, making the respective 'coefficient of relationship' 1/2, 1/8, 1/32, and so on.)

Hamilton's published two-part 1964 paper on social behaviour is in fact an expanded version of the last part of a huge manuscript. That part (number 8, in fact) was called 'Selfish and Altruistic Characters Affecting Relatives'. There, just as in his later published paper, in a two-column table Bill distinguishes between four types of social behaviours, classified as to whether an actor A is inflicting harm or benefit on another actor B. Behaviour causing benefit to self and harm to other is called selfish, behaviour causing harm to self and benefit to other is called altruistic, behaviour causing benefit both to self and other is called cooperation. Behavior causing harm to both self and other Hamilton originally called 'stupid' in his manuscript, but later identified with 'spite'.[20]

Note that social behaviour here is classified strictly in terms of the *effect* that one individual has on another individual's fitness. This is a point that is important to consider especially when it comes to the category of altruistic behaviour. Altruism in Bill Hamilton's model world is purely behavioural. It involves the consequences of an action, not its motivation. In regard to humans, this becomes especially difficult to contemplate, and Bill's theory (and later 'sociobiology' in general) was criticized for using common language in an easy-to-misunderstand, purely technical way. Of course, Bill was not particularly thinking about humans—he may typically have had plants in mind—because he wanted his theory of inclusive fitness to be as general as possible.[21]

Inclusive fitness was the central concept of Hamilton's 1964 paper. Reading his own description of his struggle with his calculations it is not easy to appreciate just what a feat his derivation of inclusive fitness was. He complains about the tedium of calculating coefficients of relatedness between pairs of relatives (calculated as the probability that they had inherited a copy of the same gene) and about the many errors he was making. But his work with trial models of pairs of relatives was ultimately aiming for a general model that would present the whole situation of social behaviour all at once, with each individual acting both as a potential donor and potential recipient. In other words, Hamilton was engaged in a difficult process of mathematical model building.[22]

What was it he wanted to achieve? As Hamilton formulated it himself, he wanted to produce something similar to Fisher's Fundamental Theorem of Natural Selection. Fisher's famous theorem dealt strictly with individuals and was based on a maximization principle: individuals were seen as acting to maximize their own fitness (reproductive success). The theorem also said that the average fitness in a population will always increase.[23]

Deliberately following in Fisher's footsteps, but taking into account the fact that animals often live around relatives, who have the ability to affect one another's fitness, Hamilton wanted to find 'a quantity,..., which...tends to maximize in much the same way that fitness tends to maximize in the simpler classical model'. (Here he meant Fisher's model.) That quantity he called 'inclusive fitness' as a contrast to Fisher's focus on traditional individual fitness. Inclusive fitness took into account two types of effects: an individual's effects on others, and the effects of others on the individual. Also, both types of effects could be either beneficial or harmful.[24]

In early November 1962 in an upbeat letter to his parents (who were in New Zealand at the time), Hamilton described his achievement. By now he was convinced that he had achieved something important:

> 14 Hadley Gardens, W 4 5/11/1962
> Dear Mother and Dad,
> Thank you for your letters. I am afraid I am rather long in sending you one, but I am pretty busy at the moment writing a paper. I hope that this one paper will get further than the last which was considered unsuitable by my supervisors—I hope it will get as far as an editor anyway. It is an expanded & generalized rewrite of the last part of the previous essay (of which you had a copy), much more elegant mathematically I think. I am pretty sure it will get published. Dr Smith is advising me to try the *Journal of Theoretical Biology*, a very new journal & I don't know much what sort of theory it does publish, but I do know that the first paper in the first number a year or so ago was a contribution by Fisher so it is good enough for me. It seems to me that my principle of social selection is very original & has implications which should be interesting to biologists concerned with evolution. In fact to tell the truth I think it is the most fundamental contribution to evolutionary population genetics since the classical theory was founded by Fisher, Haldane & Sewall Wright in the 1920's—but that is only my opinion! If it is accepted for publication I shall go on to write up my recent work in about 2 or 3 more papers, hoping that the incredibly frustrating and tedious business of writing these short and simple bits of English will get easier as I go on. Two or three good papers published should make up for the Ph D which it now seems I will not get, not having, I am told, enough of the right sort of 'results' to make a thesis, and having left the writing up too late anyway. An M Sc is much easier to get, a much rougher thesis being required & judged mainly by your own college, but whether I shall trouble to take this I don't know—probably not if I can get my work published. What I will do after this year I don't know yet. Mr Hajnal suggests I should go to America where there is a much wider field in university teaching, & he offers to me his influence with colleagues—but presumably that would get me a post under someone primarily interested in human problems, which I am rather anxious to avoid after my disappointing experience of the Galton Laboratory... What will turn up I don't know & I am too busy to give the matter much thought at the moment.[25]

We see that his advisor Cedric Smith once again tried to help, alerting Bill to the new *Journal of Theoretical Biology*. It is not clear what Hamilton means with his statement that 'two or three good papers published should make up for the PhD'. Make up in regard to what? Academic reputation? Job opportunities? Is this what he had been told by his advisers? And why is it that he believes he will not get his PhD? It seems to have to do with more than not having 'the right sort of "results" to make a thesis', because he mentions having 'left the writing up too late anyway'. Too late for getting the doctorate that year? But surely there would be the next year, and other sources of funding, at least in principle. And does the fact that his advisers were suggesting places to publish, or going to America, mean that they were at the same time advising him 'away' from a PhD?

In any case, Bill followed the advice and was planning to submit his long manuscript to the *Journal of Theoretical Biology*. But before that, he decided to write a brief version and send it somewhere for rapid publication. That paper was to highlight his main new contributions, especially 'inclusive fitness', 'Hamilton's Rule' (as it was later to be called), and 'the gene's eye view'. *Nature* was the obvious choice. It was the most prestigious and most widely read journal for short communications.[26]

The editor's response came back immediately. The topic was too specific for *Nature*, Hamilton was told. Instead he was encouraged to try 'a psychological or socialogical [sic] journal'. Bill took this to mean that 'there existed a prejudice against my topic'. His next attempt at publication was *The American Naturalist*, where it was accepted at the end of 1962 and published in 1963 as 'The Evolution of Altruistic Behaviour'. At least he now had something in print.[27]

It still bothered Bill that he had not succeeded in getting his short elegant piece into *Nature*. It would have been such a perfect forum for his great new insight. 'Hamilton's Rule' calculates under what conditions one individual is likely to behave altruistically towards another. (This happens if the benefit to the recipient is greater than the cost to the donor, taking into account their genetic relatedness.)[28] Later he reflected on the fact that he may have provided an unfortunate address for his article: The London School of Economics, Department of Sociology. Why did he choose that address anyway? One reason for the choice, according to Hamilton, was that he felt grateful to that department for supporting him academically.[29] At the same time, he may not have wanted to associate himself with University College or the Galton Lab—the paper was the product of his long, *personal* struggle, and the credit, if there was credit to be had, should be his alone.

But the *Nature* story is more complicated. In fact, it was none other than Hamilton's main adviser Professor Cedric Smith who had suggested to Bill that he submit his letter to *Nature*. In fact, it was part of a publication programme that Smith developed for Bill after seeing his latest draft and finding it unsuitable for publication as one single paper. As Bill told his parents:

'[H]e thinks it is again too long and diffuse for a paper, he wants me to make it into: 1. a letter to Nature, 2. A longer paper to J.Theor.Biol. 3. A paper for Ann. Hum. Gen.' Still, Professor Smith did not deviate from his original position that his advisee's lengthy paper was not appropriate for a PhD. (Meanwhile, Bill's proposed short *Nature* letter was based on work that he had already done during his first year at LSE, pointing out the shortcomings of Haldane's group selectionist model of altruistic behavior.)[30]

A question arises: if Cedric Smith recommended Bill to publish in *Nature*, he must have had a sense that such a publication would be possible. Advisor and advisee alike may have been further inspired by attending a luncheon talk given at University College in January 1963, a certain Dr Blest's presentation on 'Moths and Predators' on life expectancy in cryptic and warning coloured moths. Dr Blest's research in Panama showed that cryptically coloured moths died soon after reproduction, while warning coloured moths lived much longer. This empirical observation, he argued, could be seen as support for the evolution of altruism, because the warning-coloured moths needed to be around long enough for the predator to learn its bad taste and in this way spare its relatives.[31]

Bill was captivated by this talk. So far his best example of altruistic behaviour had been alarm calls in birds. Here was now one more empirical case that he would be able to quote. He went to talk to Blest after his lecture and learnt that he was just about to submit a paper to *Nature*. That same evening Bill finished his own paper, hoping that Blest and he would be published in the same issue of *Nature*.[32]

Did others, too, learn about Bill's planned submission to *Nature*? Because a few days later there is the following piece of news from Bill to Colin:

> Today has been a particularly exciting day as I received two requests for my not-yet-published paper to 'Nature' which I hurriedly finished & typed this evening. It won't be anything new to you since it merely presents the most elementary aspects of my approach to altruism which I worked out in my first term at LSE... [Here he describes Dr Blest and his lecture]... He was very pleased when I explained my work based on Sewall Wright's Coeff. of Relationship and asked for offprints when I published anything about it. I was very pleased because this

may turn out one of the first sound examples of my principle: till now I have only been able to think of the warning calls of birds as an almost irrefutable example. He is the first person I have heard talking about this sort of thing who has obviously had my own doubts about the facile explanation in terms of benefit to the group or species... We looked at each other pretty narrowly I assure you.[33]

We know Hamilton's own reconstruction in regard to what happened next: *Nature's* editor mistook his Letter called 'The Evolution of Altruistic Behaviour' for a sociological contribution and dismissed it. When Cedric Smith learnt about this from Bill, he offered to contact *Nature* pointing out that the letter dealt with genetics, not sociology but Bill declined. In a way this incident was encouraging:

On the whole this recurrent adverse reaction to my ideas encourages me to think they really are important, for there is no doubt that they are correct and that everything that appears in lieu of them in the literature on evolution is more or less nonsense. It all bodes ill for my main paper on the same subject which I am still trying to finish off and which causes me much difficulty; I think one writes worse and not better for feeling all the time that this or that sentence is likely to be considered offensive or not comme il faut.[34]

Dr Blest's paper faced no adverse reaction and was duly published in *Nature* within a month.[35]

On 13 May 1963 Bill posted his large manuscript to the *Journal of Theoretical Biology*. By now he was restored from his earlier despair. Many things were going well. Obviously at least some people believed that he had something important to say—his short paper had been published in the *American Naturalist*. The other thing that boosted Bill Hamilton's self-confidence at this time was a discovery that he had just made in the journal *Evolution*. That was two papers by George Williams (one written together with his wife, Doris) from the year 1957. He had found a scientist who seemed to be interested in similar problems as he was, and to reason in a similar way. Throughout his life Hamilton would have a strong feeling of intellectual kinship with Williams—indeed, 'twin brother' was a term that he would use in relation to him. Williams was some 10 years older, but the age difference didn't appear to matter.[36]

What surprised Bill was that Williams had already addressed two of the three topics that Bill had himself been working on up to this point (the evolution of social behaviour, or of altruism, was not the only thing that Bill had been focusing on). In addition Bill had been doing preliminary work on the sex ratio and the problem of senescence, two Fisherian themes that he saw as next in line once his work on altruism was completed. The very day after he had sent off his manuscript, Bill wrote a letter to Williams at the State University of New York, Stony Brook. Bill's letter to George Williams read as follows:

Dear Professor Williams:
If you can spare them I should like to receive reprints of the two papers which you published in 'Evolution 11'. (1957).

By a curious coincidence the subjects of my post-graduate research at London University during the past 2 1/2 years have been almost exactly the subjects covered in these papers of yours. And I was, I am ashamed to say, quite unaware of your work until I found it referred to in A. D. Blest's article in 'Nature' recently (197, 1183).

At the moment I am particularly interested in your theory of senescence and would like to be referred to anything further you may have published on the subject (and to receive reprints if possible). The theory I had evolved is essentially the same as yours; I had not worked out its points of verification on such an impressive scale as you have, but I think I can add a little in having framed my argument in terms of demographic mathematics and also in allowing for a wider range of types of genetic effects on mortality.

The discovery of your work has been a considerable encouragement to me. My own has been carried out in the face of widespread scepticism and even contempt, making me doubt at times whether my approach to these problems could really be as fundamental and as correct as it seemed to me. As regards my progress on the problem of the genetical selection of altruism, a 'letter' giving the central idea I have worked out will be published in the 'American Naturalist' very shortly. A fuller paper on the same subject has just been sent off. In it my approach differs considerably from yours, and arrives at what are I think more general and useful principles (I attempt to treat relationships in general and not just the full-sib situation); also it seems to me that you must have overlooked, as I did for some time, the fact that an analysis of the full-sib situation cannot be applied directly to the social Hymenoptera on account of their peculiar system of sex-determination. The relationship between sisters in a Hymenopteran colony is even closer than that of normal full-sibs (provided the mother-queen mated only once, which unfortunately for the theory is not always the case) and therefore potentially more favourable to the evolution of altruism. Nevertheless it is evident that we are quite united in considering that some proper argument is needed to account for the evolution of self-sacrificing behaviour in animals—something more rigorous at any rate than the usual vague appeals to the 'benefits to the species' and what not.

Yours sincerely
W. D. Hamilton

P.S. I should like to send you a reprint of this when and if I get it published. W.D.H.[37]

What Hamilton refers to here is George and Doris Williams' 1957 paper on how sibling relationships and mate choice in birds are conducive to the spread of altruism. We learn from this letter that his work on inclusive fitness was developed independently of Williams and Williams' paper. (However, Bill duly cites Williams and Williams' paper in his 1964 manuscript.) While they had the same interest in demonstrating that the trait of altruism can in fact spread under natural conditions, Hamilton was looking for a universal principle, while Williams and Williams concentrated on a particular case, from which they then generalized. Later, this difference in approach between Hamilton and Williams—'universal principle with examples' versus 'generalizing from case'—was to repeat itself in regard to other topics, too, especially the evolution of sex.

Much later, George Williams was to write about Hamilton's 1964 paper:

> It was the most important for me, not only conceptually but also personally, because my wife and I had published a clumsy treatment of a related topic, natural selection among nuclear families. It was a relief to have our ideas replaced by Bill's simple proposal of selection among individuals for the adaptive use of cues indicative of kinship with any conspecific.[38]

The above case shows a general feature of Bill Hamilton's scientific career. He often found himself working in parallel with someone else without being aware of the other person's work. This would happen again and again. But Hamilton typically worked independently, and the problems that interested him were usually not the current vogue. He himself was aware of his 'out of date' interests, which he attributed to the particular literature that was available to him in secondhand bookshops. But he defended his taste: 'Nevertheless, because topics in science often fade for no good reasons—certainly not because of error—it may sometimes help inspiration to be reminded of forgotten facts or to see new ones from an old point of view.[39]

Here, then, Bill Hamilton had finally worked out the answer to the puzzle of altruism, partly 'despite' his advisors. No doubt they believed their criticism was useful (and it probably was), but for Bill, it often felt as they were totally against him and his project. But why was there so much resistance to young Bill Hamilton? What he did not know was just how undesirable his topic was at this particular time and place that he happened to be. Bill was playing with a bigger fire than he realized.

* * *

Why was it so hard for Bill Hamilton to gain acceptance to the various places that he approached for graduate study? Let's put this whole thing in context and see if a pattern emerges.

Why was he rebuffed by academics first in anthropology, then education, and finally genetics? At least in the first and the last case, it had something to do with his explicit wish to pursue a programme of researching the genetics of altruism. But shouldn't a student in general be allowed to pursue a topic that he or she is passionate about, and shouldn't exactly this kind of independent-minded student be encouraged by the academic system? What was wrong with studying altruism scientifically? If there was nothing to the genetics of altruism (as some seemed to believe), wouldn't this become clear soon enough? The matter would be settled and the student would move on to something more sensible.

It was not that simple. As it turns out, even this kind of educational 'experiment' was too much for the academic establishment at the time. The reason for the resistance to Hamilton lay in the general intellectual climate after the Second World War, combined with a particular pre-war connotation of the term 'altruism'.

After the Second World War any work associating genetics and human behaviour had become uncomfortable. In the social sciences especially there was something of a post-war taboo on the earlier studies of the genetic basis of human behaviour. Also, with the UNESCO statement on race in the early 1950s, the genetics community had officially shifted to a state of agnosticism about the genetic basis of race.[40]

But there was an additional problem. Altruism had in fact once been quite explicitly associated with genetics, and very recently at that. That happened in 1939, on the eve of the Second World War. To be sure, the precise word 'altruism' had not been used, but instead a number of very similar expressions such as 'fellow feeling and social behaviour', 'cooperation', and 'self-sacrifice' had been mentioned. The leading scientist making this connection was the American geneticist (and later Nobel Laureate) Hermann Muller, who in 1939 at the

International Congress of Genetics in Edinburgh launched the famous so-called Geneticists' Manifesto. This was signed by some 20 luminaries in biology and later published in *Nature*. The signatories included Julian Huxley and a number of well-known left-wing scientists such as JBS Haldane, Lancelot Hogben, and Joseph Needham.[41]

Muller's Manifesto was progressive in spirit and addressed one of the great scientific topics of the time: the genetic improvement of mankind. And it did so with a left-wing twist. According to the Manifesto, if the goal was to genetically bring about the best possible human individuals, then what was needed was clearly better and more equal social conditions. It identified three key goals: improved health, improved intelligence, and 'those temperamental qualities which favour fellow-feeling and social behaviour rather than those... which make for personal "success"...' It would seem that the last criterion is readily translatable as altruism or cooperation. And, indeed, one of the leading themes in Muller's popular book *Out of the Night*, published four years earlier, was the promotion of a more cooperative society by genetic means.[42]

The big scare before the Second World War, among scientists and the general public alike, was the seemingly inevitable deterioration of the human genetic potential. The accumulation of deleterious mutations was a phenomenon that Muller, the geneticist, had discovered and documented, and for which he had invented the term 'mutation load'. (Muller was later to receive the Nobel Prize for his work on the problems of mutation caused by radiation.) In his *Out of the Night* (which sold very well in England) Muller argued for humans taking evolution in their own hands in order to counteract the deleterious force of increasing mutations. After the war, however, the pre-war fear of deterioration of the human stock ceased to be part of the open scientific and popular discourse, though for some scientists it still remained a worry.

The pre-war concern about deleterious mutations involved the whole political spectrum. In his *The Genetical Theory of Natural Selection*, R.A. Fisher had worried about the long-term consequences of the lack of reproduction in the educated classes. In protest against Fisher's conservative-sounding views, and in keeping with the general genetic discourse, left-wing scientists extended this concern to all classes. Their argument went as follows: non-optimal—unequal—social conditions would not be able to produce the best possible individuals. This was why a long list of social reforms was needed: support for women, birth control, and promotion of a 'scientific and social attitude' to sex and reproduction. More explicitly, the idea would be 'to have the best children possible, both in respect of their upbringing and their genetic endowment'. In his book Muller even suggested the establishment of sperm banks with sperm from Nobel laureates, which would be made available to women, married or unmarried, who wanted to have intelligent children.[43]

And in a would-be dialogue with Haldane, who in his famous popular book *Daedalus* had advocated 'ectopregnancies'—the growing of children outside the body, which was later to be fictionally explored in Aldous Huxley's *Brave New World*—Muller also theorized about a number of futuristic-sounding, now well-known, reproductive technologies. For left-wing scientists of the 1930s, social progress went together with science and rationality—and the creative imagination. Science and the scientists were in the driving seat at the time, solving the problems of mankind in a rational and socially progressive way.[44]

But after the Second World War, with the Nazi abuses of genetics, the left-wing position radically changed. The link between genetics and human behaviour was cut. Any suggestion for the genetic improvement of mankind, even involving individual voluntary measures, was absolutely out. Genetics of behaviour or personality was not for discussion. The only acceptable genetics involving humans was strictly medical genetics.[45]

Still, some die-hard scientists refused to abandon their original vision of improving the human race. One manifestation of this was the 1962 CIBA conference in London, with key speakers Hermann Muller, Joshua Lederberg, and Francis Crick (all three by then Nobel laureates). At that conference, Muller once again reiterated the original three goals from the pre-war Geneticists' Manifesto: the improvement of human health, human intelligence, and 'fellow feeling and social behaviour'. The trio treated the idea of the deterioration of the human stock as an established fact and suggested various countermeasures, such as genetic engineering. In other words, in the changed post-war political climate they continued promoting the pre-war human improvement programme.[46]

Someone who was painfully aware of this was Professor Lionel Penrose, Bill Hamilton's nemesis at the Galton Institute at University College. Penrose had made it his life work to save his field, genetics, from the contamination that had taken place because of its pre-war association with the genetics of human improvement. He was set on stamping out any trace of association to eugenics, both in his own Genetics Department and elsewhere. Penrose was on the left himself, but the programme of the left had changed. Consequently, Penrose's task was to uphold the post-war left-wing contempt for the pre-war left-wing enthusiasm about human improvement. A strong believer in the power of culture, he had written militant books, one of which questioned IQ measurement. He was a Quaker and a leading figure in international organizations dealing with peace research.[47]

At University College Penrose had been changing what could be changed. There was the unfortunate name Galton Institute, but what could be cleaned up had been cleaned up. So the Galton Institute had changed its name to the Institute of Genetics and the *Journal of Eugenics* had been transformed into the *Journal of Genetics*. For someone as deeply aware of the contemporary situation as Penrose, having Bill Hamilton walking into the lab, proclaiming an interest in the genetics of altruism, must have been the epitome of what was *not* desirable. It is not clear what Penrose's exact suspicions were about Bill, but his attitude was such that Bill felt himself regarded as 'a new sinister sucker budding from the recently felled tree of Fascism'.[48]

Of course, the actual spirit that Bill embraced was that of the 1930s progressive scientists, such as Haldane or Muller, with all their creative schemes for human improvement. Their ideas had appealed to Bill, just as they had to many other scientists, because of their rational and progressive nature. The ideas of the left-wing biologists of the 1930s fitted Bill's self-conscious commitment to a rational scientific world view like a glove. But just as was often the case with Bill's science, his taste was such that he was typically out of sync with contemporaneity, preferring to tackle unresolved problems of earlier generations. In this case he had become intrigued with the futuristic vision of the pre-war scientists.

In any case, now it was the early 1960s, a sensitive time, with Penrose adamant that his institution devote itself to respectable, post-war genetics. And there was something else. Penrose had taken over the Galton Lab (as it was still called unofficially) in 1945 from none other than Ronald Fisher, who had left to become a professor at Cambridge. This had presented an extra difficulty when it came to transforming the image of the Galton Lab in accordance with the new post-war line. Penrose had, however, been successful in changing the title of the professorship from Professor of Eugenics to Professor of Genetics.[49]

It would seem, therefore, that exactly during the time of Bill's studies, Penrose would have had a professional and personal interest in controlling the image of the Galton Lab and the type of doctorate degrees in genetics that were being issued. The genetics of altruism or social behaviour could simply not be associated with the lab's new image. Penrose had succeeded in discouraging Bill during the latter's earlier inquiries in late 1959, going so far as to say that he doubted a problem such as the genetics of altruism even existed. But now young Hamilton had sneaked in nevertheless—or so it would appear—and was even being advised by the respectable Cedric Smith. Absolute anathema! How had that happened? No wonder Bill did not feel welcome in the Galton Lab the few times he visited. Even the secretaries seemed to dislike him. Bill's altruism project, by triggering undesirable pre-war associations to formulations by Muller and others, threatened to undo all Penrose's hard clean-up work. And the whole thing was currently becoming extra sensitive as those three eminent speakers at the 1962 CIBA conference in London felt free to invoke the pre-war Geneticists' Manifesto.[50]

What about those other rejections that Bill had experienced? It is not surprising if in keeping with the political line, the Cambridge Anthropology Department was especially sensitive to attempts to link human behaviour and genetics. As for the negative reaction from Moray House, again, the pre-war connection between eugenics and genetics is probably the answer. Bill the 'geneticist' may have appeared a dangerous candidate for the teaching profession. What would he be telling innocents? (It is not clear what was actually said in the letter exchange. Hamilton tells us he got so angry with the negative response from Moray House that he tore up the correspondence.)[51]

In the next chapter we will take a closer look at the content of the large manuscript that Hamilton sent off to the *Journal of Theoretical Biology* and the circumstances surrounding its publication.

6

Altruism through the Looking Glass

'That story about Bill sitting in Waterloo station is absolute crap!' said Colin Hudson cheerfully to me one hot morning in Barbados. We were sitting in his home-made outdoor living room, a light roof supported by columns and surrounded by beautiful Anetto trees with flowers like red bottle brushes. I had also spied a richly fruited grapefruit tree close by, and further in the background towered a colossal fig tree. Colin, Bill's best friend from Cambridge, had moved to Barbados soon after graduating, invited by the government to develop the island's sugar cane industry. Later he became the leader of an ecological movement on the island, teaching people how to grow a variety of food plants in an environmentally clever way. Part of the experimental garden in front of us sported neatly organized car tyres used as planters. The sheltered place where we sat was also used as a meeting room, I learnt that same night at a potluck party arranged by Colin for the movement's members.[1]

Now what on earth did Colin mean by that statement? There was no reason to believe that Bill was making the whole thing up. The Waterloo story was easily imaginable. Here was a lonely chap struggling with the mathematics of what was later to become his famous inclusive fitness paper, with little academic support, and so lonely that he needed to sit on park benches and in a train station just to have people around. Bill had vividly described his plight in his autobiographical notes.[2]

It turned out that Colin did not question the Waterloo story as such—his dispute was with Bill's romanticizing of his situation. Because according to Colin, Bill was not lonely in London at all—he had plenty of support. Both his sisters lived there and regularly invited him to meet their friends (Mary ran an intellectual salon of artists and others). Bill himself had a small circle of friends, with whom he did things. Last but not least he had his home with his ever-supportive mother and father barely an hour's train ride away. And later he even had a girlfriend in London.

So Bill was not lonely, in the sense of being abandoned and without human contact, Colin told me. Bill *wanted* to be lonely! He needed it for his work. Colin explained how Bill belonged to the type of people who needed distance from other people to be able to concentrate and do creative work. This, in turn, made him feel miserable and lonely—especially when things didn't go well. But it was all his own doing, Colin insisted. Bill may even have enjoyed feeling miserable, Colin mused.

Colin himself was a very different type of person, outgoing in the extreme and naturally charismatic. He was also a practising Christian, and in general a man of causes. The way he saw his own potential to do good was at a concrete level: developing the sugar cane industry to provide jobs for women, and later promoting the environmental cause (he was proud that Prince Philip and other dignitaries had once visited his centre, called Treading Lightly).

Colin was able to refer to his own correspondence with Bill during this time to back up what he was telling me. Let's take a look. At the time of the correspondence, Bill is working on his altruism models but there are lots of distractions. For instance, Bill spends weekends at Oaklea building a greenhouse for his mother and, interestingly, observes that building a physical model takes as much energy as building a mental model. He goes to see 'The Cherry Orchard' with friends, hoping that his favourite Chechov play will not be destroyed by the actors. But he also has 'diversions', many of which are actually the beginnings of various new research projects. Most of the time he feels terribly guilty. 'I have whole days of idleness', Bill writes Colin in 1962. He oscillates wildly between various scenarios for employment. He seems on the verge of emotional collapse.[3]

Colin's verdict that 'Bill was not lonely' appears too harsh. There is a difference between being alone and being lonely. Bill felt certainly lonely, although he may have given the further impression that he was also abandoned by the world. It was the loneliness and agony felt by someone who was in the process of creating something entirely new. Was his creation real, or was it a figment of his own imagination? Would others acknowledge its existence and its significance? If it was something that only made sense to himself, that was not good enough.[4]

It is possible that during 1960–63, Bill—with the enormous pressure he felt bearing down on him—was quite close to a nervous breakdown. He seems to have felt guilty all the time for not working faster, and for not living up to all the trust placed in him. Of course, it is not unusual for scientists working intensely to suffer a nervous breakdown. One good example is Francis Galton, one of Bill's heroes, who experienced an early breakdown as a Cambridge undergraduate while he was taking the mathematics Tripos, feeling the eyes of all his family upon him. It is true that Bill had a tendency to drive himself into a state of near-depression. But he also seems to have had an inner sense telling him when enough was enough and distraction was needed. This was evident throughout his life, as he interchanged intense theoretical work with collecting expeditions. And although Bill felt that his parents expected great things from him, he also knew that he had their unqualified support.[5]

And importantly, during this crucial time he also had the support of Colin, and this was not limited to moral support. Colin Hudson was, in a sense, his long-distance, intellectual sparring partner. Between Colin and Bill there was something approaching the collaboration between Watson and Crick, the big difference being that Colin was in Barbados, and that they did not really work on a common project. The project was exclusively Bill Hamilton's. Colin, however, as a very close friend, took a deep and natural interest in helping Bill, and commented on his manuscript as well as he could. Colin understood the mathematics and occasionally criticized Bill's reasoning. He raised questions, challenged assumptions, asked for clarifications, and required examples. 'You need an introduction,' he told Bill in one exchange, explaining that he could not get the basic idea of the manuscript until he had gone a good bit into it, because of all the initial calculations. Colin's questions and challenges were so much to the point that many of Bill's carefully formulated handwritten answers to them later ended up appearing verbatim as parts of his 1964 paper, 'The Genetical Evolution of Social Behaviour'.[6]

The typed manuscript that was going back and forth between them, with comments written in Colin's handwriting and responses written back in Bill's more spindly handwriting, was a long manuscript entitled 'Genetical Models for the Evolution of Competitive and Social Behaviour'. It covered much more ground than what was later to become Bill's inclusive fitness paper. It devoted a lot of space to competition, for instance. As mentioned, it was a much expanded version of the last part of the manuscript, called 'Selfish and Altruistic Characters Affecting Relatives' that was later to become Hamilton's published 1964 paper.

Although Hamilton was working with a strictly technical definition of altruism in his model building efforts, this did not prevent him from sustaining an interesting discussion with his friend about the actual nature of altruism. As a devout Christian Colin had strong things to say about the true nature of altruism, and did not quite approve of Hamilton's callous-seeming strictly biological treatment. Throughout his life, Bill would not shy away from deep religious and metaphysical questions, although he was not religious himself. He wanted to penetrate as deeply as possible into the human existence, using different kinds of methods. One of them was probing the world's great literary treasures. Later, he was to become close friends with another devout Christian, George Price, with whom he engaged in lengthy Biblical exegesis in addition to scientific collaboration.

As we have seen, the road to what was to become Bill Hamilton's famous paper on altruism was not easy. His theoretical project was not one that could easily be completed within a limited time, but that was the way the university system worked. He had to satisfy the requirements of his supervisors and this often caused tension and conflict for him. He worried about being able to finance his studies by grant support. His shifting situation and moods can best be seen though his letters to Colin during this time. Colin served not only as a scientific and moral commentator but also as a long-distance sounding board and would-be therapist for Bill. The biggest surprise that emerges from these letters is the enormous *resistance* that Bill had to doing his altruism project, finding all kinds of 'diversions' to avoid working on it.

9 October 1961

'It is very odd—looking back over last year's work one of the things that stands out most clearly is how cleverly I diverged from what are [by] far my most important scientific projects, because, I can only think, I subconsciously felt that where I had most to gain I also had most to lose, & most shame to bear, if I found myself incompetent ['stupid' has been substituted]. (However, I have told Mr Hajnal—& let this letter also hold me hostage—that I am just about to tackle Altruism again!)'

30 June 1962

Bill has received Colin's notes on his long essay entitled 'Genetical Models for the Evolution of Competitive and Social Behaviour' and returns it 'with my notes in defense and further explanation'. He will also be sending him the next section of the essay. The concepts there 'are the most important of all those I have evolved so far'. He feels that he is not 'mentally on top'. He has a heavy cold.

He has come across Wynne-Edwards' huge book, *Animal Dispersion in Relation to Social Behaviour*, 'which is quite an encyclopaedia of fascinating facts and problems all united under an interesting but shaky theory (at least in my opinion) where group selection plays a large role. However, the author's detached philosophic approach seems to me admirable, that of a true naturalist... [The author] gives me more examples of biological nastiness than I could have hoped for...—in fact if you still don't believe that nestlings in a nest compete with one another read chapter 22.' This had been an issue of great disagreement between Colin and Bill and related to the general discussion of competition, the original paper's main topic.

The rest of the letter is largely an interesting exposition of Bill Hamilton's views on altruism and morality and a critique of the limitations of Christian morality.

11 September 1962

'There is no immediate prospect of my becoming unemployed as my MRC grant has been renewed (again thanks to LSE and Professor Glass)—and it so happens that its value has gone up by [pound sign] 50. By the end of this ac. year I am expected to produce something in the way of a thesis, which may mean some rather tedious reworking over some of the old stuff, computation on the Taiwanese etc, but if I do complete it and manage to scrape my PhD I might be fairly free...' Here he is referring to possible travel together with Colin, a long-standing plan of theirs.

Note that Hamilton is here considering *some* PhD, not necessarily a thesis on social behaviour or altruism.

8 October 1962

Bill tells Colin he is designing a conker (horse chestnut) experiment. He has promised to give Professor Smith a quantitative summary of the situation (has to do with seed competition within and between plants) and write an outline for his project, but when he weighs the conkers, he finds that quantitatively the result is not as he expected, and the situation is much more complicated. He complains that he is 'in poor mental shape' and that 'there is a new factor interfering with my work...or rather an old factor is assuming crisis proportions'.

'[I]t almost appears that I am some sort of success with women and almost in danger of becoming a "ladies man".' This worries him a lot and he wonders why that is so. Still, he continues, 'one can't brush aside what appears to be love, either in oneself or someone else, on the grounds of one's suspicions because it might really be... So those inexorable judges of my work that I have so carefully instituted—Fisher, Darwin and the rest—are finding themselves jostled by a lot of women!' He worries about losing his critical judgement of his own work.

Bill returns to discussing his conker experiment: 'This investigation which I originally intended as a light diversion from mathematics is turning out rather too demanding of both time and thought. I have made a very large collection of conkers and these are now waiting to be weighed and I have committed myself to laying out a competition experiment with them in the UC Botany Dept's garden... I intend to plant them as close-packed as is possible but nevertheless it might be several years before a yard square plot has been reduced to one or two survivors—but long before that I would have got fair impression of how often and how fast big conkers beat little conkers under such conditions... I have been trying to think whether any of the data could be used to discriminate between my two basic models of competition, what I have arbitrarily called "pressure" and "encroachment" competition.'

[5 November 1962 Bill's upbeat letter to his parents, see chapter 5.]

18 January 1963

'Dr Smith who has had it [the paper] for the past month at last has now indicated that he thinks it is again too long and diffuse for a paper, he wants me to make it into 1. a letter to nature, 2. a paper for J.Theor. Biol, 3. a paper for Ann. Hum. Gen... and he is only half way through it yet! However, when I last saw him he told me he didn't think my work substantial enough to earn a PhD, this is encouraging in a way. While waiting for his comments I have been writing up some of my work on the demographic side and I have a paper half finished there too now. This incidentally has necessitated a return to Mercury with my old program to calculate the female Taiwanese... reproductive value (to go with my male one) from some rather faked-up (?) male fertility rates.'

Bill has listened to a lecture on altruism in moths by Dr Blest and immediately sent in his own short paper to *Nature*. He has learnt programming, enjoys it and the people. He wonders if he could make a living as computer programmer?

14 February 1963

The *Nature* paper has been returned by the journal. Bill has sent off his paper instead to the *American Naturalist*. He is still trying to finish his main paper which causes him much difficulty. 'Altogether I am so fed up of endless rewritings of the same theme—I have spent almost a year at it now and although the ever improving mathematical generality has at last made some impression on Dr Smith I think even he is inclined to agree that the simpler earlier versions were really more useful—I think I must make a firm resolve to finish this business by the end of the Summer term.'

7 March 1963
Bill's short paper has been accepted and he has just discovered that the social Hymenoptera can serve as a great example for his theory. He is in good spirits:

> My "letter" has been accepted by the "American Naturalist", which is very encouraging...I am still working on the last section of my main paper—the real manifesto—and have stayed up here at my digs this weekend in the hope that I might really get it finished. This section concerned with the social insects has been giving a lot of trouble and going very slowly. This is partly due to the distractions of conker-planting but partly it must have been my guardian angel's guidance since I have long been on the edge of handing in a lot of rubbish. I was under the impression that the relationship in a bee colony (at least, between the diploid females) was that of full sisters. The day before yesterday it suddenly dawned on me that it was a special relationship and one which must be very favourable to the evol. of social instincts. You no doubt remember the peculiar sex-determination in the Hymenoptera, by male-haploidy. Well, if the queen is mated by only a single male all the sperm she gets will be genetically identical so that the paternal gene-sets of all here daughters will be identical too. Therefore take any two daughters and the expected fraction of genes they have in common will be ½ (1 (for paternal genes) + ½ (for mat.genes)) = ¾ [the right part of the closing parenthesis has been added here, missing in the letter]. For normal full-sibs the fraction is ½. For mother and offspring it is ½ with bees just as with any other sexual organism. Anyway this extraordinarily close relationship between daughters of singly-mated queens means that according to my theory they will be more keen, other things equal, to remain at home in the nest and help rear their mother's eggs than to go off and rear their own! Hence, I believe, the strong tendency for social life to develop in the Hymenoptera; it does in fact seem to have evolved repeatedly. The daughters are related to their brother drones by the fraction ¼, but to their potential sons by the normal fraction ½. Thus they should be much more reluctant to give up their drone-producing capabilities than their female-producing in favour of the queen's, and on the whole the evidence supports this since a certain amount of worker laying (unfert. eggs which hatch into drones) does seem to go on in most social Hym. Colonies.

14 May 1963
Bill sends off his long manuscript to the *Journal of Theoretical Biology*.

Having finally sent off his manuscript, Bill realized how utterly tired he was of his own theoretical work. He was looking forward to travelling and finally being able also to see Colin. After finding that one could apply to the British Council for scholarships to Brazil he had written to the legendary entomologist Warwick Kerr in Brazil and quickly received a warm invitation to come to his laboratory at the University of Rio Claro in the province of São Paolo. With Kerr's letter of recommendation Bill succeeded in getting one of the two scholarships available, and, on Kerr's suggestion, stayed for an entire year rather than a shorter period.[7]

This trip to Brazil can be seen as part adventure, part a concession to his supervisors. Bill had met with enough scepticism to realize that he needed to satisfy his professors if he ever wanted to get his PhD. And at this point, a PhD was what he thought he should aim for. His confidence had been regained and he had stopped imagining himself in various non-academic professions, which had been the case only a year or so earlier.[8]

However, it was obvious that his supervisors wanted a more traditional kind of thesis, which involved not mathematical models, but real facts of Nature. He had certainly been made to feel that he was doing the wrong thing:

> [A]s an aspirant of True Science I was acutely aware—was made to be so by scientists around me—that I had somewhow by-passed (or as I said tunnelled under) an essential stage of scientific development. I ought now to make up, to fact-pile something somewhere, test something, become respectable.[9]

So they needed evidence? The best place to collect proof to show his advisers that his theorizing described true phenomena in Nature was clearly some place where there would exist plenty of examples. The tropics seemed ideal for this purpose. Bill himself at this point was no longer in need of such factual support. His ideas were already crystallized:

> I was pretty sure that they were right—that is, that they were correctly argued. If right in this way, it was clear that no amount of evidence from nature could make them wrong; or, if it did, then at least for my comfort, Darwin's and Fisher's evolution versions would have to crash along with mine.[10]

On 16 August 1963 Bill embarked on a one-month long voyage to Brazil as a passenger on a cargo boat. He was looking forward to being around Warwick Kerr, learning various techniques and discussing his findings and theories with a friendly expert. Moreover Kerr had developed a particularly interesting theory about caste determination in bees that Bill wanted to know more about.[11]

***

Let's take a look at Bill Hamilton's monumental achievement, his 1964 paper 'The Genetical Evolution of Social Behaviour'. Unlike the original manuscript sent to the *Journal of Theoretical Biology*, the published paper consists of two parts, with the first part devoted to a mathematical derivation of the concept of inclusive fitness and the second to examples and discussion. For practical reasons, I will deal with the paper in its published form.[12]

Hamilton had been deliberately looking for a general principle similar to Fisher's Fundamental Theorem of Natural Selection. Fisher's famous theorem had stated that the average fitness in a population will always increase. Hamilton had instead introduced the new concept of inclusive fitness and, after much hard work, had achieved a similar fitness maximization principle for his inclusive fitness:

> For an important class of genetic effects when an individual is supposed to disperse benefits to its neighbours, we have formally proved that the average inclusive fitness in the population will always increase.[13]

Part I of the paper was devoted to a mathematical derivation of the concept of inclusive fitness. The concept of inclusive fitness was actually quite complicated (a factor not always appreciated by biologists).[14] Here is Hamilton's exact formulation, as he was later to express it in words in Part I of his paper:

> Inclusive fitness may be imagined as the personal fitness which an individual actually expresses in its production of adult offspring as it becomes after it has been first stripped and then augmented in a certain way. It is stripped of all components which can be considered as due to the individual's social environment, leaving the fitness which he would express if not exposed to any of the harms or benefits of the environment. This quantity is then augmented by certain fractions of the quantities of harm and benefit which the individual himself causes to the fitnesses of his neighbours. The fractions are simply the coefficients of relationship appropriate to the neighbours whom he affects; unity for clonal individuals, one-half for sibs, one-quarter for half-sibs, one-eighth for cousins, . . . , and finally zero for all neighbours whose relationship can be considered negligibly small.[15]

In other words, Hamilton's inclusive fitness was a sort of correction to Fisher, an attempt to make Fisher's Fundamental Theorem even more fundamental, by recognizing that socially living animals typically affect one another's fitness. In his article Hamilton even had the audacity to criticize Fisher on this point. After noting that there did not seem to exist any comprehensive definition of fitness in the classical model, he continued: 'And, perhaps in consequence of this lack, it rather appears that Fisher's Fundamental Theorem of Natural Selection has yet to be put in a form which is really as general as Fisher's original statement purports to be.'[16]

Part II of the paper reintroduces Hamilton's concept of inclusive fitness and highlights its two revolutionary aspects: (1) the disconnection of a gene from the individual whose body the gene happens to inhabit (because relatives, too, may carry copies of the gene); and (2) the fact that a gene's 'interest' is not necessarily the same as its carrier's interest:

> In brief outline, the theory points out that for a gene to receive positive selection it is not necessarily enough that it should increase the fitness of its bearer above the average if this tends to be done at the heavy expense of related individuals, because relatives, on account of their common ancestry, tend to carry replicas of the same gene; and conversely that a gene may receive positive selection even though disadvantageous to its bearers if it causes them to confer sufficiently large advantages on relatives.[17]

Hamilton is here using Sewall Wright's coefficient of relatedness (r) between two relatives as a rough measure of the likelihood that they carry replicas of the same gene which they have inherited from a common ancestor (genes 'identical by descent'). He warns that relationship alone does not mean *certainty* that a person carries a particular gene that the relative is known to carry. Still, he argues that Wright's coefficient of relatedness closely approximates the probability that a replica of the gene will indeed be carried.[18]

In his paper Hamilton is now prepared to 'hazard the following unrigorous statement of the main principle that has emerged from the model'. Here he is leading up to the cost-benefit analysis of what was later to be called 'Hamilton's Rule':

> The social behaviour of a species evolves in such a way that in each distinct behaviour-evoking situation the individual will seem to value his neighbour's fitness against his own according to the coefficients of relationship appropriate to that situation.[19]

One of the examples Hamilton discusses is the evolution of alarm calls in birds. The bird making the call is at greater risk of losing its life. The question therefore becomes: how large must the gain be for the benefit to others to outweigh the risk to self in terms of inclusive fitness? Here is the answer:

> For a hereditary tendency to perform an action of this kind to evolve the benefit to a sib must average at least twice the loss to the individual, the benefit to a half-sib must be at least four times the loss, to a cousin eight times, and so on. To express the matter more vividly, in the world of our model organisms, whose behaviour is determined strictly by genotype, we expect to find that no one is prepared to sacrifice his life for any single person but that everyone will sacrifice it when he can thereby save more than two brothers, or four half-brothers, or eight first cousins...[20]

Or as Bill puts it elsewhere in the paper: 'more than 1/r units of reproductive potential or "fitness" must be endowed on a relative of degree r for every one unit lost by the altruist if the population is to gain...more replicas than it loses'.[21]

Bill was aware of the fact that his concept of inclusive fitness was not perfect. There was actually a would-be paradox inherent in his whole enterprise. The measure that he was using, Wright's coefficient of relatedness, was in fact defined for situations in which *no* selection was going on! But the idea of maximization of inclusive fitness had to do with a population undergoing selection. What to do? Hamilton calmly reasoned that the weaker this selection was, the more the situation would approximate the situation of non-selection required for a proper calculation of the coefficients of relatedness. Therefore, he argued, under a condition of weak selection, his rule could be defended.[22]

This reasoning, incidentally, is typical of Hamilton. It appears that he came to many of his insights through various types of approximations. Although his aim was always to try to find the most general formula to describe a particular theory, his mind was not that of an abstract modeller. The final form of a theory was typically obtained after long experiments with plugging in numbers and seeing what the result would be. (Later, Bill would substitute this kind of exercise for simulation experiments with the help of a computer.)

For a biologist at the time, this way of proceeding was quite unusual. And, indeed, nobody had taught Hamilton how to go about these matters. Fisher's book was extremely condensed and Fisher himself (the little Hamilton had seen of him) not particularly helpful in practical matters. Population genetics had not been part of the basic biology training at Cambridge, with the result that

Hamilton had had to teach himself this new field. There was no doubt that for Bill, the Fisher book was a 'minor grail'. The problem, as noted earlier, was that Fisher almost never explained how he arrived at his theories.

Fisher was an intuitive mathematician. Bill, too, had an intuitive grasp of mathematics, but was no match for Fisher. So how did he actually proceed? It seems that he developed working models and plugged in values, aiming for a pattern to emerge, a generalization which would hold under certain conditions (for instance, as in the case above: under weak selection). This was not the way a typical mathematician would have worked. In fact, for mathematical purists, Hamilton was not general enough.

But this was instead how an engineer would proceed. Bill had transferred his early physical model building skills to a new abstract realm. He had also learnt something else about models, which would come in handy as his career unfolded. The important thing with models was that they did what they were supposed to do; they didn't have to be simple or pretty.[23]

Because of the success of the concept of the selfish gene, and the credit Richard Dawkins gives to Bill Hamilton in his book of that name, some may believe that this was Hamilton's own 'working concept', that is, that Bill was typically thinking in terms of selfish genes. This is not so. Hamilton actually followed a more traditional pattern of thinking in terms of individual organisms. At the same time, though, he had adopted Fisher's treatment of individuals in a population as gene packets of sorts. Bill was naturally thinking in terms of gene frequencies and changes in them—the canonical Neo-Darwinian interpretation of evolution by natural selection.

If anything, Hamilton had an almost emotional attachment to individual selection, especially as a sort of antidote to group selection, to which he declared himself 'allergic'. Fisher was his hero. He associated group selection with all kinds of totalitarian systems—be these systems characterized by fascism or communism. For Bill, individualism was strongly associated with human freedom, and so for him, Fisher represented the voice of reason against the prevailing post-war view that animal social behavior was 'for the good of the group'.[24]

But, even if Hamilton did not typically think in terms of genes, on a few occasions he did indeed play with the concept of the selfish gene, in the sense that he was deliberately taking the 'gene's eye view'. He did this particularly in his short 1963 paper sent to *Animal Behavior* where he launched Hamilton's Rule (and later in his 1967 paper on 'unbeatable sex ratios' and a couple of other places). For

Hamilton, the gene's eye view was a heuristic device that could help one assess what it would be in a gene's 'interest' to do. He used it especially when he wanted his readers to be able to follow him along in a new piece of reasoning. If Fisher in his theorizing was exercising what Alan Grafen has called 'licensed anthropomorphism' (Fisher was thinking of organisms as maximizing agents), then Bill Hamilton's anthropomorphism had an additional, unabashedly 'unlicensed' aspect as he actually identified with his research objects.[25]

Some have believed that Hamilton's theory about altruism basically said that relatives should stick together and help one another. In other words, Hamilton would be advocating a type of generalized nepotism. This was obviously not so—Hamilton was operating at a strictly scientific level. (This tendency automatically to interpret factual biological statements or theories as moral/political messages became particularly blatant during the later sociobiology controversy, see chapter 14.) But what is of interest here (and not always noted) is that kin altruism for Hamilton could mean a number of different things.

In his 1964 paper Hamilton worked with two separate approaches to kinship theory. One was 'direct' kin altruism, the other was what he called '*a superkinship trait*'. Kin altruism required some mechanism for knowing who was or wasn't kin. Rules of thumb could include such matters as familiarity of appearance or proximity. (Wynne-Edwards, for instance, had given examples of such matters among birds. There was also evidence that individual animals recognized each other.)[26]

But for Hamilton altruism was not only something that could evolve within a group of closely related individuals. What about 'foolishly helpful friends'? (Hamilton's own term). Clearly, there were altruistic and helpful individuals outside the family—Bill needed only to think about his godfather Charles Brasch, his mother's friends the De Beers, and some of his own close friends, such as Colin Hudson. What kind of mechanism was operating in such cases? Bill theorized that just as there exist ways in which kin is usually known, there must exist *some type of phenotypical recognition mechanism,* making it possible for individuals who share altruistic traits to identify one another. This is what he meant with a 'superkinship trait'. Richard Dawkins was later, in his *The Selfish Gene,* to call this a 'green beard effect'—a particular (hypothetical) additional trait that all altruists would have in common and would be able to recognize each other by. The basic requirement was for there to be a sufficient concentration of altruists in the same place to allow altruism to exist and spread without being overwhelmed by egotists.

For many it was Hamilton's example of the social insects, Hymenoptera, that conveyed the gist of his theory. This was clearly a stunning and concrete example of the working of Hamilton's Rule, but it lead to a widespread belief that Hamilton's inclusive fitness theory had to do particularly with Hymenoptera. In other words, rather than having provided a theory applicable to all life on earth, Bill Hamilton was seen as having solved the specific altruism puzzle for the social insects. This was especially ironic, since Hamilton had brought in the Hymenoptera as a mere example to support his theoretical model. Bill himself explained this phenomenon in terms of people not having actually read his original paper. Perhaps, but then, why this systematic bias? Well, obviously Hamilton's contribution was being reported on by various scientific and popular 'opinion leaders', and some of them may not have got things right. Others, again, may have had their own reasons for sustaining a mistaken impression.[27]

***

It is clear that there was an important moral dimension in Bill's whole being, one that did not always manifest itself in his more technical writings. His idea of altruism was not just an answer to a scientific puzzle, it also satisfied a deep need in him to know that unselfishness was possible in Nature. But it was the extreme case—self-sacrifice—that was most central to Bill. It came up again and again throughout his life.

One of the initial inspirations for his theory of altruism came from the idea of the honeybee worker sacrificing her life by stinging the intruder of a hive (something that he learnt while tending to his mother's bees). Later he suggested that suicide as self-sacrifice and self-annihilation was pleasurable, arguing that the male honeybee was 'aching for a glorious death' (as he inseminated the queen by exploding his genitalia into her and dying in the process). In both the case of the worker and the drone, Hamilton was convinced that suicide was not a design flaw of Nature, but quite the opposite: it was designed in by natural selection. For Bill, the scientist, scientific proof was needed that there would be an evolutionary point to self-sacrifice. (The gratuitous-seeming attribution of pleasurableness is actually a recognition of the eagerness of the self-sacrificing individual to do what evolutionarily it should do, in Hamilton's highly anthropomorphic language.)[28]

Indeed, when the theme of self-sacrifice surfaces in Bill's writings, it always has a positive connotation. Talking about the babassu nut and its six embryos, of which only one can become a full-grown individual, or discussing the imagined point of view of one of many embryos, or even of his own brother Jimmy (born with a congenital problem, allowing him to live only two weeks), it is remarkable how the theme of self-sacrifice surfaces. We see this more clearly in his autobiographical essays to his collected works, but it is clear that the theme has been there from the beginning. Bill's sister Mary tells of her brother's strong fascination with conkers (horse chestnuts), and the fact that one nut typically contains three embryos, of which only two can survive. One of Bill's favourite stories from the world of art and legend, meanwhile, was the tale of the Burghers of Calais, involving the voluntary sacrifice of a group of citizens to the enemy in order to save their town. He compared their sacrifice to that of the honeybee.[29]

Where might this talk of sacrifice and suicide come from? Note that for Hamilton, self-sacrifice is typically connected to resistance to authority as well, and to telling the uncomfortable truth. We are here going from intellectual to deep-seated emotional connections. The sacrifice talk may be attributed partly to Bill's general romantic vision and heroic imagination (fuelled by childhood stories about knights in armour), but also to early moral teachings, such as those readings by his mother from *Pilgrim's Progress.* At an early point in the story, however, one of Christian's close travel companions is sentenced to death and executed because of his faith. This shocking development may have left a deep impression on little Bill Hamilton.

Bettina herself was not overtly religious. She had been brought up a Baptist in New Zealand, but had later moved away from this faith. Still, as we have seen, she and Bill's father often went to church on Sunday, and sometimes took the children with them. Bettina's own interest in evolution and the way she conveyed this to Bill and the other children may in fact have taken the place of her earlier religiosity. (As a student of medicine, she had been exposed to these ideas in some courses.) Nevertheless, Bettina was strongly morally oriented and had wished to become a missionary before she married Archibald Hamilton, I learnt from Bill's brother Robert in New Zealand. Although the Hamilton home was not religious, knowing the Bible was part of general education at the time, and Bill, with his phenomenal memory, soon became quite expert on Bible quotations. Here a couple of short 'Bill stories' may be relevant. Bill once opened the door to a Seventh Day Adventist, and to his daughter Rowena's amazement—and pride—soon found himself engaged in a spontaneous game of Bible quote-counter-quote with the poor fellow. On another occasion in the United States, he wanted to volunteer for a public debate with a creationist, and was sure that he would have won. He was however discouraged by his colleagues, who saw this as more than a matter of Bible knowledge.[30]

There is still another explanation for Bill Hamilton's attraction to the idea of altruism, and especially his later development of the point that altruists need to find one another. Perhaps he absorbed the general spirit of British war resistance, with its emphasis on solidarity and sacrifice. He may have adopted a vision of the need for good people—altruists—to stick together against a looming evil. At the time the British war pamphlets were conveying ideas of the invisible enemy and internal spies—for instance, one flyer told people to watch what they said in public. The question became: who can you trust? And one answer was, clearly, family and friends.

During Bill's early years men fighting for their country were viewed as fighting for the survival of the group. But as we have seen, Bill didn't like the idea of group selection. Unlike many liberal or left-wing proponents of group selection, who identified it with cooperation, small-scale socialism, and the like, he associated it with totalitarianism. (A decade or so later Bill found another reason to dislike group selection on scientific grounds in his development of a theory of levels of selection. Selection at the group level was no panacea, because due to the layered structure, traits that were advantageous at one level became disadvantageous at the next.)[31]

Here I have tried to reveal a deeper layer of Hamilton's interest in altruism than the intellectual challenge that he obviously saw in solving Darwin's longstanding puzzle. His wish to take up that challenge was already clear from his Cambridge efforts, and his interest increased as he realized that the pioneers of Neo-Darwinism really hadn't solved this problem yet.

But Hamilton is also looking for avenues of escape. It is as if he would not like altruism to be exhaustively defined by his own theoretical efforts. There is a romantic realm of 'true altruism' somewhere on the horizon. Bill also leaves room for mysterious happenings in regard to humans. At the beginning of his collected works he says defiantly that he does *not* want his own theories to apply to himself or to his friends, he expects far more mysterious forces to be operating.[32]

Much of Bill Hamilton's academic work involved finding ways to escape the consequences of his own theorizing. This was to drive him to ever new efforts to find acceptable avenues for the development of altruism. He so badly wanted to find a scientific foundation for altruism—it was as if a phenomenon was not real if it was not scientifically demonstrable. Or rather, by demonstrating that altruism could indeed be derived from Nature, he was able to validate a deep conviction that Nature could be trusted as a model for humans, too. Bettina had told her children that Nature was a precision engineer. For Bill the scientist, this meant that its features were adaptive; there was a meaning to Nature's various strange-seeming organisms and behaviours. Bill never lost his faith in the adaptiveness of Nature's diverse phenomena—he took them as personal challenges to his intellect and explanatory ingenuity.

7

Brazilian Break

Brazil was a totally new experience for Bill Hamilton. His natural habitat was the English woodland and the chalky lands and tangled banks of Kent, where he knew so well the lives and habits of his various 'model organisms'—the limited set of plants and animals that he mentally consulted for a first, rough check of the plausibility of some new insight of his. He had an almost personal relationship with the cinnabar moth and its corresponding flower, the ragwort, and many other insects and plants, and by now he had collected most of the notable moths and butterflies in Great Britain. He had even made a few serious forays into the European landscapes and their insect populations.

Brazil sent Bill's collecting zeal into overdrive. The sheer vastness of the country, the many different types of habitat, and the overwhelming number of new and exotic plants and animals almost overstimulated him. There was so much to investigate, and everything in principle lent itself to comparison with something that he already knew. Moreover, he was in a place where he could obtain assistance from local experts, who would lead him to wasps' nests and introduce him to new fascinating phenomena. He was fortunate enough to be connected to the lab of one of the entomological gurus of the world, Warwick Esteban Kerr, who could personally help him to make sense of his observations. This was El Dorado! (Or, rather, O Dourado, its Portuguese equivalent.)[1]

This was exactly the break that Bill had hoped for after his hard theoretical work. He admits that he started collecting social wasps and wasp nests 'almost compulsively' wherever he went. Moreover, he put his mind to doing this 'properly', labelling everything. Bill had the exhilarating feeling of being a pioneer. He realized that much of what he collected and observed at this point must in fact be new to science. Bill's old 'hunting instinct' had now been transferred from butterflies to wasps and wasps' nests. There was the excitement of finding 'some magnificent new nest structure, which I thought perhaps had never been seen by scientists before'. (Of course, he realized that most of it was not new to the Brazilians who led him to the nests.) He succeeded in collecting a number of different nest forms.[2]

But wasn't this rather dangerous business? Bill did realize that before him he had, in principle at least, 'an army of Lilliputians that might, by their extreme numbers or subtlety of venom, leave me unconscious or even dead in the forest or savannah'.[3] No problem. Bill explains how he became practically 'immune' to wasp stings:

> My efforts to obtain nests or even just to get a standard sample of the wasps as specimens usually resulted in one or two stings from each nest, but luckily I never developed any of the venom allergy that not uncommonly impedes students of social insects. Rather, in my case the opposite happened: I became less sensitive to stings. It was lucky, because the wasps were often very fierce.[4]

As he travelled around, returning to Kerr's lab after his expeditions, Hamilton kept notebooks to take down all his observations. Of course, what he was also supposed to be doing was to find more supportive evidence for his theory of altruism, in order to convince his supervisors. As he later explained: 'I badly needed examples...where both self sacrifice and the limits to it were indisputable'.[5] But that would emerge as part of the process. Bill was used to having many parallel projects running at the same time.

One of the particular problems that intrigued him at this time was the phenomenon of multiple egg-laying queens in Brazilian social wasps. How did these social wasps manage to keep their colonies cohesive and cooperative despite having several queens?[6] This question could not just be investigated by observation. What was needed was detailed knowledge of the physiology of the individual insects. In September–October 1963 Bill set up an index card system and noted there the results of his dissections of the various species of wasps, bees, and ants that he studied.[7]

David Hughes, a student of Bill's, who followed some of his naturalist trails, and did some detective work comparing Bill's notebooks with his collected specimens in the British Museum of Natural History, describes Bill's working method:

> Through a number of dissections he sought to find out; how many of the females on a nest were queens (as evidenced by developing ovaries), how many were inseminated and what state of development had the fat bodies attained. In some cases he also measured wing size and the number of hamuli (small hooks) on the wings; both features indicating caste. He did not restrict himself to social species but also examined solitary wasps and bees. These cards contain the details of 8 genera of wasps, 8 genera of bees and 2 genera of ants that he encountered and dissected. This amounted to a very large number of dissections. Concomitant notes in his journal of the time . . . detailed behavioural observation on *Polistes* wasps upon nests as well as results of staged encounters between queens taken into the laboratory. In one case he performed wing-clipping experiments to determine the effect this had upon dominance order.[8]

Although Bill went on field trips with staff members, he didn't need to go far from Kerr's lab in Rio Claro to find interesting phenomena. For instance, he started wondering about the extreme sluggishness of some aggregations of *Polistes* wasps (*P. versicolor*) that were living on the century plant (*Agave americana*) surrounding the university buildings. Those wasps seemed to be just hanging around in an idle way. Why did they not follow the example of their fellow wasps, nesting on the buildings? Out with the dissecting scalpel! Bill discovered that in some of the sluggish wasps the ovaries were not developed, and that some were parasitized. It was these kinds of observations that Bill was to store for further use also in his immense personal data base—his own big head—and put to good use many years, even decades, later. Indeed, in 1998 Bill, who had a phenomenal memory, remembered this very incident and compared it with a case of another species of idle-seeming *Polistes* wasps, this time in Tuscany, Italy.[9]

Fond as Bill was of *Polistes* wasps, he was also fascinated by their parasites, particularly the genus *Strepsiptera*. This meant that he was already keeping the theme of parasites on a back burner in his mind, constantly accumulating information until, from a number of scattered observations, an underlying pattern would finally emerge. That was to happen some 15 years later when he would more systematically start exploring the role of parasites in evolution. At that time, his careful note keeping in journals and on index cards would stand him in good stead.

Warwick Kerr was a specialist on social bees, and especially stingless bees. He was the kind of man that Bill immediately felt comfortable with: immensely knowledgeable, generous, and fun. Kerr told me about the discussions they used to have together—about socialism, religion, politics, in addition to their discussions about Bill's findings and more specific laboratory techniques. Kerr's English was fluent, but Bill insisted on speaking Portuguese in order to learn the language better, Kerr laughingly told me.[10]

Bill loved everything about Brazil: the landscape, the language, the people, the food, the fruits, the interesting-tasting drinks (especially those made of the fruit guaraná, which he describes as looking like bloodshot eyes). It must have felt like an excursion, an escape from his usual self, to be in a sunny place, surrounded by fun loving and friendly people. Serious scientists, yes, but with an appreciation for life, laughter, and occasional strong 'cafezinhos'. Hamilton especially loved the uncomplicated life far away from the city and enjoyed chatting with the Amazonian natives, gleaning from them an abundance of information about the region and the properties of various plants. It felt empowering to be able to communicate in a different language. The first time Bill had encountered this sensation was in 1957 when he had worked in Lille and learnt 'workman's French' which he later used in Paris (of all places). Bill had studied Latin at Tonbridge, and was fascinated by language in general—the fact that he had no ear for music did not seem to impede his ability to learn new languages.

Bill may even have got somewhat carried away with his newfound powers. Perhaps it was his wish to combine many things—learning Portuguese, travelling, and doing some theoretical work at the same time—or maybe just a test to see if his ideas would fly, but I have it from Kerr that Bill soon tried to teach his idea of inclusive fitness to a group of Amazonian natives. They seem to have gotten the gist of it.[11]

It was the Amazon that was the biggest experience for Bill on this first Brazilian trip. Bill had known about the Amazon since early childhood when his mother had read aloud to him and Mary the adventures of Henry Walter Bates, the early explorer of the Amazon. This was to remain one of his favourite books. It was here, among the immense diversity of species, that Bill would develop many of his deepest insights and most audacious comparative flights of fancy.[12]

It is hard to say if it was animals or plants that most fascinated Hamilton. On later trips, he seems to have focused more on plants. He was especially intrigued by the adaptation of the plants of the so-called flooded forest to being regularly submerged under water. There were also the various species of insects, especially ants, that somehow survived the regular flooding of their nests. How did they do it?[13]

The Amazon provided exactly the whiff of danger that Bill seemed to crave in order to feel fully alive. His hero Bates had encountered a jaguar at close range during one of his walks, and had described the encounter in a lively manner. It is not clear that Bill ever came face to face with a jaguar, although he certainly saw some. Piranhas were a different matter, and he somehow avoided being attacked by them, although he insisted on swimming in uncharted waters. There is also no obvious anaconda or other snake story that I know of. But then again, all 'Bill' stories are far from being known.

An animal that seemed to inspire fear, at least at the sound of it, was the howler monkey, typical of the Amazon region. In fact, the sound was so awful that when Bill first heard it, he believed that he was listening to two jaguars in battle.[14]

Hamilton was proud to report that he was 'immune' to insect bites, but there were all kinds of nasty parasites that could get into the body. Indeed, he had an experience with an intestinal parasite, which he writes he caught by just drinking from 'an innocent-looking clear stream in the forest'. On another occasion he got infected by Leishmania (a parasite). Of course, he learnt from these incidents and developed counterstrategies. In a lengthy footnote to one autobiographical essay in *Narrow Roads of Gene Land,* Hamilton shares with his readers a hot tip when it comes to fooling a particular biting beetle. The trick, apparently, is to lure it out with a piece of bacon![15]

If anyone was directly associated with danger, though, it was Bill's host, Warwick Kerr. And here we get into stories about African killer bees. Indeed, it was none other than Kerr who in an effort to raise the domestic honey production in Brazil had originally imported Africanized bees in the hope of their mating with domestic ones to get a better honey yield. The African bees were wild, living in trees (or rather, they were ordinary honey bees who had gone wild). As a defence against 'harvesters' of various kinds, especially humans, they had developed a very sensitive disposition, prepared to attack at the slightest sign of disturbance.[16]

It was some of these bees that were let loose in an accident in 1952, and that wreaked havoc especially on smaller farm animals, as they spread throughout South America and even over the border to the United States. One big factor was people's lack of familiarity with these bees and the proper way to handle them. Later on, when better guidelines were established, their higher yielding honey production came to be appreciated, and Africanized bees were often preferred to domestic bees. One basic safety measure was to keep the hives far away from people and domestic animals—and from other hives.[17]

When it comes to interactions with killer bees, we have a good 'Bill' story, and this one told by Bill himself. He was helping take honey from what a farmer described as a 'strong' hive. The farmer had been impressed with the professional looking veil and leather glove that Bill typically used for approaching wasp nests and took him for an obvious bee specialist. The hive was in an upturned wooden packing case, which Bill now approached with his machete. He had not been sure what type of bees these were. He was soon to find out:

> Within about thirty seconds... everything had clarified beyond any doubt: everyone anywhere near to the hive was going to have to flee for their lives. My hands, in so far as I could still see them through the brown fog of bees, had become boxing gloves: they appeared as large brown moving balls... I was wearing only tropical clothing. The farmer rushed to his house, I to my jeep. A mile up the mountain road when there were more bees inside my vehicle than were accompanying it outside, I jumped out and ran further on foot into the forest, and finally, when there seemed to be more bees inside my veil than outside, I flung that off as well and ran further.[18]

He had been really worried about the fate of the farmer and his family. But after half an hour he came back to find the farmer and his family unharmed. The farm's pigs had also survived, however, the hens and the ducks were dead.[19]

Bill later reflected on this incident, comparing it to his earlier encounter with explosives. That for him had marked the end of his love for chemistry. But with bees, he noticed, the result was different. He found that despite the killer bee event, he still enjoyed honey bees, and still 'felt the need to walk up to any nest of any newly discovered social insect as close as he could'—which of course he would continue doing throughout his life.[20]

In addition to collecting wasps and wasps' nests, Bill continued his old habit of rolling over old logs and looking under old bark. But Brazil was also giving him ideas for new projects. 'I was beginning to find, in the tropical strangeness of the fauna..., one of my next main insect interests. This came to me in the lives, the sex ratio, and finally in the remarkable major evolutionary potentials, of all kinds of tiny arthropod inbreeders'.[21]

In other words, while he was finding examples for his existing theories in Brazil, his observations in that country helped to stimulate new research ideas and theories. It was a fruitful combination of deduction and induction, theory and nature, a system of feedback and multidimensional comparison between phenomena in England and Brazil.

Despite all distractions and side projects for Bill in Brazil, he knew that his main task was to revise his manuscript for the *Journal of Theoretical Biology*. The journal editor had informed him that the reviewer of his manuscript wanted the paper to be divided into two parts. The first part would be a derivation of the concept of inclusive fitness. The second part would show how inclusive fitness played out in practice in some concrete cases and give examples from nature. This looked like a lot of work to Bill and with all his collecting activities, he knuckled down to it rather slowly (this was the time before word processors and involved plenty of tedious typing and retyping). But he also knew that he had already been able to amass some convincing examples from nature.[22]

The result was a two-part paper, just as the referee had requested, with the derivation of the principle of inclusive fitness in the first part and the second part devoted to examples and discussion of extensions and problems. Also, just as the referee had requested, the second part now started with a quick review of the principle of inclusive fitness, referring back to Part I. The paper had now become more readable and contained new evidence. Bill had also profited from conversations with Kerr and his colleagues. The first part of the paper, however, retained some rather unusual mathematical notation. On 24 February 1964, whilst still in Brazil, Bill put the manuscript in the post to the *Journal of Theoretical Biology* and hoped for the best.[23]

Bill had made rather ambitious plans for his return journey. This South American trip was his golden opportunity to go and visit Colin in Barbados. In their letter exchange, Colin had suggested that they go on a wasp collecting trip together, working their way through parts of Mexico and the United States. After that the idea was to travel together to England via Canada. One stop on the way would be Chicago, where Colin knew a geologist friend who could put them up. Bill thought this was an excellent plan.[24]

With this in mind, Bill had made additional complicated travel arrangements for himself (this was typical Bill—throughout his life his professional trips often consisted of a chain of mini-trips of various kinds). This meant that before meeting Colin in Barbados, one of the other trips Bill made was a brief collecting expedition to Curaçao in the West Indies. It did not seem to bother him that this was in fact a considerable distance from Belém on the northern coast of Brazil, a place which he was planning to return to by travelling by jeep overland, taking the newly constructed Brasilia-Belém road from Brazil's capital to the Atlantic coast.

Looking at the map, the distance between Brasilia and Belém may not seem that huge but Brazil is an enormous country, and the road conditions are often quite horrendous. The potholes are notorious. (When I visited Brazil in 2004, the country's central government itself posted online warnings about the condition of the roads.) Bill could of course have taken an airplane but, ever the explorer, he preferred to 'weave his way' around the potholes, partly for the chance of collecting interesting specimens along the way, among other methods by sticking his hand into other holes at the side of the road. As already noted, Bill prided himself of having been stung by over 1,000 different wasp species. This was not mere bravado—one of his gathering techniques was deliberately to let an insect sting him and then catch it and examine it more closely! (With this in mind, that earlier Greek car accident story should probably be re-examined in favour of the 'bee inspection' version rather than the 'bee flicking' one.)[25]

Hamilton's companion on this long journey was Sebastião Laroca, a colleague of Warwick Kerr. The jeep itself was an old American type, and up to the task. Bill's ever thoughtful and responsible father had sent Bill a sum of money and persuaded him to buy an appropriate, used car for this expedition. (Bill, as usual, had been in close letter contact with his parents, during his trip.) Archie certainly remembered the incredible-looking wreck that his sons had brought back from their 1959 Anatolian adventure and realized that that trip had probably been more touch and go than his nonchalant sons were letting on. He did not want something similar to happen again. At the same time, he knew that his son was always operating on a tight budget.[26]

Collecting insects in South America, Bill Hamilton may have been partly reliving what he called 'his childhood mania'. But now he also had a professional framework for doing this. Before his trip he had entered into an agreement with OW Richards, the famous entomologist at London's Imperial College, to bring him South American insect specimens. Richards had also been advising Hamilton on Part II of his big 1964 paper, for which Hamilton was grateful. Knowing that he would still be travelling when his paper got published, Bill had been sending his father a list of people who were to be sent a reprint of his paper immediately. OW Richards was on top of that list.[27]

This is the way Bill describes the day of his major scientific triumph: the publication of 'The Genetical Evolution of Social Behaviour'. It was now already July:

> I was travelling up mainly overland from São Paulo towards Canada on my way home to Britain. Very probably the sun of the day that witnessed my paper going into the post from the offices of the *Journal of Theoretical Biology* would have seen me weaving my old American jeep between the corrugations, stones, and potholes of the Belém-Brasilia road (first of Brazil's transcontinentals, just 2 years old). At midday it would have blazed near vertically on the top of my head as I stopped at the roadside and collected wasps from some nest; later at sunset, if still able to pierce the haze, it would have seen me and my Brazilian companion, Sebastião Laroca, slinging our hammocks between low cerrado trees not far back from the stony or sandy piste where occasional lorries still groaned on into the night. For sure, both that day and night I was blissfully untroubled about the finer points of measuring relatedness.[28]

The agony and obsession of some four long years was over.

What was over, too, alas, was Bill's relationship with his girlfriend. In January 1964 Bill had surprised Colin by informing him that his girlfriend was going to come over and join him. She would then be driving with him to Belém, following along to Barbados and then coming with Colin and Bill back to England. But she did not come over as soon as expected, and appeared undecided whether to join them or not in her letters. In early March Bill told Colin that she would not be coming after all. She had decided to break up with him. The reason for this (as she informed Bill and also later explained to Bill's mother) was Bill's strange idea of marriage. Bill had told her that they should have no more than two children, of which one child would be his and another her child with another man. His girlfriend thought about this and realized that she could not agree. She also wondered what lay behind this suggestion. (She probably did not recognize Bill's experimental scenario as coming straight out of some book like *Out of the Night*.)[29]

Bill finally reached Barbados and met up with Colin, but now it was already late summer, and they almost immediately set off on their planned wasp collecting trip. In Mexico, they visited an active volcano, something that Bill had always wanted to see. It was after the volcano expedition that Colin realized that Bill had some kind of odd fever. It was not malaria, they thought, since the symptoms were different to those experienced by Bill when he had caught malaria on his last expedition to Eastern Europe. The fever seemed to come and go, and Bill just waved it off. Continuing by bus, they decided, on another whim, to get off in the middle of a desert and spend the whole day collecting there, planning to catch a night bus later on. Colin later reflected on this as a rather crazy idea, but he explained that this was the way they were at the time.[30]

Where was the bus? Colin and Bill were waiting in a small shed-like bus shelter in the middle of the desert in the middle of the night. Bill was by now feeling increasingly ill and needed to stretch out on the bench in the shelter. The air was stuffy, so they propped up the door with one of Bill's suitcases. They waited and waited. Both may have dozed off for some time. The air had got stuffy again. They soon realized the reason: the suitcase propping up the door was gone. Someone had stolen it.

Unfortunately, the case contained a number of Bill's carefully collected and labelled specimens of insects from all over Brazil, though not all of them. But much, much worse: that suitcase had contained a box of cyanide. Who in his right mind carries around cyanide? Colin explained to me that when he met Bill, Bill had run out of killing bottle material and had asked him for some suitable poison. Colin had then presented this request to one of his friends, a researcher who had been planning to get rid of the box of cyanide in any case.

The terrible problem that Bill and Colin saw immediately was that the substance in the box looked just like ordinary salt. The amount in the box would be enough to poison a whole community! What to do? Word had to get out to the thief himself, he needed to know what he was dealing with. And people had to be warned. Finally, the bus came, and Bill was able to get himself to a radio station, where his warning was duly transmitted. One can just imagine Bill feverishly trying to explain in Portuguese to confused Spanish speakers his unlikely-sounding story. Colin said he believed that the message got out and was effective.[31]

As one might guess, this was not the end of Bill and Colin's adventures. Having reached Chicago, with Bill now quite weak from his fever, they set out to look for Colin's geologist friend. Bill was now so frail that he had to lean heavily on Colin as they walked. They must have looked like two drunks, because before they knew it, they were picked up by the police and deposited in the back of a police car. Luckily, Colin was able to explain their situation to the police. The police's response was to give Colin and Bill police car escort to Colin's friend's house in Chicago.

***

Back in England Hamilton set about catching up on his journal reading. What had he missed during his year in Brazil, when he had only sporadically been able to follow the scientific literature? Now here was something interesting! He was gratified to see that *Nature*'s editor had finally changed his mind about accepting articles on evolutionary biology. He remembered only too well how his own brief submission in 1962 had been rejected almost by return post, and he had been told to send his brief ('Hamilton's Rule') altruism paper to a sociological journal. Well, perhaps the editor had not understood that his theory was expected to apply to all of life. In any case, in November 1963, there was a nice summary article by Vero Wynne-Edwards of his recent book, *Animal Dispersion in Relation to Social Behaviour*,[32] a book that had received considerable attention as a serious attempt to formulate a 'good for the species' explanation for animal altruism.

Although Bill thought that Wynne-Edwards' particular explanation for animal altruism did not really hold water (according to Wynne-Edwards, bird populations regulated their reproduction based on information about their numbers that they gleaned from yearly mass gatherings), he much admired the naturalistic observations and general knowledge of this man. In his own 1964 contribution, Bill did not criticize Wynne-Edwards; instead, he cited the Scottish naturalist a number of times as a source of useful naturalist examples.[33]

But what was this? In the 24 March 1964 issue of *Nature* Hamilton saw a couple of letters to the editor. One letter was unusually long and included a mathematical model. It was entitled 'Group Selection and Kin Selection'. The author was John Maynard Smith. 'Kin selection' was a term that Maynard Smith had coined here and introduced as a contrast to group selection.

Bill sensed that something was going on because he vividly remembered the complete lack of interest shown by Maynard Smith in the early 1960s when the two had been introduced by Bill's genetics supervisor Cedric Smith and Bill had tried to explain to Maynard Smith his interest in the genetics of altruism. Cedric Smith had probably hoped to get some assistance for Bill from his colleague at University College. Bill seemed to be bogged down in a potentially fruitless pursuit and was also having trouble with the mathematics. Perhaps Maynard Smith could help? That did not happen. Somehow at this meeting Bill was not able to convey to Maynard Smith what he was after, and Maynard Smith had not been interested in inquiring further. In fact, the meeting was a total disaster.[34]

But now here was Maynard Smith, seemingly fully engaged in an area that he had earlier shown no interest in whatsoever. This was odd. Bill read on. After first commenting on Wynne Edwards' article, Maynard Smith went on to offer an alternative to that author's group selectionist 'good for the species' argument: 'It is possible to distinguish two rather different processes both of which could cause the evolution of characteristics which favour the survival, not of the individual, but of other members of the species. These processes I will call kin selection and group selection. Kin selection has been treated by Haldane and Hamilton.' (Here Maynard Smith gave references to Haldane (1955) and Hamilton (1963).)[35]

Reading further it became clear to Bill that the gist of what Maynard Smith called 'kin selection' was exactly the idea that altruism can spread if the recipients of altruism are related—an idea that he regarded as his own. Maynard Smith wrote:

> By kin selection I mean the evolution of characteristics which favour the survival of close relatives of the affected individual, by processes which do not require any discontinuities in population breeding structure. In this sense, the evolution of placentae and of parental care (including 'self-sacrificing' behaviour such as injury-feigning) are due to kin selection, the favoured relatives being the children of the affected individual. But kin selection can also be effective by favouring the siblings of the affected individuals (for example, sterility in social insects...) and presumably by favouring more distant relatives.[36]

This was all very upsetting to Bill. His life's work so far, his long *Journal of Theoretical Biology* paper, was newly published (in July that year), having been in the pipeline for several years, but during the time that he was revising and resubmitting his paper in 1963 and early 1964, Maynard Smith had managed to get in this brief communication in *Nature*, of all journals. He had effectively scooped Bill by coining his own term 'kin selection' for what was essentially Bill's idea. Certainly, he had referred to Bill's brief 1963 paper but at the same time he had given priority to Haldane.

This was especially irritating because Bill's own 1963 paper (which he had submitted to *Nature* without success) had in fact contained a discussion of Haldane. But Bill's reason for introducing Haldane at all had been precisely to show that Haldane *failed* in his modelling attempt to derive altruism from group selection (in his 1932 book). Bill had used Haldane's failure as a foil for his own 'Hamilton's Rule' which he introduced as 'an extension of the classical theory' based on individual advantage. But here was Maynard Smith acting as if Haldane had 'said it first' (he was referring here to a later 1955 article by Haldane). Bill was convinced that Haldane had not done so.

Bill felt doubly robbed, by Maynard Smith and by Haldane, of something that he had considered his scientific priority. Nobody before him had been able to show in serious mathematical detail just how the counterintuitive feature of self-sacrificing behaviour could actually make evolutionary sense—that is, before his 1964 paper, where he introduced the seminal concept of 'inclusive fitness'. (In his 1963 paper he had mostly given a quick sketch of 'Hamilton's Rule', the general condition for the evolution of altruism.) But Maynard Smith's 'Group Selection and Kin Selection' piece was published in March, *before* Hamilton's *Journal of Theoretical Biology* paper, which did not come out until July. To top it off, Maynard Smith's piece even sported a mathematical model (his 'haystack' model), illustrating how, given certain artificial assumptions, kin selection might operate in principle. So, Hamilton had tried to use Haldane's shortcoming to boost his own theory, but Maynard Smith, Haldane's favourite student, had put his master back on the throne. Also, there was no doubt that 'kin selection' was a catchy term. Maynard Smith's further moves would have to be closely watched.

This episode would create a deep psychological wound for Bill which would take a long time to heal.

Meanwhile, Bill still had not attained his PhD. By now he had published what he thought was an important paper, but he felt academically vulnerable. One older colleague who was able to come to his direct assistance was EO Wilson from Harvard. Wilson was visiting London in 1965 in order to give a paper at a meeting of the Royal Entomological Society in London. He was interested in talking to 'young Hamilton', whom he had 'discovered' in 1965. In his autobiography Wilson tells a story about how he underwent what he called a paradigm shift during a long train ride as he was struggling with Hamilton's 1964 paper, going from incredulity to conversion. Wilson regarded himself as the world expert on the social insects, and here was this newcomer explaining the evolution of their social behaviour! Finally, he tells us, he had to give in.[37]

Walking with Wilson in London before the conference Hamilton told him about his advisers and his difficulties in attaining a PhD. Wilson thought this was a great pity. He decided on the spot to devote a significant portion of his talk to presenting Hamilton's ideas to this audience, which he knew contained the top figures in British entomology. The result was that a third of his hour-long presentation was spent on Hamilton. Wilson had also prepared responses to potential objections. 'It was a pleasure to answer them', he recounts. 'When once or twice I felt uncertain I threw the question to young Hamilton, who was seated in the audience. Together we carried the day.'[38]

Memories such as these, of friendly support from colleagues from the United States while facing difficulties at home, were to slowly start tipping the balance when it came to Bill's decision on the ideal place to work. But nothing drastic was to happen for more than a decade.

8

Sex and Death

Silwood Park is the Field Station of Imperial College, located to the northwest of London, in Berkshire, near Ascot. Originally a country estate, its grounds offer a splendid environment for all kinds of field explorations and experiments, encompassing as it does environments ranging from woodland to marshland. Impressive huge cypresses and other planted trees surround the red brick Victorian mansion, whose halls and living quarters have been turned into spaces more appropriate for research scientists. A large plaster beetle now occupies the place of the family coat of arms in the large common room on the ground floor—a mischievous indication of the takeover by entomologists. The huge conservatory and the lawns seem to cry out for tea parties, and afternoon tea indeed forms an important part of the overall atmosphere of this pleasant environment.

It was here that Bill Hamilton spent the beginning of his academic career, from 1964 to 1977. Actually, to be more precise, he did his research and advising of graduate students at Silwood but travelled to Imperial College's London lecture halls in South Kensington for his undergraduate teaching. During his stay at Imperial College, Bill was to publish some of his most important new ideas after his first breakthrough with his inclusive fitness theory: his theory of senescence, his paper on extraordinary sex ratios, his 'geometry for the selfish herd', and his and Robert May's paper on dispersal in stable habitats. It was also at Silwood that he clarified and expanded his original ideas about the evolution of altruism and developed his new theory about the origin of social evolution in insects.[1]

In 1964 Hamilton had not yet received his PhD. What he had achieved, though, was his huge *Journal of Theoretical Biology* paper and his shorter 1963 summary in the *American Naturalist*. Still, those were only two papers, and both very recently published. Bill's scientific standing was by no means clear. It took the imaginative director of Silwood Park, entomologist OW Richards, to realize Hamilton's potential despite his lack of a doctorate, and what was more, lack of formal teaching experience. He was, however, familiar with Bill's scientific contributions and, as mentioned, had even advised Bill on his 1964 paper. Another feature that made Bill attractive to an entomologist was surely Bill's enormous knowledge of insects, especially wasps. By now, Bill had made valuable observations on insects in the Brazilian rainforest and collected some rare species. (Indeed, many of these were later to be exhibited in the Natural History Museum in London and written up as part of Richards' major taxonomic work on South American wasps).[2]

Bill's own story of how he got the job is rather entertaining. Times were very different in academia then. In this case it took only a letter of recommendation and an interview to secure the position of Lecturer in Genetics at Imperial College, London. Bill had only one rival for the position, and this man was in fact offered the position first, but declined. Probably he got a better job elsewhere. According to Hamilton, 'jobs for academics almost fell from the trees'. But it may also have been that his relative lack of self-presentation skills sent something of a warning signal to those looking to hire teaching staff. Although they were willing to take a chance on his academic promise, this was, after all, also a teaching position. Would this shy young man be able to handle hordes of unruly undergraduates?[3]

The teaching turned out to be a bigger challenge than expected. Bill was by no means a nonchalant instructor—according to his sister Mary her brother wanted desperately to be a good teacher. Before his first official teaching job, Bill practised on Mary and her friend David Harris, giving them a short course in statistics in Mary's London apartment.[4]

One problem was that Bill's model for teaching came from classes that had typically not been very big. Most recently he had attended small graduate lectures. Moreover, his particular course of study had offered him no opportunities to give presentations—or at least he had avoided such opportunities. In any case, he had no experience of how to present science to a bigger group. Also, at the time, no formal teacher training existed and new teachers were totally on their own.

Despite Bill's good intentions, his classes were a disaster. Here were young students oriented toward learning the basics of genetics, and here was their new genetics instructor, who wanted to teach them everything that *he* thought was important, including his own brand new theory of inclusive fitness.

Let's hear a testimony of one of Bill's students at the time:

> Bill first shambled into my life in 1964 when I turned up for the first presentation by the new genetics lecturer at Imperial College, London. I was a third-year student at the time. Bill was a lecturer and it is no exaggeration to say that Bill's lectures were models of everything a lecture should not be. They were disordered, not leavened by anecdote or theatricality, and they juxtaposed the relatively elementary with the very cutting edge of the subject.[5]

Bill taught these students 'male arrhenotoky' (at the time Bill used this term instead of haplodiploidy) and other esoteric-seeming concepts. But, says our informant, 'like generations of students before and since, we taught ourselves such elements of the topic as we decided we needed to know.' Still, there remained a mystery. How was it that the omniscient OW Richards had chosen this young scientist as their lecturer in genetics? Speculation abounded. The students believed they could find the answer in the pubs in South Kensington. And there they encountered a totally different Bill Hamilton. It soon became clear, as one student formulated it, that 'Bill lived on an intellectual planet that few of us could see, let alone share.' It was especially Bill's deep knowledge of natural history that became the key to their respect for him in other matters too, such as his insistence that research on social behaviour could and should have a mathematical basis. For serious students, Bill was an inspiration, and a model to follow.[6]

Bill had clear plans for what he wanted to do in his new position. His primary interest was research. There was no doubt that the best place for an office would be at the Field Station in Berkshire, close to the edge of Windsor Great Park, not in Imperial College's London quarters. This meant that his office would be located in Silwood House itself, surrounded by a research park (the former estate), and he himself would be surrounded by scientists from the departments of entomology and plant pathology, housed in the same main building. For Bill, the park provided a daily distraction as he walked to work from the modest room he rented some distance away.[7]

This was the ideal place for Bill to continue the work on wasps and bees that he had started in Brazil. He also had some new ideas. But he soon realized that he needed to start finishing off some earlier projects first.

Bill, as we have seen, habitually worked on several projects at once. These were typically overlapping. Being an enthusiast and getting new ideas from all kinds of sources—as he walked in Nature, travelled, browsed in books, heard about some evolutionary oddity, read a piece of literature, or even sometimes as he lectured himself—he typically became excited by something new before an older topic had been fully explored or written up. But Bill did not run away with his new obsession, his scientific professionalism showed itself in disciplined efforts to 'tidy away' and write up his older findings before embarking on a new adventure.

This is how it was with the topics of sex ratio and senescence, the two big projects immediately following his 1964 paper. He had, in fact, started researching both earlier and had made some good progress before he went to Brazil in 1964. Bill saw sex ratio and senescence as closely interwoven with his thinking on inclusive fitness. 'They represented the most difficult and most interesting side issue of my main theme, those issues that needed special care and that had philosophical interest in their own right.'[8]

Senescence was one of the many big and intriguing topics raised by Fisher in *The Genetical Theory of Natural Selection*. In typical fashion, Fisher had treated senescence in a very cursory way. What did he mean when he wrote that he found it 'not without interest' that 'the death rate in man takes a course generally inverse to the curve of reproductive value.'[9]

Bill was sure that this sentence had some deep meaning. This was his general experience of Fisher—after all, he had successfully been able to engage with Fisher's thinking before when he developed his inclusive fitness theory. Fisher was always onto important things, and usually right, but this did not necessarily mean that he presented the most general possible case. It had been a major coup for Bill to use the idea of inclusive fitness to extend Fisher's own Fundamental Theorem to include social behaviour. So, by analogy, it seemed that what Bill would need to do with any other Fisher statement would be to think deeply about what exactly might be going on evolutionarily speaking and then try to use mathematics to capture the phenomenon and explore its implications.[10]

The problem at this point, however, was that Bill had almost lost interest in the senescence problem. This happened as soon as he found that his ideas were not new: he had formidable forerunners, such as PB Medawar and George Williams. Typically, rather than reading the literature on a subject, he was working from scratch in an independent manner, only to be bitterly disappointed to discover later that others had been there before, reaching similar conclusions.[11]

Still, Hamilton had his unique approach to fall back on. Unlike many others, who were satisfied with partial solutions, Bill was always aiming for maximum generality. As we saw from his letter to George Williams in 1964, he was aware that Williams had been there before him, but still thought that he had something additional to offer. And obviously he had. The result was his 1966 theory of senescence, published in the *Journal of Theoretical Biology*. That paper mostly agreed with the ideas of his two forerunners, but it had the advantages of being internally consistent, identifying the most important factors, and not 'bowing to authority'. (The last expression referred to Medawar's apparent need to quote Fisher despite the fact that his own theory did not support Fisher. Bill, on the other hand, had a tendency to be rather cocky sometimes in his early writings.)[12]

Fisher saw a direct connection between reproductive value and mortality. He believed that the reproductive value of an individual started diminishing as mortality increased. But Hamilton found that there was no necessary connection between sex and death. In fact, as he later noted, he got a large psychological boost out of finding that his hero was wrong on this point. Instead, Hamilton went on to develop models to demonstrate that senescence as such is inevitable, quite independently of reproductive value. No organism can escape it. One reason is that genes that are good for us at an earlier point in life become detrimental in later age (this point was also made by Williams).[13]

Senescence was of course a topic of particular interest to people. And for humans there existed demographic data that could be used. Bill had started looking into this topic already at the LSE under the guidance of John Hajnal. He had originally wanted to find data on mortality and longevity of populations as close to the conditions in the Paleolithic as possible. But that was difficult. The closest data that Bill had found was clearly off, but still of some value: information on Taiwanese late-19th century near-subsistence families. He had already written up part of this demographic research in late 1962 when he stalled in his altruism project, and even considered it a possible topic for his PhD. Now this work could finally be published—it came in handy as his empirical study for 'The Moulding of Senescence by Natural Selection'.[14]

Silwood Park was a stimulating place for Bill in yet another respect. He was surrounded by enthusiastic entomologists, exchanging insect lore of various kinds. His colleagues appeared to have an inexhaustible amount of almost unbelievable, scandalous gossip stories about these small creatures. Not only did some insects in their relentless incestuous pursuits appear to violate all moral laws, but their patterns of inbreeding stretched the most creative imagination.

The examples he was able to collect at the Field Station 'would have delighted even Fabre's taste for the bizarre and for the quasi-moralistic reflections that they could cast toward humankind', Hamilton mused.[15] Take for instance the button beetle, found in date stones or buttons of 'vegetable ivory' made of palm seeds. A female virgin beetle, after having excavated a hole, first uses parthenogenesis to produce four or five males. She waits for her first son to mature and mates with him, whereupon she eats him as well as her other sons. At this point she is ready to produce her main brood in the chamber that she has first enlarged. There she now lays her eggs fertilized by her first son plus about three other males. The result is a brood of about 70 females.[16]

Or what about 'incestuous love-making by babies within the womb', Bill's terse labelling of one of his discoveries:

> I had pulled a piece of white-rotten wood from a dead branch and had noticed some odd pearly droplets adhering to the velvety broken surface of fungal mycelium. Under a microscope the drops proved to be even more strange than I had imagined: each possessed a head and legs... Each drop was the swollen body of a female mite distended with eggs and young. Tiny fully formed adults were bursting from eggs inside their mother... The males... spent all their time searching for new sisters in the crowd and right there, within their mother's womb, copulated with them.[17]

The main point of all this for Bill, however, was the scientific challenge these inbreeders presented. There had to be an explanation for all this. This type of inbreeding in insects had to be adaptive. And the overall explanation would also have to fit with his earlier findings in kinship theory.

It was clear to Hamilton as he was observing some of these behaviours under a microscope and identifying males and females, that the optimum ratio between the sexes was not the expected 1:1—one of Fisher's main conclusions, which by now everybody took more or less for granted. Instead what Hamilton saw was a clear deviation from Fisher—to the great, persistent advantage of females. In fact, as he continued assessing the male-female ratio (in insects in the field or bred in the laboratory), he found that the condition of inbreeding often resulted in a strong female bias—sometimes up to as much as 9:1. (If it took just one male to inseminate a number of females the optimum rate of females to males could be way off Fisher's expected 1:1).[18]

Note, however, that in these cases the reproductive pattern had to be one of inbreeding. Now, where in nature was inbreeding especially likely to take place? Bill's answer was: in closed environments, such as those within rotten wood or under bark, or in discarded pupae shells, or in certain types of closed fruits or flowers. Bill felt there was a meaning behind his life-long interest in chiselling bark, kicking tree trunks, and digging into holes in search of insects. These closed spaces housed valuable examples of his emerging theory of 'extraordinary sex ratios'. (Later he was to develop a theory suggesting that the probable evolutionary origin of whole new taxonomic orders of insects may have been the exact result of this kind of cloistered incestuous life style.)[19]

The image Bill had of himself was that of a hunter. He felt excitement as he was going off into the woods: what insects and what sex ratios would he find today? Would Nature oblige him, as he counted the males versus the females? Was he onto one of Nature's hidden patterns? But it was not enough to collect material in the field or in the laboratory—Bill's real hunting instinct expressed itself predominantly in libraries. He needed to find literature documenting the existence of extraordinary sex ratios. And so he combed through all kind of books that just might give him the desired information, in library after library, most of them conveniently located very close to where he was teaching in South Kensington.

'I think I may have found something close to being another eusocial insect...' he reports in a letter to a colleague, immediately adding—for effect—'in a library'. What was special about Bill was that it didn't seem to matter if he had found something himself in Nature or if he had encountered it in a dusty volume, often devoted to some very different purpose than the one he was using it for. Bill's love for and comfort level with books, even quite outdated ones, was of great use, as was his love of language. Bill was good at reading all kinds of texts and gleaning from naturalistic descriptions or throw-away comments the kind of information he was looking for.[20]

For Bill, the libraries were stimulating in another way, too. They presented an interesting personal and professional challenge, one that he savoured. This challenge was twofold. In these quite specialized libraries he took pride in not asking any librarian for guidance. He became his own expert. Sometimes the issue was more basic: avoiding being found out in libraries accessible only to special card holders. Hamilton tells us about the strategy he used to fool librarians in such places. The point was to walk in with great confidence and immediately start one's card catalogue research as if one knew what one was doing.

Hesitation would be an immediate give-away. Hamilton's success indicates that his ethological insights about librarian behaviour were on target—or perhaps he deliberately over-dramatized the threat of these librarians to give an additional thrill of danger to the hunt. Bill knew how to make an adventure out of the most mundane occurrence.[21]

Already during his time at LSE he had been studying the entomological literature closely enough to see that there was a real puzzle about the departure from Fisher's 1:1 principle in regard to small and inbreeding insects. In addition to collecting examples of this phenomenon, he had also started developing the argument for the main model of what was later to be called 'local mate competition'.[22] But that was not enough. What Bill required to clinch the case was to be able to test his emerging sex ratio theory in some way and assess its predictive power. And for this he had a great new research tool at his disposal: computer simulation. (This was thanks to the course he had taken in programming at the University of London on his own initiative, without consulting his advisers.)

The way he used simulation, Bill explained, was to check his own mathematics or logic:

> Never having much confidence in maths or logic, or at least not in the lines of these that I generated for myself, I was beginning to use simulation repeatedly to check some of the more unexpected predictions from my thoughts or my algebra.[23]

Hamilton's earliest programs when he was still at LSE had actually dealt with demographics. It was in fact from running simulation programs on the development of the sex ratio of a population under various genetic and ecological assumptions that he realized that a determinant of the growth rate was the sex ratio itself. That made him re-examine Fisher's idea on the sex ratio.[24]

This was the time of enormous mainframe computers and nightly runs—the heroic early stages of computer use in science. Undaunted, Bill conducted the runs on his university's large computer in Gordon Square in London—sometimes assisted by his younger sister Janet and his friend (and childhood neighbour) Marian Luke. The use of the mainframe computer presented its own challenges. It was almost to be expected that one's programs had 'bugs' in them, and this typically meant another long wait. Bill's response was simply to treat this as yet another game of outsmarting authorities (and lightly breaking the rules). To speed up the process, therefore, Bill corrected his reel of punched tape by covering up some holes and punching in others, before he got quickly back in line. This action was absolutely forbidden, which added to the thrill. (The reason was that the sellotape used to cover the holes often clogged up the paper-tape-readers). Another strategy of Bill's was to keep his programs extremely short, which resulted in his perpetually bad programming style.[25]

But what was the result of all this research, forbidden procedures or not? Hamilton's paper 'Extraordinary Sex Ratios' was published in *Science* in 1967. To his delight he could actually declare that when it came to his sex ratio theory, 'its predictive power was almost comparable to what is standard in the hard sciences'. Indeed, Bill Hamilton's paper on the sex ratio is considered by many to represent one of the best demonstrations ever of the actual power of natural selection in its ability to provide quantitative predictions. It was once again a Hamiltonian tour de force. Hamilton later wrote that he was 'more proud of this paper than of almost any other I have written. It helped sex ratio theory well on its way to being the section of evolutionary theory that best proves the power and accuracy of the Neodarwinian paradigm as a whole.'[26]

Still, Bill knew whom he really had to thank for his marvellous results:

> [T]he insects continued to provide the best examples of all; indeed it was as if in return for the affection and curiosity I paid to them they kindly agreed to fit my quantitative theories more closely than a population or evolutionary biologist normally dares hope for.[27]

Cooperating insects, indeed. And yet another thought occurred to him:

> But then perhaps it is just that there are so many insects so diverse that if you just look far enough you can find examples to fit any theory, however bad ...[28]

Hamilton's paper about extraordinary sex ratio turned out to be very rich in content. He had seen this phenomenon before and reflected on this.:

> As happened with the ideas about kin selection I had worked on earlier, identification of the evolutionary advantages that determine sex ratios turned out to have much wider applications than the insect examples that I had first been puzzled by and had used to illustrate and develop them.[29]

Later on he was able to identify at least five new ideas in his 1967 paper. He listed them as follows:

1. The levels-of-selection debate
2. The idea of conflict within the genome

3. The 'evolutionarily stable strategy'
4. The initiation of game theoretic ideas in evolutionary biology
5. More indirectly, by emphasizing the costliness of male production for females and for population growth, as well as the ever ready 'option' (among small insects, for example) of parthenogenesis, the paper helped to initiate debate over the adaptive function of sex.[30]

Hamilton is correct in tracing the origin of these various strands of thought to his 1967 paper. Although he usually operated with individuals, just like Fisher, on a couple of occasions he did play with the idea of the 'selfish gene'. He was taking the gene's eye view as a methodological device just as he had done already in his earlier 1963 paper where he launched Hamilton's Rule. Hamilton's anthropomorphic approach served him well in his biological imagination. The most surprising idea launched in this paper was probably the notion of 'intragenomic conflict'. Hamilton could explain intragenomic conflict as resulting exactly from the fact that the Y and X chromosomes would have different 'interests' and 'want' different things (this was later developed further by Robert Trivers and David Haig). And it was indeed Hamilton's sex ratio paper with its 'unbeatable strategy', a game-theoretical concept that was to inspire the biologically handy concept of 'Evolutionarily Stable Strategy' (ESS), which John Maynard Smith and George Price were to introduce five years later. It is also not surprising that it was at least partly this work on skewed sex ratios, with the seeming dispensability of males, that led Hamilton on to his next major project: the evolutionary rationale for having sexual reproduction at all or put in another way: what use are males?[31]

The 1967 paper also had a personal, psychological meaning for Bill. Having identified intragenomic conflict, he felt that he now had the scientific answer to something that had been bothering him, his sense of an inner split. As he later reflected:

> Seemingly inescapable conflict within diploid organisms came to me both as a new agonizing challenge and at the same time as a release from a personal problem I had had all my life.... Given my realization of an eternal disquiet within, couldn't I feel better about my own inability to be consistent in what I was doing, about my indecision in matters ranging from daily trivialities up to the very nature of right and wrong?... As I write these words, even so as to be able to write them, I am pretending to a unity that, deep inside myself, I now know does

> not exist. I am fundamentally mixed, male with female, parent with offspring, warring segments of chromosomes....[32]

At Silwood Park Bill much enjoyed being the only one pursuing evolutionary questions. This was the kind of situation he really liked. He could go about his own theorizing in a data-rich environment, which in this case encompassed both Silwood Park and Windsor forest, the library, and the wonderful stories told to him by his entomological colleagues. As before, it appears he liked to have people around, as long as they did not bother him. 'All was just as I chose' nicely sums up his contentedness with his early years at Silwood Park.[33]

But of course no pleasure could be greater than talking about evolutionary puzzles with a kindred soul. In 1967 when Bill was in charge of inviting lecturers to Silwood he asked George Williams if he would like to come and give a talk. And Williams came, all the way from Iceland, where he was in 1966 and 1967 on sabbatical (moreover functioning in Icelandic, a language which he had taught himself). Williams later reflected on his visit to Silwood Park:

> This was just two days after visiting David Lack at Oxford, UK. Thus, on this one brief trip I met two most pleasant and immensely important scientists of widely different ages. That the youthful Hamilton could be so far ahead of almost all living biologists on such important issues was mighty encouraging.[34]

Williams had been in correspondence with Bill not only about kin selection and senescence, but also about his own 1966 book, *Adaptation and Natural Selection*. Here and there in the book, Williams acknowledges his debt to Hamilton and his 1964 paper. (The main thesis of Williams' celebrated book, though, was a methodological exhortation to his fellow evolutionists: do not postulate that evolution happens at a higher level than is necessary. This was a way in which Williams wished to discourage any unreflexive use of group selectionist arguments among his colleagues. He professed to have been provoked to write this book by having read Wynne-Edwards.)[35]

Although Bill's desire to see Williams was surely genuine, of course, the fact that Bill had been able to get such a leading biologist to visit would also reflect positively on him. Note that at this point Bill had not yet completed his dissertation. Rather than picking a topic suggested by an adviser—something that he had already contemptuously dismissed in Cambridge—he had opted for the difficult route of publishing original papers, in a sense using the judgement of the scientific community as a tool against the scepticism of his doctoral committee. He had also had the good fortune of having the support of Ed Wilson at the London entomology conference. Perhaps Williams could back him up further, legitimizing his approach as a sound one? After all, Bill was being evaluated on his performance as a lecturer at Imperial College, and it would not hurt to have some support from a big name.

It is probably not a coincidence that this peak performance of Hamilton's in his sex ratio paper came about during a time when he was in a state of heightened emotion, seeing the woman who would soon become his wife, Christine Friess. Christine was a medical student in London (she later switched to dentistry), and a friend of his younger sister Janet. Christine used to travel to see him in Silwood Park on weekends in the flat he shared with Yura Ulehla, a research colleague from Czechoslovakia. Christine was the daughter of Herbert Friess, a Lutheran pastor who later also cared for Anglican congregations. At this point he and his family were living in County Mayo in Ireland, where he was in charge of a small congregation. Canon Friess had left Germany around the time of the Second World War when the Nazis had tried to prevent him from practising his profession and found refuge and work in England. It was through Christine helping Bill with a German translation that the two had initially met. A mutual friend said that there was a 'timeless' quality to both of them.[36]

By this time, Bill was in his early thirties and all his closest friends were already married. Bill and Christine were married in Ireland in the summer of 1967. The wedding picture shows the newlyweds standing just outside the church. Christine is very pretty in a small cloche hat, with a dreamy look in her eyes. Bill is staring straight into the camera—his typical behaviour in front of a camera ever since he was a child. The best man is Colin Hudson. It is a small wedding, with only the closest family present. (Bill's sister Mary is missing from the picture.)

This happy event was soon to be overshadowed by family tragedy. In October the same year Bill's younger brother Alex (Leco) was killed in a mountain climbing accident; he was only 18 years old. His school year at the University of St. Andrew's in Scotland had just begun, and this was the first outing of the mountaineering club. When everyone was asked about their previous climbing experience, Leco presented himself as not an absolute beginner. He had been climbing somewhat apart from the others together with an inexperienced fellow student, to whom he was tied with a rope. Suddenly Leco's partner lost his grip and together they fell down quite a distance into the water below. Before anyone could rescue them, they had both disappeared from sight. Their bodies were never found. Later, a funeral was held on the cliffs over St Andrew's Sound.[37]

This was a great blow for the Hamilton family, and a personal one for Bill. Leco is commemorated by a memorial wall, forming the back wall of the Nissen hut in the Oaklea garden (now used largely for storing of archives). The atmosphere of the whole place is akin to a chapel, with the light coming in through what looks like a large stained glass pattern. On closer inspection, it turns out that it is a clever mosaic of the bottoms of glass bottles, many of them cobalt blue. There is also a plaque to commemorate the incident. This was one way Leco's family and friends could do something concrete and beautiful together to express their grief.

9

Challenges of Social Life

In 1968, Bill Hamilton had arranged to go to Brazil again for a year, supported by the Royal Society. He was to do his own research, but also be part of the official Royal Society and National Geographic Society's nine-month expedition to Mato Grosso, a huge area and the last virgin land in Brazil. His wife Christine travelled with him in the role of the expedition's official dentist. (Christine had attained her dentistry degree by the time of their marriage. Bill, being something of a Victorian in his attitude to sex roles, did not particularly want her to work for a living, but when money was needed, Bill was happy for Christine to take on temporary dentistry jobs.)[1]

To allow for his travel, Bill had arranged to cover his teaching duties by doubling up on teaching courses during the previous term and making other necessary arrangements. Also, in 1968 Bill had finally been awarded his PhD, mostly a compilation of his four published papers with a covering essay. His strategy to publish directly in journals had borne fruit. Colin Hudson told me of the great anger and frustration that Bill had felt in regard to his supervisors. It is true that they were critical of his abstract and mathematical approach. At the same time, they obviously executed what they saw as their advisory role and asked for the type of evidence that they believed was required. (Interestingly, it appears that Hamilton's advisers actually helped him with both his sex ratio and senescence papers. At least Bill thanks them for their guidance in the acknowledgements sections of these papers.)[2]

Hamilton seems to have had a hard time in general acknowledging superiors or supervisors who exercised their authority just because of their position. He himself tended to operate in an ideal space according to how things 'ought to' be. (They usually weren't.)

There are some interesting details about Bill Hamilton's dissertation. His supervisors were Cedric Smith and John Hajnal, but the external examiner was John Maynard Smith. This last fact is mentioned nowhere by Hamilton himself, but we find it mentioned in passing by Maynard Smith.[3] What we do not learn either is that Bill is said not to have spoken a word to John Maynard Smith during his *viva* (the formal defence of a student's dissertation). There is yet another interesting detail. Bill is said to have requested a DSc rather than a PhD for his degree. A DSc is an extraordinary type of degree that is usually given for especially meritorious work beyond a PhD. This was not seen as applicable in Bill Hamilton's case, but it surely demonstrated his keen sense of the value of his own contribution.[4]

After Hamilton had obtained his doctorate, his genetics adviser Cedric Smith was impressed enough to offer him a position as lecturer at University College, a position that Bill declined.[5] He already had a position, a fact of which Smith appears to have been unaware. Moreover, at that point he simply wanted to go off to Brazil.

In a letter to his good friend Yura Ulehla Bill describes the expedition and the general area. As the last virgin land in Brazil, Mato Grosso needed to be explored before any development took place. He describes his own work as rather specialized, and not related to the rest of the expedition. He writes that he has found some new species to add to his collection, and he has been stung by wasps. In general, he is having a very pleasant time. Christine, also an avid letter writer, reports on the daily life in the camp and how they arranged a system for taking showers. She tells a story of a tame capybara, which used to follow people around and took pleasure in nipping them as they were swimming.[6]

It is hard to collect wasps' nests without being stung. Anyone doubting this can check the book about the Mato Grosso expedition written by the expedition's leader, Dr Anthony Smith, and the photographs there of Bill and Christine with badly swollen and distorted faces. But Hamilton was undeterred, he was a professional. (Richard Southwood told me he thought Christine had possibly been even braver.)[7]

Undergraduate students from Imperial College who remembered Bill Hamilton as a mumbling, though knowledgeable, lecturer, would certainly have been surprised to see this naturalist side of him. This was another Bill Hamilton altogether, and one with a big smile on his face. Just as he relentlessly hunted in the libraries back in England, so he tackled the fauna and flora around him in Brazil. His eyesight was extremely good, and he could spot apparently interesting phenomena at quite some distance.

Sometimes Hamilton had to use a good dose of ingenuity in order to get to the nests that he wanted. And here some knowledge of engineering and construction principles could come in surprisingly handy. Consider the following description by Bill himself:

> This nest of the wasp *Agelaia angulata* was in the central chamber of a large disused termite mound in Mato Grosso, Brazil. The mound had been previously excavated and its colony destroyed by an armadillo, whose burrow the wasps later adopted for their entry. To reach the nest I had to dig a 4-m trench (covered with tarpaulin for protection), followed by a 2-m tunnel that enabled me to break into the chamber from below and behind. The worker wasps are extremely fierce and leave their stings in human flesh like honey bees: even after losing them they continue to cling, buzz, and mimic the motions of stinging, so contributing to a general terror; where unable to sting due to a bee veil, they jet venom and this often reaches the eyes. In spite of bee clothing (veil and gloves) I was incapacitated for two days after my first observation of this nest.[8]

Bill took his scientific task very seriously. Once again he was collecting specimens for OW Richards. A crowning achievement of all Bill's careful collection work was the insect that was later named after him: *Stelopolybia hamiltonii* OW Richards.[9]

For maximum efficiency, Bill needed to have his equipment handy:

> He was the complete field naturalist. There was always a magnifying glass around his neck, and binoculars, a camera, and a rucksack of killing bottles, notebooks, and the like.[10]

Bill's methods in the field were idiosyncratic, but they produced results:

> To be with the biologist Professor WD 'Bill' Hamilton... in the field, particularly in Amazonia, was exhilarating, educational, and usually terrifying. To find wasps—with hymenoptera a favoured group—he would thrash wildly at passing vegetation. Any co-walker, sanely situated some 40 yards astern, would suddenly see Bill stop, place a hand where he was being stung, carefully pinch off the insect, examine it closely, recognize the species, and then know where to look for its nearby nest.
>
> Even if the nest was very high in a tree, perhaps resembling a three-foot fir cone, Bill would fashion a club to bring its occupants nearer. He would smite the tree and observe what he called the 'bombing'.

> Clusters of disturbed wasps would drop from the hole at the bottom, hanging together in a compact mass, and then separate to search a particular level. With these horizontal groupings each seeking the cause of their disturbance, there was an inevitable query. 'Why aren't they searching where we are standing?' 'Probably,' he answered, 'because most of their intruders come from higher up.'[11]

There is no doubt that Hamilton was enchanted with Brazil, especially the Amazon jungle. He brought back with him not only wasps' nests but also other interesting items, such as lianes, which when they dried up looked like porous cricket bats with holes. Later on, Bill and Christine put up a huge photograph of the Brazilian rainforest as a backdrop for their living room. The Hamiltons had effectively brought the jungle back home with them.

They brought back to England something else from this expedition too: two foster children, Godofredo (Godo) and Romilda, brother and sister from the Goias province. Godo was 12 and Romilda 14 at this time.

On the face of it, this whole initiative appears somewhat out of character for Bill Hamilton, if we take seriously some statements in his autobiographical notes. Didn't he declare that he was basically uninterested in humans, preferring the natural world instead? What was the reason for his interest in these two children? Could it have been that, at the age of 33 and married for two years, he had started feeling a strong desire for a family? There may have been something in that—in a congratulatory letter to his younger sister Margaret who had given birth to a daughter in New Zealand, Bill expressed his wish to have children himself. But Bill's thinking behind this whole initiative was much more complicated, and indeed rather 'Bill-ish'. Here is the story.

In 1964 when visiting the Goias province on his first trip to Brazil, Bill had been impressed with the very detailed knowledge of the natural world that one particular little boy seemed to possess. This little boy soon became his guide to the local flora and fauna. Little Godofredo seemed to have expert knowledge about the habits and habitats of insects and the properties of regional plants. And he was only eight years old! If he was this clever at such an early age (and no doubt Bill here compared Godo to himself at the same age), what might become of him when he grew up? This was a young boy in need of a good education.[12]

Bills original plan, therefore, was to pay for a good regional school for Godo as he got a little older. Probably Bill had something like his own Tonbridge in mind, except he was looking for its Brazilian equivalent. Surely there existed schools of that type there as well? It turned out, however, as Bill and Christine talked to various school officials in the area that this was not possible to arrange. This was a disappointment to Bill, because it was one of the things he had hoped to accomplish during his stay in 1968–69. Now the plan seemed impossible to realize.

At this point Bill came up with Plan B. What if they brought Godo to Britain and had him attend school there? He would get a solid British education, something that would be a great foundation for his future career. And he wouldn't have to live at a boarding school (Bill had not liked the idea of boarding himself). Where, then, would he live? Bill had the perfect answer: at Oaklea with his mother and father. He reasoned that his mother, for whom her youngest son's recent death had been a great blow, would be cheered up by having a youngster in her house again. (This was typical Bill—he often planned for people in a well-meaning way without checking with them what they wanted themselves.)

The term 'foster child' is somewhat misleading. Godo was by no means a child in need of a home—he already had one, including parents, and a number of brothers and sisters. This was a special arrangement. As Bill discussed with Godo's father the possibility of taking his son with him to England, the father agreed, but only if Bill and Christine were willing to take Godo's older sister Romilda as well. This then was how Godo and Romilda ended up becoming the Hamiltons' foster children. However, the Brazilian authorities insisted that the foster children would have to be returned to their real parents when they turned 18.

Soon after this Bettina and Archie once again had youngsters around at Oaklea. Both children attended school, quickly picked up the English language, and made good progress. Romilda's homework books are written in a neat handwriting, and are often marked 'good'. Godo was more of an impatient boy, not really the scholarly type, although clearly a fast learner. There was a lot of genuine affection between the children and their new 'grand' foster parents, but no doubt the situation put a strain on the latter.

In the early 1970s Godo and Romilda moved in with Bill and Christine. The Hamiltons had by now bought a house in the vicinity of Silwood Park, abandoning the apartment they had rented after their return from Brazil. The children continued with their schooling and Bill and Christine took an interest in their homework and hobbies. They enjoyed their family life but were somewhat disappointed that the children did not make more academic progress.[13]

In late 1970 Bill realized that although the education had been clearly beneficial for the children, it had not really led to the revolutionary results he had hoped for, especially in the case of Godo. Romilda was already becoming an older teenager, close to marrying age in her own society. And there was the additional happy news that Christine was now pregnant with her first child, due in May 1971. Under these circumstances the best solution for all parties seemed to be for the children to return to their parents, and this was arranged at the end of the year.[14]

In May 1971 Helen was born. Bill had hoped for his firstborn to be a son to keep the Hamiltonian line going (he admitted that this was an aspect of traditional male pride). (Later Bill and Christine were to have two more daughters, and Bill wisely changed his attitude. In fact, he used to boast that he had three girls.) But there was some worry about the Hamilton name. Archie had been the only male in his generation (he had two married sisters, each with several children), so already he may have felt some pressure. Bill, the oldest son, may have felt similar pressure in his turn. Leco had died, but fortunately for the Hamilton name, Robert in New Zealand had produced three sons. Still, some ingenuity could be employed in regard to the British branch of the Hamilton family. Janet, Bill's younger sister, kept her maiden name when she married and was able to arrange for her three sons to use Hamilton as their preferred surname.[15]

Over the years, Bill and Christine continued to keep in touch with Godo and Romilda, and Bill used to visit them in Goias whenever he was in Brazil. Romilda became an English teacher and a community leader, and Godo an engineer, though later health problems made it hard for him to work. The English educational experiment had been made in good faith, though Bill had overestimated the adaptability of the children to the British system. It had also been a successful experience in family life, and fond feelings persisted between 'grand foster parents', foster parents, and the children.[16]

***

Back from the Mato Grosso expedition, Bill had not been in England long before he took off again, this time to attend a major conference at the Smithsonian Institution in Washington, DC. The conference, held in May 1969, was called 'Man and Beast: Comparative Social Behavior' and was arranged by the then secretary of the Smithsonian, S Dillon Ripley, an eminent ornithologist. The plan was for the proceedings later to be published as a book. Hamilton was one of the invitees and one of the main speakers. He felt very honoured, but at the same time rather bewildered. Here he was recently back from the wilderness, suddenly transported to an almost unimaginable world of grandeur and lavishness, with enormous receptions, speeches by Nobel laureates, and politicians everywhere as far as the eye could see.[17]

Was Bill impressed? Not really:

> I recall names dropping round me like warm snowflakes (a special Washington style I was to decide after a second visit), and just a few of those names being known to me and the rest potentially officers of Xerxes' Persian army as far as they mattered or impressed me.[18]

Just as he had earlier entertained himself by imagining the streets of London as actually representing his favourite walks in Kent, now he amused himself by comparing the luxurious hotel with the Brazilian cerrado: the carpet, the floor, the sounds . . . the cerrado won.[19]

But what was the real aim of this kind of conference? Why was all this money spent? What was expected from the invited speakers? With an unfaltering ethological eye, Bill took in the nature of the conference attendants:

> I recall talking to senators who seemed proportioned in size to the halls where they held court . . . I recall scientists of every type but most of them larger both physically and, seemingly, in spirit than those I knew in England. Here were the laureates, the pompous, the hirsute, the fantastical, the unbelievably industrious, the funny, pugilistic, queer . . . ; Tolstoyans like Alexander, Rabelaisians like Chagnon. Perhaps I would have been less amazed and less entertained if I had been to a scientific meeting in England, but this was my first.[20]

The aim of the conference soon became clear. The urgent topic was inner city violence. The conference was convened to invite ethologists and biologically minded anthropologists to formulate solutions to current social ills. The Great Society was being threatened by unleashed aggression. The United States was still in shock from the Little Rock riots and the 1968 Chicago democratic convention, and in the background loomed the Vietnam War. This was not the place for mere abstract theorizing, as Hamilton soon found out:

> I recall a more European-sized wife of one senator pinning me at once with her chin and her fierce eyes as she asked me how my theory could help to reduce violence and crime in America.[21]

One of the books that had stimulated the convening of the conference had been Konrad Lorenz's *On Aggression*, which in the late 1960s 'culturist' climate was polarizing opinions about human nature and its biological foundations. The Man and Beast conference came down on the side of biology: 'Should the study of man be based on man alone?' Ripley wrote. 'Do genes control how we hunt, protect our young, affiliate, cooperate, fight or claim territory...The study of the inheritance of traits which affect culture is much needed.'[22]

Hamilton had taken the conference's invitation very seriously. He felt that he was invited especially to try to apply his idea of inclusive fitness to humans. Although he himself later downplayed his paper as largely an overview of various recent themes 'combined with an artificial effort to draw out anything that appeared relevant to an understanding of Man', this was his opportunity to present to a large audience his basic ideas and to develop some new ones.[23]

In fact, Bill used the occasion for experimenting with a game theoretical explanatory approach, applying it to pair-wise interactions, gregarious behaviour and inter-group hostility. This was the first time he used this type of analysis in a scientific presentation. He especially wanted the conference participants to become aware of the fundamental problem of the ubiquitous 'Prisoner's Dilemma' situation and the obstacle it presented for social cooperation. The problem in the Prisoner's Dilemma model situation is that what seems like the most rational course of action to each individual in the short run will not produce the best solution in the long run.[24]

Hamilton approached the Prisoner's Dilemma following a typical game theoretical set-up. Interactants (or 'players') can choose to be either selfish or cooperative, and there will be a 'payoff' for each participant depending on his/her own actions and the actions of the other party. In regard to the 'game' of evolution Hamilton explained: '[I]f the Prisoner's Dilemma is played between individuals meeting at random and if the payoffs are fitnesses, we have seen that it "pays" in natural selection to take a selfish course consistently, because the type which does so gets greater than average fitness when associated with any other type, in no matter what ratio.'[25]

But this was not the whole story, Hamilton protested. In the formal development of game theory, 'the possibility that the payoffs might be other than as they are represented in the matrix of a game is carefully precluded...the payoffs must show what the players really prefer'. But the formal theory assuming rational actors missed an important point: 'It is quite unrealistic, for example, to represent the payoffs of Prisoner's Dilemma in terms of years in prison, or as sums of money expected from a partnership, *without making allowance for the degree of fellow feeling of the players*' [my italics], Hamilton noted. Because in real life there did exist certain conditions under which cooperation rather than selfishness would actually be the preferred behavior, exactly because it would result in a better evolutionary payoff. 'Assortation', that is, interaction among individuals sharing the cooperative (altruistic) trait, was the key. Although, as Hamilton had explained, it 'pays' in general in natural selection to take the selfish course consistently.... '[i]f assortation occurs, however, this outcome is not certain'. Indeed, he suggested that his idea of inclusive fitness provided one such possible route of escape from the 'hydra' of the Prisoner's Dilemma. The incentive to 'let the partner down' could be expected to vary with different degrees of relationship, for example, it would fall to zero for partners that were sibs. 'The concept of inclusive fitness provides a simple test for the resolution of games in this way', he concluded.[26]

Still, it was very difficult for Bill to address the issue of humans.

> It is hard even to feel and harder still to write in a way that runs counter to a current world view, especially a moral one, and it is all the harder when the way is re-shaping a plane of perfection to which all civilized cultures are thought to be striving. A scientist or philosopher with a programme of such heresy has to be tough if he or she is to communicate it and, while doing so and for long after, must endure the tortures of Orestes.[27]

In regard to the topic of aggression there was very little data available at the time, for humans or for primates. (For instance, the group conflict and organized warfare that takes place among chimpanzees had not yet been described by Jane Goodall.) But Bill believed that on theoretical grounds war-like tendencies would most likely have evolved:

> For me, it was the discussion of the 'innate aptitudes' that I believed must exist in all human populations and most individual humans, and of the selection our forebears must have undergone through competition between populations, creating their warlike inclinations... that caused me most pain to write.[28]

But he felt that, despite the pain, he had to tell the world the truth. And this was an attitude that Hamilton would stick with throughout his life, whatever the cost:

> Theory concerning the nature of the 'beast within' was why I had been invited and I continue to believe that only from a basis of honest description can there be hope of taming what we have and may not like.[29]

Later on, Hamilton occasionally wondered if *not* telling the truth would actually sometimes be more conducive to social peace. But in such internal debates his scientific conviction typically won out. And if need be, he could get support from Aeschylus himself. Had not Aeschylus told the hounded Orestes: 'Then fly and do not weaken'? If Orestes could do it, so could Bill Hamilton. Hamilton saw himself as a serious messenger of scientific truth, and he was fully prepared to suffer the consequences.[30]

The need to write up the results from the conference led Bill to some interesting insights, which he was to explore further in later papers. One was the formulation of relatedness in cases of inbreeding, the typical situation in 'viscous' populations (populations with low dispersal). This was something that he had deliberately not addressed in his 1964 paper in order to keep things simple.[31]

This was also the point at which he started thinking about the connection between kinship theory and dispersal, which he was later to address more explicitly. In 'viscous' populations organisms do not leave their original habitat, and the result is typically inbreeding. Now inbreeding would appear to be obviously favourable for the evolution of altruism (since in this way high relatedness and high concentration of altruistic genes can be easily achieved). But this is not the case: inbreeding also has its disadvantages. In a non-dispersing population offspring will be competing for resources with individuals of their own kind. Therefore, Hamilton realized, 'to be effective, altruism must put offspring into competition with non-altruists, not bunch them in a wasteful competition with their own kind.' But if this is not possible, then the whole group will need to expand and replace other groups, a tendency associated with xenophobia, he reasoned.[32]

The Man and Beast conference paper, 'Selection of selfish and altruistic behaviour in some extreme models', published in 1971, demonstrated Hamilton's deep struggle with the consequences of his own reasoning. He devotes a whole Appendix to the discussion of how altruism may in fact be able to evolve. A clue seems to be the idea of subdivided populations. Altruism can be favoured in subdivided populations, because altruistic genes can rapidly concentrate by chance due to genetic drift. Hamilton admits that within each of these subdivisions the frequency of altruists typically tends to decrease. But he is looking for processes by which it will be possible for altruists to find one another and the overall number of altruists in the population as a whole can increase.[33]

Another way to establish altruism would be if altruistic migrants were to leave a group and these would be accepted in other groups. Yet another method for the altruistic gene to spread would be for vigorous altruistic groups to grow in numbers and then divide when a certain size was reached. This is similar to Haldane's 1932 model, Hamilton notes:

> Haldane discussed the selection of altruism with this sort of process in mind, but accompanied his verbal discussion by a strangely irrelevant mathematical analysis in which he considered numbers of genes rather than gene frequencies.[34]

But there is the same well-known obstacle in all these cases, be they genetic drift, emission, or fissioning:

> As in all cases the altruistic mutation still faces the difficulty that at its first occurrence it tends to waste its altruism on individuals unrelated at the locus in question... But... provided that the altruism is not suicidal, the recurrence of a mutation plus the possibility of a local increase by drift can overcome this barrier. Eventually there will occur a clump of altruists of such size and purity as to reverse the local counterselection, and with further growth increasingly determinate progress can take hold.[35]

This sounds like an almost Biblical prediction of the way in which Good can finally overcome Evil.

At the same time, for Hamilton, the struggle for altruism is a tale out of some knightly storybook. The altruistic mutation is the hero. There are difficulties in its way, but if it joins in with more of its kin (others who carry such mutations), together they will be able to form a righteous band that can beat the local power holders and open up the way to general progress and prosperity for all.

At the Washington meeting 'young Hamilton' met some key scientists involved in the research of the nature of the biological underpinnings of human nature and formed personal friendships that would last a lifetime. Many of these colleagues he was going meet again at future conferences. Little did he know at this first meeting, where he awkwardly and tentatively applied his theorizing to humans, that in less than two decades he would be elected the first President of the newly founded international Human Behaviour and Evolution Society.[36]

The Washington conference marked a turning point for Bill Hamilton. From now on, he stopped seeing himself as a potential crank. He had been taken seriously and was starting to get recognition—at least in the United States.

***

Hamilton had another memorable experience during this American trip. He had been invited to visit Harvard and give a lecture there. (It is not clear if it was before or after the Smithsonian conference in Washington, DC, there are different reminiscences.) The most interesting thing for Bill during that visit was the opportunity to meet Robert Trivers, who was to make such important contributions to human sociobiology and take many of Hamilton's ideas further.

This was the first time that Bob Trivers met Bill in person. Trivers was at the time a graduate student in biology at Harvard, having taken an interest in Hamilton's early papers on kinship theory and realizing that there was an important type of social behaviour not touched on by Hamilton: 'reciprocal altruism'. Trivers had written a brief paper on this topic, which he gave to Bill to read, and they had some discussion about it.[37]

Hamilton remembers from their early meeting that he was very excited about Trivers' willingness to tackle this topic. He himself had been concentrating on kinship altruism and had more or less relegated other kinds of cooperative behaviours to something that he called 'all those other kinds'. He had assumed that they had been more or less covered by Adam Smith whom he had read at Cambridge. But what Trivers called 'reciprocal altruism' was certainly an important category. It was definitely an evolutionary force on a par with kinship altruism ('kin selection'). It was just that in Bill's mind, this type of behaviour could not be classified as altruism at all. For something to be called altruism, the behaviour had in principle to be imagined to have suicide as one of its extremes, Hamilton insisted. And there was no way in which this type of interchange of mutual benefits could have anything to do with suicide.[38]

This was why Bill tried to dissuade Trivers from using the term 'reciprocal altruism'. He pointed out to Trivers that his examples were actually not examples of reciprocal altruism, but rather of something that could be called 'return effect altruism'. But Trivers stood his ground and refused to change his term. (This more detailed discussion took place in later correspondence.)[39]

Trivers was especially grateful for 'the gentle tone' in which Hamilton commented on his paper. Indeed, his general approach was to encourage Trivers to continue, while suggesting that he perhaps not use mathematics (Trivers later realized that his early calculations had been littered with errors). Trivers just wanted so much to be like Hamilton: 'I was very much trying to mimic, however feebly, Hamilton's way of thinking: try, if at all possible, to get the formulation down to the level of genes, to be more sure that you have got the argument right, and to give a quantitative form to it if possible'.[40]

This was the beginning of a long and fruitful friendship between Bill Hamilton and Bob Trivers.

But what about Bill Hamilton's Harvard lecture? It was to be a general lecture to the faculty and students. Trivers guessed that Bill was probably covering the same material here as in his talk at the Man and Beast conference. From Bill's point of view, however, the situation was far from optimal. Here he was, barely back from Brazil, and with the cerrado on his mind. In fact, this may have been his first invited public lecture at a university. He was clearly out of lecturing practice in general. He was probably unprepared. He may have had jet lag. And he may have been sensitive to heightened expectations in the audience.[41]

In any case, here is Trivers' description of his first experience of a Hamilton lecture:

> There were perhaps eighty or ninety people, almost filling a lecture hall, most of us with eager anticipation. Hamilton got up and gave one of the worst lectures, as a lecture, that I have ever heard. There was an emeritus Harvard professor who occasionally used to give a lecture, widely appreciated, on how to give a poor lecture. W. D. did not need any teaching in this regard and had generated some wonderful tricks of his own. I say this as a man who loved W. D., but his early troubles in this regard were sometimes very funny. For one thing, he lectured for a full forty-five minutes without yet getting to the point. It was abstruse and technical; he often had his back to us while he was writing things on the board; you had difficulty hearing his voice; you did not get any overview of where he was going or why he was going there. When he realized that he was five minutes over time and still had not gotten to the point or indeed very near it, he looked down at Ed Wilson, his host, and asked him if he could have some more time, perhaps ten or fifteen minutes. Of course Professor Wilson granted him some more time, but he also made a rolling 'let's-try-to-speed-this-up' motion with his arms. Hamilton then called for slides. The room went dark, and there was a rumble and a roaring sound as about 90% of the audience took this opportunity to exit the room for some fresh air. Some students were nearly trampled, I am told.
>
> I remember walking home from the lecture with Ernst Mayr, both of us shaking our heads. It was obvious that the man was brilliant, a deep thinker, and his every thought well worth attending to, but whoa, was he bad in public![42]

There would be many Hamilton lectures very similar to this in the future and many audience comments very similar to that of Trivers. But there would also be good Hamilton presentations, and occasionally excellent ones.

10

The Price Effect

Things were going well for Bill Hamilton professionally. He continued his work at the Imperial College Research Station, now under new directorship. 'Silwood Park is thriving under Richard Southwood', Bill told his friend Yura in a letter in 1971. Richard Southwood ('Dick' to his colleagues) was one of those rare academics who was able to appreciate immediately Bill's scientific genius. My own couple of interviews with Southwood gave me the impression of a man with great social intelligence, open minded, and with a sense of fun. It was in fact Dick Southwood who had irreverently introduced that big beetle semi-relief in the place of the former coat of arms in the great hall of Silwood House.[1]

A very curious naturalist, Southwood appreciated young Hamilton's audacity and ability to bring things together theoretically. The two thought alike in a number of ways, and their interests overlapped. Already such shared notions as the importance of knowing about both insects and their preferred plants made for a tacit understanding and a lot of good naturalist conversation. On the other hand, Southwood was an administrator and had to keep an eye on all aspects of his area of responsibility. One of these involved teaching performance among his staff. And there, regrettably, Bill was not making visible progress.[2]

At Silwood Bill continued his studies of insects in rotting wood. He also took up breeding butterflies, just as he had done before at Oaklea. His butterflies livened up both the conservatory at Silwood Park—Bill sometimes put on butterfly shows at tea time—and his new home, surprising dinner guests. (No doubt Christine, who was also very good with plants, helped create the right habitat.)[3]

Hamilton's main interests at this point, however, were theoretical. His first and foremost task was to update his 1964 inclusive fitness paper in the light of new empirical findings and to respond to a number of challenges. The result was a long review in *Annual Review of Ecology and Systematics*. An update of the ideas in the 1964 paper had become even more important since George Williams had in 1971 made Hamilton's original article available to a broader audience in his small edited volume entitled *Group Selection*.[4]

Bill's central point in his *Annual Review* paper was to improve Hamilton's Rule by introducing a more satisfactory expression for r, the relationship between donor and beneficiary. In 1964, his discovery that he could use Sewall Wright's r, 'the coefficient of relatedness', represented a great achievement for Hamilton, who had long been struggling to find a general expression for this relationship. But later he had found an even better tool with which to capture the donor-beneficiary relationship: 'the regression coefficient of relatedness'. This was good news that he wanted to share with his fellow scientists.[5]

Hamilton was eager to show how by introducing the regression coefficient he could clarify the relationship between Fisher's 'classical theory' and his own inclusive fitness theory (which, as we know, he saw as a generalization of Fisher). To allow his readers to better follow his reasoning Hamilton first took them on an imaginary trip using a 'gene's eye' perspective. Indeed, he had already introduced the gene's eye view in his earlier sex ratio publication, but this was the first time that he demonstrated how useful the gene's eye view could be as a tool for thought.[6]

In this review article he also took up several challenges that had been raised by colleagues in the meantime. One of these was the topic of inbreeding. He also showed how his general theory could cope with such things as rival queens and egg-laying workers. Moreover, he discussed more closely the role of the males. Further themes that he addressed, with references to new empirical research, were the origins of eusociality, as well as cooperation in the absence of kinship. This article now gave him an opportunity to state that he was not insisting that eusociality in the social Hymenoptera was necessarily connected to their special family relationships, although he had presented this as a hypothesis in his 1964 paper. He accepted that recent empirical evidence appeared to point away from this suggestion. In fact, he noted that there could very well be cooperation in the absence of kinship. And finally here Hamilton took the opportunity to briefly introduce Robert Trivers' recently coined term 'reciprocal altruism' and the conditions under which it can take place, and to expound on the idea of high mobility versus viscosity in regard to antagonism and mutualism.[7]

All in all, Hamilton's 1972 *Annual Review* paper was a thorough job with plenty of examples. It introduced some important new ideas and competently engaged with the many findings or commentaries that had accumulated since his original 1964 contribution. Hamilton was later, and with justification, to refer to this paper as his 'paradigm for kinship theory'.[8]

More fun for himself and his readers was surely Hamilton's short 1971 paper, 'Geometry for the Selfish Herd'. That paper was originally written for *Scientific American*, which explains its more relaxed style. Bill loved puzzles and had sent in his piece to appear in Martin Gardner's special section on mathematical puzzles. But alas: just in the same way as Bill had earlier run into trouble because of his contributions being 'too new' and therefore difficult to get published, now he was told instead that *Scientific American* didn't publish original work. Very well—Bill sent off his piece to the *Journal of Theoretical Biology*, where it was much appreciated, later to become one of the most cited papers ever listed in *Current Contents*.[9]

'Geometry for the Selfish Herd' captured the interesting general question as to why animals that were pursued (or were expecting to be pursued) by predators 'bunch' together rather than doing something else—say, spreading in all possible directions. Bill was able to show that for each animal, from a selfish point of view the optimal place to be was a point where it was surrounded by other animals—hence the persistence of rounded cattle flocks.

Bill himself had observed some early thoughts on this puzzle in an unexpected place, an obscure article written by Francis Galton on African cattle. (African cattle are not the first thing one associates with Galton!) Galton, an accomplished and intrepid traveller and with a great zeal to measure all kinds of natural phenomena, had encountered this social behaviour but had not treated it more systematically from a mathematical point of view. Looking more closely into others who had approached problems of this kind, Bill was not surprised to find that his 'twin brother' George Williams again had got there before him. Williams had been dealing with schooling in fish, putting to fish the same question as Galton and now Hamilton put to cattle. Williams had also observed deviations from the rule—sometimes, instead of schooling, the fish actually preferred a swift spreading in all directions, something that Williams called a 'spark'.[10]

The main intellectual figure in Bill Hamilton's life from the mid-1960s to mid-1970s, however, was George Price. Price was the type of eccentric, brilliant person to whom Bill was naturally drawn. In the first place, Price, although he had a doctorate, was absolutely outside the system. An American and an engineer, he had relocated to London after a difficult divorce and having been awarded a large insurance sum after an injury settlement. It seemed that he now wished to pursue some intellectual interests and had found London a congenial place for this purpose.[11]

Bill must have felt particular curiosity and responsibility towards Price when he learnt about the impact that his 1964 paper had had on this man. Price was one of those people who took everything very seriously and literally—including Bill's idea of inclusive fitness. Price had studied Hamilton's inclusive fitness paper closely and become worried about that paper's social implications. In Hamilton's words, Price was 'as deeply shocked as the Victorian lady and her friends had been by evolution itself'. If Hamilton was right about the workings of natural selection, then the prospect for humanity did not look good. Could this limited nepotistic altruism be the best, the most 'humane', that evolution could achieve? There must be a flaw somewhere! Price was determined to find it. But to find the error in Hamilton's calculations he would first have to learn genetics.[12]

Here we have, then, an example of a major intellectual effort undertaken with the motivation to prove someone else wrong—for moral and social reasons rather than scientific ones. (But actually, for Price these were the same, as will soon be apparent.)

After some time, Price concluded that Hamilton was basically right—nepotistic altruism seemed indeed to be the best that evolution could achieve. He was extremely disappointed. He had hoped for a better result coming from science, because it was to science that he looked for moral guidance. For Price, who was an atheist, there was no God or other moral authority. At the same time Price was a 'scientific fundamentalist' of sorts: if science had come to a particular conclusion, this would have to be accepted as a fact. Moreover, the moral and social implications of the fact would need to be identified, and taken as a guideline to follow. It seems that Price was an unhesitant believer in the so-called naturalistic fallacy, which for him was not a fallacy at all. The combination of atheism and this particular type of fundamentalist reasoning gave Price no way out. Soon he was pushed into a state of despair.[13]

For science, however, there was an unexpected consequence of Price's avid attempt at error finding. In his efforts to prove Hamilton wrong, Price had tried to develop a different formula himself. This he later sent to Bill. (There was some time lost because Bill was in Brazil.) Looking at Price's equation, Bill immediately realized that the result was correct. At the same time, what Price had sent him did not at all look like a re-derivation of Bill's own formula for inclusive fitness, starting from the same premises. Rather, Price had come up with 'a strange, new formalism that was applicable to every kind of natural selection'. And this new formalism, Price's 'covariance formula', was something that Price, who had never studied population genetics at all, had derived all by himself.[14]

Bill instantly realized the value of Price's work. George Price had found a way to mathematically reformulate and rewrite Fisher's insights in his classical theory in such a way that they could be made compatible with regression analysis. Strictly speaking, Price's formula, the Price Equation, was a tautology. It was not saying something new about the world; rather, it reformulated what was already known. It can be described as simply a notational convention. But what this unusual new notation did was to force a problem to be formulated in a statistical way, which in turn made it easier to analyse with the help of Neo-Darwinian evolutionary theory.[15]

Introducing the regression coefficient instead of the correlation coefficient in his 1970 paper opened up new possibilities for Hamilton. In contrast to Sewall Wright's correlation coefficient r, the regression coefficient could take on a negative value. More immediately and temptingly, Price told Bill that he could see how his new formula could be usefully applied to spiteful behaviour. Spite was one of the four types of social behaviour that Hamilton had outlined in the original 1964 paper, but he had so far done little to explore it further. Spite was there defined as behaviour that harms both the donor and the recipient, although the recipient more than the donor. This was certainly an idea that appealed to Bill, who began working on this new problem with gusto. Meanwhile, they agreed that it was important for Price to publish his valuable formula in the best possible journal—*Nature*—and the sooner the better.[16]

For Hamilton *Nature* was his standing challenge, a beautiful, glittering lure. This was the way that this journal would appear to him throughout his career. For each submission, his optimistic hope of triumph over *Nature*'s editor and referees would win out over the potential pain that would follow in case of a rejection. Bill had by now been rejected once: his very first paper. Well, he knew that *Nature* as a journal was not very keen on findings in evolutionary theory, and everybody knew that it was dominated by cliques. But now *Nature* had a new editor. Perhaps this signalled a new era for Bill too? A submission would at least be worth a try. Hamilton encouraged Price to go ahead and publish his formula as soon as possible, while he was finishing up his new spite paper inspired by Price and using the Price Equation.[17]

What was the outcome? Price's piece on his covariance formula was rejected by *Nature*. But Hamilton did not think this should be the final verdict. Examining the referees' reports and having learnt by now that one does not just accept what the referees say, but goes on fighting, Hamilton wondered if the reviewers had actually understood Price correctly. Maybe they were more confused than really negative? Bill therefore suggested to Price that he resubmit a revised version. That proved over-optimistic: *Nature* had not regarded this as an invitation for resubmission at all, and simply returned Price's paper without comments. But by now Bill had finished his spite paper and sent it in to *Nature*. This time, his paper was accepted.[18]

What *Nature* did not know was that this was actually part of an experiment conducted by Hamilton and Price as co-conspirators and with Bill as the chief strategist. Bill described the next step as follows:

> I now wrote to *Nature* regretting that I must withdraw [my manuscript] because the powerful new method I had used and cited had been recently refused by *Nature* and I could not proceed until that method was published somewhere.[19]

This was true, but it also presented the journal with an interesting conundrum. *Nature*'s editor broke the impasse by telephoning Bill, asking him to contact Dr Price who could not be found. The end result was that *Nature* took both papers. In this round of his game with *Nature*, Hamilton had won. Together, Price and he had outsmarted this top journal. Bill remained convinced that special measures were needed to get published in *Nature*. Following normal procedures, he and Price would most likely have been rejected:

> [I]f in 1970 George and I had combined instead of splitting, if we had sent a single paper deriving the covariance formula, re-establishing kin selection with it, and finally exemplifying spite, all in one glorious synthesis, I am fairly sure that the joint effort would have been rejected.[20]

Subsequent events were to prove Hamilton right. What he had said about Price's paper would also apply to himself, as he submitted his co-authored parasite avoidance theory of sex some 15 years later. When it came to *Nature*, Bill Hamilton would never really be able to penetrate the citadel. Perhaps what was needed was a back door of sorts—or maybe he should simply have submitted manuscripts of a less innovative and controversial nature?

During his time in London, Price was also collaborating with John Maynard Smith, resulting in a joint paper in which they developed the idea of an Evolutionarily Stable Strategy (ESS). In fact, that collaboration came about because Maynard Smith, too, was eager to get Price's ideas published and thought this would help the matter. Price himself was making inexplicably slow progress on this. In an apologetic note to Maynard Smith, Price expressed his gratitude, saying it was not often that a referee of a paper became the joint author of the paper. That was true: Maynard Smith had discovered Price when he was refereeing (for *Nature*!) Price's very long paper on alarm calls in animals (quite unsuitable for that journal because of its length). He ended up urging Price to submit it to the *Journal of Theoretical Biology* instead. Meanwhile he invited Price to work with him on a new joint paper, which they would make short enough to send to *Nature*. In this case no author collusion was planned, since Maynard Smith was a *Nature* insider. (It was, incidentally, Maynard Smith, acting as a referee, who had approved the revised Price paper, and probably Bill's spite paper too).[21]

But there were complications. As George Price was working on his various papers, things were happening in his personal life. On his arrival in London, he had been an atheist. Before long, however, he underwent a total religious conversion. This seems to have happened practically overnight. Being the serious and uncompromising man that he was, Price now completely changed course and started behaving like a true Christian, as he saw it. From his earlier apartment he moved to a telephone-less room in Soho and started taking care of homeless people, helping them with money and jobs. The money he had from his insurance settlement was rapidly disappearing. To make matters worse, some of the people he was trying to help blatantly abused his kindness. This deeply depressed Price, who regarded this as his own failure.[22]

How were his journal articles coming along? Price, surprisingly, had started working on an exegesis of the Bible, especially the New Testament. But wasn't he supposed to write scientific papers? What about his real work that was being delayed? Maynard Smith was still to learn that Price had concluded that his Bible exegesis *was* his real work. For Price, that work was as important as his science. Or rather, for him scientific work or biblical exegesis were just different approaches to the same truth. He saw a direct personal connection between his science and his religion.[23] Probably nobody understood George Price's quest better than Bill Hamilton, and he describes it beautifully:

> George believed that the discovery he had made in evolutionary theory was truly a miracle. God had given him this insight where he had no reason to expect it.... Because it was a formula missed by the world's best population geneticists throughout the past 60 years, it was clear to him that he had somehow been chosen to pass on a truth about evolution to a world that was, somehow, just now deemed ready to receive it. How was he supposed to do it? How much was he expected to tell and how? He... decided it was right to treat the matter in just the way he saw divine Truth being handled in the Testaments—that is,... perceived only slowly by disciples... Through such first interpreters, through such glass, so darkly, along with religious truth, evolutionary truth was supposed to filter outwards. In this process I believe I was chosen to be his first initiate.[24]

Indeed, soon Bill himself was involved in George Price's religious work, specifically, a piece on the Twelve Days of Easter, which Bill thought had real merit. Price thought that he could demonstrate that the current calendar was incorrect and that the true Easter in fact fell on a different date. Bill, who advised Price on this paper and its publication, thought that it would have been a suitable discussion paper in the cultural pages of a major London newspaper around Easter time. Both were sorry when the paper was not accepted.[25]

In Price's everyday life his absolutist stance was becoming increasingly untenable. But he seemed to find no way toward compromise. His friends tried to help him. Maynard Smith insisted in a letter that Price accept some money. Also around this time Albert Somit, professor of political science and Price's old college roommate, visited London. Seeing Price's situation he offered him a loan of a few hundred dollars if George promised he would not give them away. Price responded that he was sorry but that he could not guarantee. Cedric Smith, Bill Hamilton's PhD supervisor, had earlier arranged for Price both a grant and an office in the Galton laboratory. Later when Price lost his apartment, he had let Price briefly sleep there too. Unfortunately a scandal erupted as unruly protégées followed Price to his new quarters, and Price had to move.[26]

In regard to money, Price had deep faith that solutions would appear at the very last minute. As he explained in a letter to John Maynard Smith, 'God usually provides for me... I expect that something will happen. I am now down to my last 15p. I will be very interested to see what happens when that is gone'. (It was to that letter that Maynard Smith responded: Please let me help!)[27]

The same was the case with Price's health. He was dependent on medicine (he needed to take thyroxine on a regular basis for his thyroid condition) and here again it seems he expected God to intervene to get him his medicine in time. He already had two experiences which he believed demonstrated divine intervention at the last minute. But there was not to be a third one. Price may have taken that as a signal that God had given up on him. He committed suicide shortly after Christmas 1974, leaving a note saying that he didn't want to be a burden to his friends.[28]

Albert Somit was not really surprised when he heard about Price's death, he told me. He thought that George Price was an example of a man with a 'death wish', as he put it. This was evident already in college and later in his life. If there was a choice, George always took the more difficult option, choosing a type of operation that had a larger risk of leaving one handicapped for life, say. Perhaps this kind of fearless high risk gamble was among the features of Price that impressed Bill Hamilton?[29]

The Price affair was very upsetting to Hamilton for a number of reasons. As he explained to Price's brother Edison in a letter: 'It was as if I had lost a second self'. He also felt enormous guilt. After all, he had gone to Ireland with his in-laws for Christmas and left Price to his own devices. Never mind that he had invited Price to stay with him and his family for a week just before they left and that he and Price had arranged to meet again after the holidays. And never mind that Price probably had some acquaintances with whom he could celebrate Christmas—he had done such things before. The thought was nagging Bill that he could have avoided Price's death if he had only taken him with him to Ireland.[30]

That this was foremost in his mind was clear when I talked to Bill in the late 1990s. Justifying his decision to leave Price, he explained that he and his family had planned to go to visit his in-laws for Christmas and that he couldn't very well have imposed Price on them. Later on I reflected on this—actually, why not? Bill was not the conventional type. Of course, he may have been considering all aspects of the situation, and felt responsibility not to disrupt a family occasion in the event of Price behaving in an unpredictable way. On the other hand, Price had stayed at Bill's own home many times and his wife and children had enjoyed his visits. Also, might not Canon Friess have enjoyed a little game of exegesis with someone who knew his Bible as well as George Price? Couldn't Price's Christian connection actually have been a plus?

That was just it. Price's exegetic fervour may have been just too much of a good thing. Price could get rather intense, Bill knew from his experience with the Twelve Days of Easter, and there was no guarantee that Canon Friess and

Price would see eye to eye on the Bible. But that was not the main problem with bringing Price and Christine's father together. The problem was the whole idea of a serious Christian Christmas celebration and what that might do to Price.

For some time there had been tension between Price's scientific and religious interests. He had been stalling on his scientific papers, explaining that he had other important work. That work had ranged from helping the homeless to doing Biblical exegesis. Meanwhile Price knew that he needed money. Initially, in fact, he had hoped that he would impress the world with his scientific work and in this way get some employment—and here he had succeeded already in getting a grant and an office from Cedric Smith. But that was coming to an end. To continue getting money for his research, he would have to procure another grant, or get a job as a research associate, or something. This meant scientific work. But Price believed that God wanted him to be a good Christian, and that meant caring for others.[31]

Interestingly, just before that fateful Christmas of 1974 Price had come to a different interpretation of what it meant to be a good Christian. It did not necessarily mean giving all you had to the poor, or spending all your time helping the needy. Instead, God had now told him that the right thing to do for him was to pursue his science. This was a great revelation. Price had important papers to write (Maynard Smith had been on at him to submit his long original paper to the *Journal of Theoretical Biology*; and their joint paper was to appear in *Nature*). And Bill had just invented for him a position as Bill's research assistant, which would bring in some money. This looked hopeful. Indeed, in a letter to his daughters Price said that he was looking forward to doing research again.[32]

So in the late autumn, Price's attitude had taken a surprising turn in the right direction, from a scientific point of view at least. But how long would this last? Price's mind had a tendency to oscillate. The obvious point, therefore, would be to keep Price on the right track and prevent a relapse—and that meant keeping him away from any religious temptations. This was the reason why Price had to be barred from the Friess-Hamiltonian Christmas celebration, for his own good.

But had Hamilton calculated correctly when it came to costs and benefits in the Price case (considering these strictly in regard to Price)? This was what he kept asking himself. Would the risk of bringing Price to Ireland have been worth taking, after all? What especially worried Bill was that he had not contacted Price immediately upon his return after the New Year. 'I wonder if George had counted with me as a kind of last chance', Bill later wrote to Price's brother

Edison. These kinds of 'if only' scenarios can be very disquieting. No wonder that Bill later wrote to a close friend that he wanted to forget the whole Price affair or that he mentioned in his autobiographical notes that he remembered the Price story as if it were a dream.[33]

Bill may also have felt guilty for another reason. He was grateful to Price, but not quite happy that Price had proven himself to be the stronger mathematician. It was akin to the feeling he had had as a youngster when he had been fascinated with the patterns on butterfly wings, but later heard that Alan Turing had taken up 'his' project. His conclusion at that point had been that this was not quite what he wanted anyway, and he had gone on to other things. Something similar occurred when it came to the re-derivation of his inclusive fitness with the help of the Price Equation. Moreover, he didn't like the implications—as little as Price had done.

There may have been yet another source of Bill's ambivalence and guilt in regard to Price. This would have related to Price's attempt to clear up matters between Bill and John Maynard Smith, treating the whole thing as a misunderstanding. In his letter to Maynard Smith on 19 October 1972 Price first discusses his own situation and the paper the two are working on, 'The Logic of Animal Conflict' but in the middle of the letter he drops a bombshell:

> However, there is one matter that has to be brought up now and on which I need to hear from you before I try doing any re-writing. This is a rather unhappy matter to mention. It involves Bill Hamilton, whom I know moderately well and probably would know better if he had not moved out of London. In general it is wrong to repeat what is said to one privately. However, I have thought it out carefully, and I think that in the present case there isn't much choice, but I have to bring it up. But please don't repeat it to anyone else (unless to Sheila). I will mark the envelope 'Private and Confidential' so that it won't get even to your secretary. Especially, please don't say or write anything to Bill about this. If you happen to feel that you want to say something to him to explain that he misconstrued what you did (as you probably will want to do), then the way least embarrassing to me would be to have me tell it to him... That is, if he learns that I told you what he has said to me, I want to be the one who tells him this.
>
> Anyway, the point is that he thinks that you wronged him on the matter of 'kin selection'. His account of the matter is that you refereed his 1964 paper for the Journal of Theoretical Biology, and required a major revision (changing it from one paper to two) that caused a nine-month delay in publication and

> meanwhile you sent Nature a letter with the term 'kin selection' that has received much of the credit for the idea. In your paper on alarm calls [a paper published 1965 in *Animal Behavior*] you don't cite Hamilton at all and in the paper you just sent me [the draft of their joint paper on the logic of animal conflict for *Nature*] you again don't cite him on kin selection (though you do cite him on group selection and on genetic strategy). Especially undesirable is that you cite not only yourself but Levins on kin selection. This is just adding fuel to the fire.
>
> ....
>
> Now, in order to develop the explanation of the Hamilton matter, I have to tell you something else, which I mention with considerable embarrassment. In 1968 I was desirous of rapid publication, since I was living on savings that were nearly gone, and I had hoped that if I published some interesting papers I could obtain financial support. Consequently I was very irritated that Nature took seven months to report on my paper, and I made complaining remarks to many people about this. Then when I learned that you had refereed the paper, I put this together with what Bill had told me, and to about four people I made rather unkind comments about how you had delayed the paper. One person I said this to was Richard Andrew. Then when I visited Falmer [the location of the University of Sussex] I found that you were not at all as I had pictured you, and you were extremely kind and considerate toward me. Therefore I remarked to Richard during the afternoon that I was sorry that I had said that about you, and you were so considerate that I thought probably it was another referee who was responsible for the delay, or if it was due to you, then you had some good reason for it. (He looked at me with a rather serious expression on his face, nodded, and said, 'Yes, that is right'.) So the last time this thing came up in talking with Bill, I mentioned my mistake and suggested that if he knew the full story he might also find that he had been mistaken about you.[34]

John Maynard Smith responded immediately. Here is the excerpt from his letter dealing with Hamilton:

> I was distressed about what you have to say about Bill Hamilton, particularly since I may be partly at fault. I did referee his paper for the J. T. B. Also, I did urge a major revision. Looking back, I think I should have asked Bill's permission to quote his '64 paper as (in press) to show that I had seen it. Actually, my paper was mainly concerned with the Wynne-Edwards, group selection issue. All it contributed to the discussion of kin selection was the term itself (I think—I don't think Hamilton used it?), and a distinction between kin and

PLATE 1 The Hamilton children at Oaklea, mid-1940s.
From left: Margaret, Mary, Janet, Robert, and Bill. Kindly provided by the Hamilton family.

PLATE 2 Archibald and Bettina Hamilton, Bill's father and mother.
Kindly provided by the Hamilton family.

PLATE 3 Oaklea, seen from the garden. Kindly provided by the Hamilton family.

PLATE 4 Family picture, early 1950s. *From left:* Robert, Bill, Archibald, Alex (Leco), Mary, Bettina, Janet, and Margaret. Kindly provided by the Hamilton family.

PLATE 5 Silwood Park, Zoology Department, around 1964. Bill: 2nd from left in the 2nd row, Richard Southwood: middle of the 1st row next to Nadia Waloff. Provided by Claire Henderson, credit: Silwood archives.

PLATE 6
Bill stung by wasps during the Mato Grosso expedition 1968–69. Kindly provided by Christine Hamilton.

PLATE 7 Bill Hamilton and Robert Trivers, joint seminar, Harvard spring 1978. Credit: Sarah B. Hrdy.

PLATE 8
John Maynard Smith (photo taken 1992 in the South Downs, Sussex, by Victoria J. Rowntree).

PLATE 9 John Maynard Smith talking to Jon Seger, late 1990s.
Credit: Sarah Hrdy/Anthro-Photo.

PLATE 10
Bill and Sarah Hrdy wondering if the giant sequoia is really parasite-free. California, 1980s. Credit: Dan Hrdy.

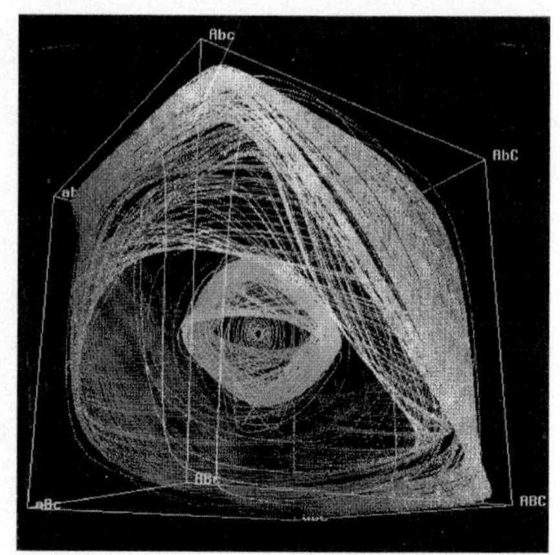

PLATE 11
'Computer art', 1986. Bill loved this simulation of parasites 'herding' hosts like sheep dogs and an eye watching (see exact quote in this book chapter 18, note 6, p 401). Credit: Naomi Pierce.

PLATE 12 Human Behavior and Evolution Society Inaugural Meeting, Evanston 1989. *From left:* Irenäus Eibl-Eibesfeldt, George Williams, Edward O Wilson, Richard Dawkins, and Bill Hamilton. George Williams gave me this picture taken by Philip Rushton.

PLATE 13 Bill in the Amazon, mid-1990s. Credit: Marcio Ayres.

PLATE 14 A group of Hamilton's Master's students at the Mamirauá Ecological Reserve, mid-1990s. Bill is in the back. Credit: Peter Henderson.

PLATE 15 Bill in the Amazon, mid-1990s. He loved the flooded forest and its potential to produce new forms of life (see this book chapter 19, p 285). Credit: Peter Henderson.

PLATE 16
Hamilton's Rule as posted on a sign board at his department at Oxford (Hamilton's Rule was formulated in his early 1963 and 1964 papers). Taken by Ullica Segerstrale.

> group selection. There is no question that Bill's 1964 paper did far more to establish the importance of the concept. I think I have done too little to acknowledge this fact since. For example, I have just looked at the second edition of my Penguin [this refers to his popularly written book *The Theory of Evolution*, the second edition of which came out in 1966], and I find that although I do quote Hamilton, I could with fairness have been more generous in acknowledging his contribution. I shall look for a further opportunity to make his priority clearer—although my impression is that his contribution is pretty well understood.
>
> As to our draft... Oddly enough (I must be Freudian) I got the reference wrong in the draft! I had intended to quote Hamilton and myself for the concept of 'kin selection', and Levins and myself for the relative ineffectiveness of group selection, and quote *only* Hamilton for kin selection.
>
> ...
>
> About your own paper—I don't think I kept it an unduly long time to referee. I don't have any record of dates of receipt, but I doubt that I had it more than a month at the outside.
>
> About the future—I don't think that Bill Hamilton feels too badly about things now. I certainly won't discuss the matter with anyone. But I will try to set the record straight without making too much of an issue of it, because Bill certainly deserves any credit that is going. I leave it to you whether to say anything to Bill or not.[35]

Presumably Price told Hamilton about the result of his inquiry, since to him it may have seemed as if everything was now cleared up. He may have been puzzled or even hurt at Bill's response, which may well have been negative or ironical. Bill tended to be stubborn when it came to the story with John Maynard Smith.

But whatever his feelings, Hamilton was tidy in his attitude to science. There was an obvious need to write up in more accessible form the Price-inspired rederivation of inclusive fitness, and show how his formula could elegantly be used to address the problem of levels of selection. And the perfect place to do that was in the context of a request that had come to him from the anthropologist Robin Fox in the meantime, to participate in an anthropology conference in Oxford in 1973. He felt, as he wrote up his presentation, that he was honouring Price. Price had seen an early draft and liked it, although he had been upset at what he saw as the social implications.[36]

In his chapter in the conference volume edited by Fox, 'Innate Social Aptitudes of Man: An Approach from Evolutionary Genetics', Hamilton suggested that culture is not left totally to its own devices; a genetic system is underlying it and safeguards it. This means that some things that we may believe to be purely cultural are actually of ancient biological origin—for instance, xenophobia. These kinds of thoughts, of course, were exactly those for which EO Wilson would be attacked a couple of years later. Compared to Wilson, Bill got away without too much criticism, although, as he put it, some 'hot coals' did appear at the 1973 Oxford conference.[37]

The person providing most of these 'hot coals' was the American anthropologist Sherwood Washburn. Ironically, Washburn was someone that Hamilton admired and whose quotation he had chosen as a vignette for his own chapter. In this quotation Washburn pointed out that learning is genetically prepared—that is, different species learn different things with different levels of ease. (This was an important point made originally by Konrad Lorenz.) The conference was organized in honour of Washburn and Niko Tinbergen—the famous Oxford ethologist who later that year was to share the Nobel Prize in physiology or medicine with Konrad Lorenz and Karl von Frisch—with the intention of bringing the new developments in ethology into the general anthropological discussion. If the expectation was for the anthropologists to react, the whole thing succeeded almost too well.[38]

What was it Washburn reacted so strongly against? From his later writings, we see him employing the usual barrage of criticisms against what was later to be called 'sociobiology'. But the basic problem may have been the same that later put him at odds with his own favourite student, Irven De Vore: he deeply resisted the shift from group selectionism to 'kin selectionism' that was starting to take place at this point. And of course, Hamilton epitomized the 'wrong' way of thinking, being the very inventor of Hamilton's Rule and inclusive fitness, which basically emphasized the value of kinship altruism.[39]

Worse, Bill was making various statements explicitly putting down group selection. Washburn on the other hand had a soft spot for group selection. He saw it as an expression of the power of social organization, which for him was an important force in human history. For him, ascribing human progress to genetic changes was anathema, and that was what he thought Hamilton's position was all about.[40]

Central to Hamilton's 'Innate Social Aptitudes of Man' paper is the idea of *levels of selection*. There is no point talking about 'the' unit of selection—natural selection operates on many different levels simultaneously. But because the levels form a nested hierarchy, there can be internal competition between the levels. Especially in regard to social traits, traits that may be beneficial at one level (say, altruism of individuals in a group is good for the group) may have different effects at the next higher level (say, this group will now be more powerful and beat another group with fewer altruists among its members). Traits that contribute to solidarity in the in-group make the in-group stronger in its competition against the out-group. Or take conscience, says Hamilton. It is good for a group to have members with a conscience, altruistic members. But this may make the group grow too fast, which in turn means that it will use up a larger part of common resources in a 'selfish' way.[41]

It is here that Hamilton introduces Price's covariance formula. It addresses exactly the question of how two successive levels—the intra group and the inter group—contribute separately to the total amount of a particular trait in a population. Moreover, if both individuals and groups can be regarded statistically as collections of traits, the relative presence of a particular trait (say, altruism) can be measured at different levels.

The point of most interest for Hamilton, of course, is how to increase the level of altruism in the population. Haldane had suggested 'tribe splitting' for groups reaching a certain size. But Hamilton had noted that this was not enough—the ensuing competition between the new groups would invariably lead to the strongest group prevailing, after which altruism in that group would slowly disappear. There had to be a way to stop altruism from diminishing in the population!

The solution was to get positive association between individuals carrying this trait. And here the simple principle was for the benefit of an altruistic act to fall on individuals more likely to be altruistic, than on random members of the population.[42]

In terms of the Price Equation this involves the ratio between the within-group variance and the between-group variance. Expressing this mathematically as the probability that a certain number of 'fitness units' are given to a fellow altruist rather than to a random member of the population, Bill arrives at a criterion for 'positive regression of the recipient's genotype on the donor's genotype'. Bringing together all the donor's effects on recipients of various relationships plus the donor's basic non-social fitness, one arrives at the donor's inclusive fitness.[43]

Hamilton has here re-derived inclusive fitness with the help of the Price Equation, and he is very pleased. He has killed several birds with one stone:

> The usefulness of the 'inclusive fitness' approach to social behaviour... is that it is more general than the 'group selection', 'kin selection', or 'reciprocal altruism' approaches and so provides an overview even when regression coefficients and fitness effects are not easy to estimate or specify.[44]

And here Bill can finally put his foot down and clarify some issues that have bothered him for a decade. He explicitly tries to put a stop to the distinction that others have made between kin selection and group selection. This is not his doing. And people seem to believe that kin selection and inclusive fitness are just alternative ways of talking about the same thing. They are not. Inclusive fitness is a far broader concept. The important issue is not *kinship* between donor and recipient—rather, it has to do with carriers of altruistic genes recognizing one another: 'assortation' in a general sense. Otherwise put:

> '[K]inship should be considered just one way of getting positive regression of genotype in the recipient, and... it is this positive regression that is vitally necessary for altruism. *Thus the inclusive-fitness concept is more general than "kin selection"*' [italics added].[45]

Assortation in fact includes a range of possibilities. '[I]n the assortative-settling model it obviously makes no difference if altruists settle with altruists because they are related (perhaps never having parted from them) or because they recognize fellow altruists as such, or settle together because of some pleiotropic effect of the gene on habitat preference.'[46]

Therefore, Hamilton reasons,

> [i]f we insist that group selection is different than kin selection the term should be restricted to situations of assortation definitely not involving kin. But it seems on the whole preferable to retain a more flexible use of terms; to use group selection where groups are clearly in evidence and to qualify with mention of 'kin' (as in the kin-group selection referred to by Brown), 'relatedness' or 'low migration' (which is often the cause of relatedness in groups), or else 'assortation', as appropriate.[47]

Here, then, we have Bill Hamilton's serious effort to sort out the 'group' and 'kin' conundrum. It is all a tour de force and a triumph of his research programme, always striving for greater generality. It is therefore so much more surprising that he should have chosen to tuck away this particular re-derivation of inclusive fitness and showing how it works with levels of selection in a conference volume—in the field of anthropology of all things. In his autobiographical notes Bill tells us that he used to amuse himself thinking about people getting into a conflict as to whether they should read his chapter or not. On the one hand it contained an important new derivation which clarified many misunderstandings about group selection, kin selection, and inclusive fitness. On the other hand, he had things to say there about human nature and human history which might be easily regarded as offensive.[48]

It seems that he succeeded only too well in maintaining a low profile for this work. No reaction followed among his biological peers. There was no big Aha! experience—and no Boo! either. Did his peers even know about the existence of Hamilton's chapter? It seems that his colleagues just continued seeing kin selection as opposed to group selection, and treating kin selection as an alternative name for inclusive fitness.

People are typically not aware of 'Hamilton II'—the group selectionist. It has fallen to such scientists as DS Wilson and Elliott Sober to try to rectify the situation, bringing back Hamilton's re-derivation.[49]

Meanwhile, could it be that Hamilton himself was conflicted about how much to publicize his re-derivation? Although he obviously had to recognize the correctness of his own re-derivation of inclusive fitness and his wish to honour Price, Hamilton may not have been particularly pleased with the whole outcome. In his heart he may have remained Hamilton I, the man who had declared himself 'allergic' to group selection, rather than Hamilton II, the reformed group selectionist. He may also have keenly remembered his own long and difficult struggle to derive a general expression for inclusive fitness. The reconciliation of kinship theory with group selection with the help of Price may have seemed to somehow belittle his own earlier effort. Or was there something more? At the very beginning of his career, in his 1963 paper, Hamilton had proudly dismissed Haldane's group selectionist derivation of altruism. And now he had been almost forced to show how Haldane's approach could be made sense of after all with the help of the Price Equation.[50]

With the Fox volume chapter, Bill had a sense of intellectual closure in regard to the whole business on the evolution of altruism. What could be said about kinship theory, reciprocal altruism, and the like, had pretty much been said. The rest was a mopping up exercise that he was not interested in—it was time to do something new. From Hamilton's own perspective the kinship paradigm was at the point of becoming exhausted, or, alternatively, the research programme around kinship theory did no longer seem excitingly progressive. Now was the time to pursue some healthy field research and take a break from the theorizing. For Bill, the circumstances were very similar to the one that drove him to Brazil the first time around: hard theoretical work, psychological exhaustion, and the craving for a total change of scenery.

11

Creativity in a Tight Spot

The bruises from Robin Fox's Oxford conference had not been that bad, save for the grumblings of Sherwood Washburn. The suicide of George Price, though, lay rather heavily on Bill's conscience. In any case, time had come for a break from Imperial College and its teaching duties. And Bill knew exactly where to go—where else, if not Brazil. Warwick Kerr, Bill's initial contact, and an increasingly important statesman of science, had by now moved to the university in Riberão Preto, a prosperous coffee plantation town in the province of Sao Paulo, and so had a number of his colleagues. It was here that Bill and his family spent some nine months in 1975–76.

Bill and Christine were lucky in that they found a nursery in a convent nearby where the sisters could look after their two daughters, now aged three and two. The girls soon became fluent in Portuguese. But what were their parents up to meanwhile? Anybody watching Bill Hamilton during this time would have seen a man almost fanatically gathering figs and slashing them open. What on earth was he looking for?[1]

The answer is: the tiny fig wasps. Bill and Christine found themselves involved in monumental fieldwork, sorting males from females and trying to classify different species. What was so intriguing, and explained Bill's determination to slash as many figs as possible, was the variety of forms that crawled out of the figs. It was a parade of extremely strange creatures, wingless and big-headed, with fantastic mandibles and other types of lethal-looking weapons. How might one make evolutionary sense of such grotesque creatures, which were so different from the winged ones emerging. What accounted for this difference?

It was not long before Bill was caught up in the excitement of a great puzzle. He opened up more and more figs, sometimes using just his bare hands. The feeling he was getting reminded him of his earlier enthusiasm for solving what he called his 'Melittobia problem', named after the little parasitoid *Melittobia acasta*, a vivid example of the deviation from Fisher's 1:1 sex ratio among small interbreeding insects.Bill's examination of the mysteries of fig wasps was to keep him occupied for the next few months.[2]

Bill could easily explain what was going on in figs in general. Pollination is species specific, each species of fig has its own fig wasp, and it is important for males and females from the same species to find each other. An obvious strategy is to have wings and fly to find your mate. But another strategy which may be more reliable in the case of a less common type of fig tree/fig wasp, is to try to have a rendez-vous already inside the fig. Since in a given fig there is often great male-female discrepancy, with males outnumbering females, the result is competition for females.[3]

The fig wasp males are of two kinds: winged and wingless. For wingless males their only chance to mate is to fight off other males—hence the various nasty-looking paraphernalia. Hamilton describes the tactics of these fig wasps as perhaps 'cowardly' looking—they come out for a quick strike and immediately pull back and hide. But, he says, consider their situation. What would you do?

> A male's fighting movements could be summarized thus: touch, freeze, approach slowly, strike and recoil. Their fighting looks at the same time vicious and cautious—cowardly would be the word, except that, on reflection, this seems unfair in a situation that can only be likened in human terms to a darkened room full of jostling people among whom, or else lurking in cupboards and recesses which open to all sides, are a dozen or so maniacal homicides armed with knives. One bite is easily lethal.[4]

The general discussion of fighting strategies in the fig wasp paper is an interesting one. Here we see the game player side of Bill. His observations range freely from insects to humans—from horned beetles to left-handed boxers, and their relative advantages and disadvantages.[5] Bill makes an important distinction between honest and false signalling and gives particular consideration to the case of bluffing. When is it worthwhile to bluff? And is there ever any point in bluffing 'down' your capabilities? (Judging from Bill's own experiences, the answer to this is yes. If one is on a mission, say, to find a particular rare insect in places such as a biscuit factory or hospital—having been given a hot tip by one's pest exterminator friend—the best role and attire to assume is surely that of a plumber with no knowledge of insects whatsoever.)[6]

Ironically, this whole rich fig wasp project was entirely unplanned. Hamilton had actually received a research grant from England for a comparative study of insects in Brazilian old tree trunks and those in British decaying wood. This was to be a follow-up to a study on species and life forms in this habitat that he had begun at Silwood. He had thought that a comparative study would be valuable, but had not taken into account the location of the place where he would be going to live and teach this time—Riberão Preto. That place featured not a single piece of rotten wood! If anything, the trees surrounding the main buildings of the old coffee plantation that had now been turned into a nice-looking university campus, were as healthy as possible. Eucalyptus trees in general belong to those trees that are practically insect free. What was more, it was the dry season and the only thing growing was fig plants—as weeds.[7]

But then Hamilton remembered something from his previous visit: a student of Warwick Kerr's in Rio Claro, Lucio Campos, had shown him some strange insects that could be found inside figs. So he looked for them in the local figs, and sure enough, there they were.[8]

And then he had an Aha! moment. The wingless fig wasps struggling to mate inside the darkness of the fig in fact very much resembled the cloistered life that existed under the bark of rotting wood that he knew so well. So Hamilton's fig project came about simply faute de mieux, because he could't find any suitable old tree trunks close by. It was a challenging situation, but Bill got creative and against all odds was able to make a connection between his current fig project and his original proposal. He realized that what he was studying in both cases was the broader question concerning evolution in closed spaces, and that the results from his fig research would actually help him theorize more generally about this problem. But first he needed to make the not-so-obvious connection between the decaying wood habitat and the fig environment. As Pasteur said, 'discovery favours the prepared mind'.

Bill made one more important discovery during his trip. That was that he actually enjoyed teaching. This was quite a surprise to him, since he had rather bad experiences from Imperial College. There his teaching duty involved big classes of seemingly unmotivated undergraduates—Bill himself reflected that the things he was teaching them were not what they were interested in (he guessed that they perhaps wanted more practical knowledge). Things had got so bad that students regularly complained about Bill's lectures. Now here he was in Brazil, teaching population genetics to a small class of six graduate students in Portuguese. And it went well, with students being polite and respectful. Clearly, the situation was pleasantly different. Moreover, as I learnt from Warwick Kerr, there was a tacit backup system of sorts—if there was something that the students didn't understand, they could always go to Kerr for explanation. It is not clear whether Bill knew this, but it probably made the students feel more secure. In any case, Bill's conclusion was that he did have a chance as a teacher, after all.[9]

Bill reflected on the difference between teaching graduate students and undergraduates:

> The course I taught was my first wholly at a graduate level and my class was just six. To my surprise I enjoyed the teaching. I had the difficulty of speaking in Portuguese but, against this, how different were these thoughtful, respectful Brazilians from the hundred-headed swarm I faced in my customary nine o'clock lectures at IC. There I was lucky if I could see one or two of the heads caring one Johannsen's polygenic bean for my Hardy-Weinberg equilibrium or my definition of linkage: during the classes most students would be good-humouredly chatting or perhaps deep in their morning newspapers. And toward the end of the course I knew they would have another topic: they would be discussing who was to lead this year's delegation to the professor to complain about the irrelevance and incomprehensibility of the lectures I was giving. In those days apparently 'inapplicable' subjects such as population genetics and evolution were not the zoology that aspiring entomologists and parasitologists had come to Britain's prime technical institute to learn. Instead they wanted insect physiology, practical parasitology, and the like—in short, knowledge to make a living with.[10]

After his Brazilian experience, Bill knew that he wanted to teach only graduate students.

***

Let us take a closer look at Hamilton's life-long fascination with 'life under bark'. This theme related to a long-standing question of his: what were the factors that pushed social insects over the threshold to eusociality? This problem deeply interested him. It would have to be something more than a mere overlap of generations 'before such situations can evolve towards an outbreak of that extreme sociality and sterility such as is found in ants and termites, both of which groups... are very likely to have originated in this environment'.[11] What we are dealing with here is in fact Hamilton's second, 'underground' social theory. This theory deals with the very origin of sociality and eusociality.

At Silwood Park Bill had been working with, among others, his student Victoria Taylor who was awarded her PhD for her study of the biology of the half-millimetre beetle *Ptinella*.[12] Her mentor had become greatly intrigued by the fact that most species of this genus exist in two different forms—one wingless, colourless, and blind and another winged, tanned, and able to see—the first adapted for underground expansion, the other for dispersal. It seemed a promising project to continue the inventory of dead-wood insects in a more systematic way, looking for patterns that could later be generalized and focusing especially on wing polymorphism.[13]

And so he did. After chiselling away at old tree trunks around Silwood Park and Oaklea and exhausting the entomological literature in the London libraries—his usual combination of naturalist and library research to find underlying patterns—he felt that he could confidently make a number of generalizations. He presented his findings at a meeting of the Entomological Society in Kensington (later written up as 'Evolution and Diversity under Bark' in the conference volume published in 1978).[14]

In his long and detailed essay Hamilton calls attention to the huge difference between the open green habitat exposed to sun and wind and the murky, closed underworlds.[15] He points out the remarkable diversity in the host of genera, families, and higher taxa that can be found only among dead-tree insects. A peculiar feature of that insect fauna is its amazing degree of *functional convergence,* despite its extremely diverse *phyletic divergence*. How is this to be explained? In his paper, Hamilton suggests that the reason is *similarities in breeding structure induced by the environment.*

He finds four such functional convergences in dead tree insects:

1. Different types of wing polymorphism, which appears to be switched on by *environmental cues,* not by alternative genotypes;
2. Male haplodiploidy—about four of the six inventions of male haplodiploidy exist, and may have originated in this habitat;
3. Advanced social life—at least two origins of advanced social life in insects (termites and ants) can be found in this habitat. Also there are many examples of sub-social life in this type of environment;
4. Numerous developments of sexual dimorphism, often connected with particular roles of males (wingless, unusual size, strange armaments, etc).

The radical, novel point that Hamilton wants to make is that *this particular habitat has the ability to force a particular kind of breeding structure, favourable to rapid evolution.* He believes the environment of dead trees and rotten wood is the place of original divergence of many of the major insect groups.[16]

So for Hamilton it is *ecology* rather than genetics that is the motor in regard to the origin of major insect groups. What is it, then, about this specific ecology that makes it so conducive to rapid evolution? It is the 'cavernous quality of the insect living spaces that dead trees offer'. In addition, argues Hamilton, dead trees tend to be uneven and scattered, which connects up well with Sewall Wright's point that rapid evolution can happen if a population is scattered into many small 'quasi-isolated' demes.[17]

Hamilton goes on to list additional 'evolutionary facilitations' that can arise due to this kind of local isolation:

1. Evolutionary change (due to drift and recombination);
2. Social evolution (promoted by the formation of kinship groups);
3. Protection from major disaster (through deme subdivision);
4. Interspecies reciprocation and symbiosis (this is Trivers' reciprocation argument for individuals, here extended to multigeneration local stocks);
5. Polymorphism (promoted, eg, through alternating selection for sedentary breeders and dispersers).[18]

What is more, Hamilton has found paleontological evidence to back up his conclusion that ecology is the motor for the origin of major insect groups. Take beetles and their ancient heritage:

> The vast order of beetles, the most speciose group of the living world, almost certainly originated in the habitat of rotting trees: its first representatives are found in the Permian period and closely resemble particular primitive rotten-wood species of today.[19]

But the same is true of the origins of other orders as well. Just look what crawls out of Bill Hamilton's cornucopia of bizarre underground life:

> Here the world's smallest beetles fold within their half-millimetre bodies sperms longer than themselves (likewise fold the delicate feather wings that experts say they should not be able to fly with, but they do); here *Xylocoris*, the prototype bed bug, lives with its psychopath male crimes all long forgiven and forgotten, its

> astonishing act accepted throughout an entire descendant superfamily; here lives *Pygmephorus*, the *Stereum*-sucking mite that balloons its body to be a gigantic and watery pen for sex-wise incestuous infants; here *Cephalonomia* the proto-ant, fitted with more castes-in-prospect than most of the real ants, moves weasel-like through tiny tunnels after doomed beetle-larva hosts; here *Heteropeza* nearly forgets the midge identity and open-air life of its ancestors and teems by bursting its stiff, fibre-like maggot offspring out of the large mothers (when grown, these are due to explode themselves in birth and so onward in an endless progression); here beetle larvae lie in soft cradles quilted with living food, no more than turning their heads to feed all their lives through 'kindness' of an organism from another kingdom; here fight syncytial amoebae; here luminous wood chambers still brighter-glowing lampyrid firefly larvae within it; here are polymorphisms for wings, harems sequestered by armoured blue-beards, horned beetles fighting, guarding or in copula... A family style of life was so common that I came to expect ever new groups to reveal it. Always I found many ages mixed together, all within small cavities or self-formed patches of other kinds. Worms, mites, woodlice, and insects of various types, were among my examples.[20]

For Hamilton, Brazil was particularly interesting for its beetles. His all time favourite beetle was *Titanus giganteus*, a dark brown beetle as big as his hand, breeding in rotten palm trees in the state of Amazonia. Another favourite and fascination of his was the Brazilian carrion beetle, which was able to bury small dead animals. (In typical flight-of-fancy Bill style—but also in typical earnestly-meant Bill style—he later included in a contribution to the Japanese journal *The Insectarium* a hypothetical description of how it would feel to be buried by carrion beetles).[21]

But how was Hamilton's theory of ecological constraints and social evolution received by his professional colleagues? Despite its radical claim, Bill's paper 'Evolution and Diversity under Bark' presented at the Entomological Society of London, did not seem to have had any earth shattering effects. Later he reflected that there may have been something wrong in the style of his paper, or perhaps it was too much geared to entomological aficionados.[22] Both points are probably true. The piece is very long and detailed. Moreover, an (over?) abundance of specialist entomological terminology makes it difficult to read for non-entomologists. In short, the article was not easily accessible. A simple explanation, therefore, might be that two things happened with this paper:

entomologists found Bill's thesis too radical for their taste, while general evolutionists, who might conceivably have been interested in the idea, never knew of its existence. They would not have felt a special urge to open a book with the title *Diversity of Insect Faunas*. Bill's thesis about the origin of eusociality and life underground got buried in the conference volume.

* * *

Some seem to believe that Hamilton in his 1964 two-part paper in fact *claimed* that a strong relationship existed between haplodiploidy and eusociality in Hymenoptera. He did not, but others have made this connection. In fact, Hamilton's exact formulation was quite cautious. And as time passed and more evidence came in, his position became even more careful.

Let us see exactly what Hamilton said in his 1964 paper. Under the subheading 'Colonies of social insects' (in the second part of his paper) he starts off as follows:

> Caution is necessary in applying the present theory to Hymenoptera because, of course, their system of sex determination gives their population genetics a peculiar pattern. But there seems to be no reason to doubt that the concept of inclusive fitness is still valid.[23]

He goes on to imagine a concrete situation for a colony. If a worker's tendency is to go back to her mother to help rear her children, rather than starting her own, how can colony reproduction ever take place? His answer is an *ecological consideration of costs and benefits:*

> As soon as either the architectural difficulties of further adding to the nest, or a local shortage of food, or some other cumulative hindrance makes the adding of a further bio-unit to the colony 1 ½ times more difficult than the creating of the first bio-unit of a new colony, the females should go off to found new colonies.[24]

At this point Hamilton brings up a suggestion by the famous entomologist Charles Michener, according to whom the development of a worker caste has been possible because of unrelated bees sharing a common entrance tunnel to their burrowing nest. Hamilton's first reaction is surprise:

> The classical theory concerning the evolution of the social insects has always posited a wide overlap of generations allowing mother and daughters to co-exist in the imaginal state as one of the preconditions for the evolution of this kind of

> sociability... That such altruism could arise through genetic interest in the offspring of unrelated bees sharing the same excavation, as Michener actually suggests, seems to me incredible.[25]

But he immediately goes on to consider another more acceptable idea, 'another important type of social behavior to which Michener has re-drawn attention which might well arise on the basis of much lower relationships.' This is a case where entrance-guarding females take turns at fending off intruding parasites. Hamilton, ever the theorist, even goes on to suggest an imaginary system for further improving the situation. There should be some way of shielding it from insects that may be shirking at their duty. Here is his suggested evolutionary remedy:

> [T]he bees could evolve an instinct which allowed them to leave duty at the nest entrance only on the stimulus of another recognized tenant coming in... By going out when supposed to be on duty, a bee would jeopardize her own brood as much as, if not more, than, the broods of the others, so that selection would tend to stabilize the instinct.[26]

In other words, it would seem that, at least in principle, Hamilton is willing to entertain the possibility of cooperation without relatedness as long as there is some self-interest involved for the cooperators. The reason he does not consider it very likely, however, is that he is keenly aware of a problem: *the evolution of countertendencies*:

> Nevertheless, even if it is possible to account for the evolution of guard-instincts without a basis of social relationship between the bees, it is hard to see how other socially disruptive practices, such as robbing within the nest system, could fail to evolve unless a bee's co-tenants are also usually the carriers of some part of its inclusive fitness.[27]

Hamilton's early concern for countertendencies is a natural part of his game-theoretical thinking. Meanwhile, counting with the evolution of countertendencies may not be typical within traditional theorizing about social insects.

Later, in his 1972 *Annual Review* paper, Hamilton revisits the haplodiploid thesis, now looking more closely at the available empirical evidence. He finds supporting facts and arguments but also clear counterexamples. It is clear that haplodiploidy is only one factor among many when it comes to the evolution of eusociality:

> [V]arious facts suggest that the social groups of Hymenoptera originated from ancestors which were effectively monogamous in the female sex.... Assuming single mating in the socially inclined ancestor, an attractive explanation for the basic feature of hymenopteran sociality, worker-like attributes in females, is at once apparent. In the sense of inclusive fitness a female prefers sisters (B=3/4) to sons or daughters (B=1/2) and therefore easily evolves an inclination to work in the maternal nest rather than start her own.
>
> Such a bias latent in the male haploid system helps to explain the multiple origins of social life in the Hymenoptera, contrasted with one origin only for the Isoptera [termites]... *Nevertheless, the occurrence of this special relatedness to sisters must not be overemphasized. Male haploidy is certainly not the only prerequisite for evolving a sterile caste. Perhaps the preadaptations of solitary nesting Hymenoptera as porters and builders are equally important.* We have to explain why other male-haploid groups have not evolved social life, and also why clonal aggregations like those of aphids have not done so. Equally troublesome, why is thelytoky [unfertilized eggs giving rise to females] not more common in social insects considering that it occurs at all? [italics added].[28]

Hamilton goes on to give examples of social insects that are thelytokous but not social, or degenerate and suggests that 'the catch lies in the danger of abandoning sexual reproduction, and this danger is still imperfectly understood'. Also, why is there not more cooperation in aphids—why have workers failed to appear? He also mentions some male haploid groups which may have worker-like adaptations, particularly ambrosia beetles of the genus *Xyleborus* which have developed elaborate gallery systems and some colonial fungus-feeding and gall-forming species in thrips.[29]

Hamilton then systematically surveys the existing empirical evidence for the possible factors that might have created conditions for the evolution of eusociality. He explicitly considers such possibilities as selection operating between and within colonies, pre-adaptations, and various ecological conditions.

Starting with termites, which have normal reproduction in contrast to the haplodiploid Hymenoptera, Hamilton notes that '[a]s with all social insects the co-operation and altruism within a colony of termites is in strong contrast to their hostility to members of other colonies'.[30] Meanwhile, termites have a basic system of monogamy, and this ensures that the relatedness within the colony will be high. Also, inbreeding is common, which rapidly increases the general level of relatedness. Still, Hamilton notes, this doesn't seem to be enough. There are groups with similar characteristics (claustral life, flightless, inbreeding) which have not been able to surpass the threshold to eusociality. Special features that could have aided termites in this could have been their additional interaction with protozoan symbionts, and the need for a regular pattern of social contact for their unusual feeding pattern (they eat each other's excrement). Later, as the colony expanded and moved deeper into its burrows, these arrangements could have resulted in the division of labour. 'These particular factors have no necessary connection with genetical relatedness', Hamilton observes, '*but, of course, insisting on the necessity of relatedness in no way precludes other factors as necessary or contributory* [italics added]'.[31]

Hamilton realizes that inbreeding is not necessarily a good thing; it lowers a species' ability to adapt quickly in a changing environment. But if one regards a colony as a *superorganism*, 'inbreeding makes the superorganism more nearly a reality. It should facilitate the peaceful coexistence of multiple reproductives (where advantageous) and facilitate the cooperation of their broods.'[32]

Hamilton then goes on to look for other groups with normal reproduction which might be showing beginning tendencies to altruism of some kind; 'pre-adaptations' to social life. One example is what he calls 'trophic altruism'—say, a cricket mother allowing her own body to be eaten by her brood. And there are more such candidates: 'The omnivorous diets, the manipulative ability (burrowing, entrance closing, food storing), and even the stridulation [movements that create a high pitch sound] of crickets suggest pre-adaptations to social life. The wing polymorphism that is frequently present...may also be pre-adaptive: local population viscosity permits the evolution of friendly relations between neighbours while some distant dispersion prevents gene fixation and preserves adaptive plasticity'. Still, he concludes, this was not enough to produce worker-like altruism.[33]

As for Hymenoptera, Hamilton warns that calculating coefficients of relatedness is actually quite complex. This can only be done when the sex ratio has been specified, which is in turn an empirical question. The issue has to do with the balance of power between the queen and the workers. Hamilton describes the situation:

> [T]he queen may be inclined to produce more males than the sterile workers regard as ideal. On the other hand a laying worker may want more males than the queen does—provided the extra males are her own (the worker's) offspring. The factor of multiple insemination also affects these issues and so do differing roles of the two sexes in gene dispersion.[34]

In advanced swarming species, such as *Apis* and *Eciton*, the complete dependence of the queen upon workers becomes important. Outbreeding, multiple insemination, and helplessness of queens should all correlate with high male production.[35]

Another complication is that depending on the amount of multiple mating, the relatedness between sister and sister may be higher or lower than the relatedness between mother and daughter.[36]

A particular issue that needs tackling again is Michener's thesis that worker-like behaviour can develop through association between unrelated females. According to Michener association takes place first and this then gives rise to the matrifilial colony. Hamilton now addresses this argument head-on. He is convinced that the required traits develop first and that this is what later enables the association between the foundresses of a new colony:

It is certainly not impossible for worker-like behaviours to develop in a group of sisters if the advantage to the colony is high enough. On the other hand, high advantage is unnecessary to arrive at the matrifilial colony: male haploid animals gravitate naturally towards this condition provided that the sex ratio or some ability to discriminate enables the worker to work mainly in rearing sisters. *Therefore it seems likely that the worker-like attributes involved in association—submission, ovary inhibition, etc. —arise during a matrifilial phase and that these attributes subsequently permit association between foundresses when certain additional conditions are satisfied* [italics added].[37]

He suggests the following conditions for the development of worker-like behaviour:

1. all adult females have the potential to behave as workers;
2. homing instincts keep groups of sisters together;
3. inbreeding further increases the relatedness of sisters and relates also occasional pairs of associates that are non-sisters;
4. an associated group of females has a much higher reproductive potential than a nest foundress alone.[38]

Association is also helped if males and females remain close to their nest of origin and mate close to it (an island or stepping stone population structure), which ensures at least some relatedness between the members of the population, which in turn helps association. Also, weak flight capability helps association by bringing about local genetic homogeneity.[39]

But are there also general examples of *cooperation without relatedness*? Hamilton is systematically investigating the limits of his paradigm. Yes, there are. A typical case is where cooperation also supports an organism's own interest (eg a female guarding an entrance to a burrowing system which also includes its own cell). Also external threats can induce cooperation (eg predation by ants on polybiine wasps induces an efficient mix of queens and workers that can quickly rebuild the nest).[40]

And then there is the newest suggested possibility of cooperation without relatedness: Trivers' theory of reciprocal altruism:

> Reciprocal altruism, as defined by Trivers, has no need of relatedness. It requires instead that interactants should remain together long enough for their roles as donor and recipient to reverse several times, and also that they should be endowed with flexible behaviour that curtails further benefits to individuals which are observed not to reciprocate.[41]

Finally, Hamilton pushes his paradigm to the extreme. What about *interaction between different species*? Here certainly the kinship factor is excluded. Trivers had invoked the case of cleaning fish. He argued that for this arrangement to work, a pair of animals needed to have to have a long-term association. An inhibition has to be established against the cleaner fish eating its cleaner. Hamilton decides to consider instead the relationship between pathogens and their hosts. What are the strategies here? Bacteria should develop benign relationships with their host in order not to get killed themselves—which will happen if they succeed in killing their host. Moreover, the best strategy for the pathogens has to do with the nature of the host population—is it mobile or immobile? By moving, a mobile population provides ever new hosts for virulent pathogens, so in this case the pathogens should keep up their virulence. But if the host population is relatively immobile, it is likely that the pathogens may become symbionts.[42]

This same reasoning about mobility versus immobility also roughly fits the phenomenon of *mutualism*, cooperation among species, which Hamilton also briefly discusses. Examples abound. In bees few symphiles (species that live together) are benign, whereas in ants and termites symphiles are often benign. There are well-known wasp-ant associations. In wasps, adults have been found to be dependent on their larvae for digestion of proteins into sugars. And just like humans, ants probably moved to agriculture and cattle rearing of their own kind, with mutual benefits for both, Hamilton muses.[43]

Here, then, we have seen Hamilton, the naturalist in action, closely observing life especially underground and theorizing about the origin of social life. He acts here largely as a creatively theorizing naturalist and ecologist, who engages with existing theories and takes data collected by his colleagues (and himself) seriously, while among these very diverse sources and types of data he is trying to identify broader regularities and patterns and even find a tentative answer to the question of the origin of sociality.

Throughout his life Hamilton continued assessing the existing empirical evidence. Fast forward to the mid-1990s. Here is Hamilton's yet again updated view of the role of the haplodiploid relatedness pattern in shaping eusocial trends:

> Elements of the actual pattern, such as the almost complete non-working of hymenopteran males in contrast to the 50 per cent of workers being male in termites, do in fact accord well with the theory. *Whether other factors that must also apply to the evolution of sterility may be more potent forces, however, so that the special pattern imposed by haplodiploidy is swamped by them and therefore hardly detectable, is much less clear.* On the side of haplodiploidy having relatively high power I take some tentative encouragement from the very recent discovery of a probable sterile caste in thrips, which is another haplodiploid group. However, in some thrips species males also develop into fighters and in at least one species have been observed attacking adventive enemies. Although males are apparently less active in this respect than females, the observation means, again, that *biases of relatedness are clearly not the whole story* [italics added].[44]

So, although he sees relatedness as an essential factor, he is willing to consider that it might be obscured by other factors. He even suggests the possibility that a *synergy* between the various factors might contribute to a sudden transition to eusociality. In very rare cases, new eusocial groups may emerge without the help of haplodiploidy at all.

> What the extra-factors might be remains a subject of very active discussion. Haplodiploidy is probably just one of them. This itself, as I indicate in the paper [here he refers back to his paper 'Evolution and Diversity under Bark'], may through microbe interference be especially apt to become originated in the seething underbark world. Other courses toward social life arise, somewhat paradoxically, from the fact that life in families is not necessarily peaceful; others arise from the temporal limitation of the dead tree's supply of food

> and shelter, leading to the split incentives...to be a stay-at-home breeder or a disperser; yet others again arise from patterns of relatedness (see Chapter 8) ['Evolution and Diversity under Bark']. *Extremely rarely, as it appears, enough of these factors may concur and synergize so as to launch, without the help of haplodiploid asymmetry or superrelatedness, a new and thoroughly eusocial group* [italics added].[45]

We see that Hamilton is more than willing to consider various factors that may contribute to the evolution of high sociality or eusociality, many of these ecological, having to do with the availability of nests, food, etc. He also realizes that sometimes the haplodiploid relatedness may be obscured by other considerations. In other words, he is very far from believing in the total explanatory power of the Hymenoptera hypothesis.

Especially when it comes to life underground, which he knows so well, Hamilton is as much an *ecologist* as he is a theoretical modeller. It is interesting to note how much interest he takes in such things as nutrition and development. Here is a zoologist who knows his beloved study objects in intimate detail. And as an inductivist in regard to empirical studies, Hamilton seems open to a complex vision of interacting forces when it comes to the origin of eusociality.

***

Hamilton himself has not been harping on about the haplodiploidy hypothesis—but others have. The point is of course that Hamilton never depended on the case of Hymenoptera as a cornerstone for his inclusive fitness. That connection has been made by others.

We have, for example, EO Wilson presenting and discussing the Hymenoptera connection in his *The Insect Societies* in 1971. He is interested, but also somewhat critical.[46] We have Robert Trivers in 1976 (with Hope Hare) declaring the case of Hymenoptera the strongest support for Hamilton's theory of inclusive fitness. (Later, Trivers' empirical study supporting the haplodiploidy hypothesis is enthusiastically cited and commended by Wilson in *Sociobiology*, chapter 20). The Hymenoptera case is popularized for a larger audience in *Natural History* (and later in a popular book) by Stephen J Gould. And John Maynard Smith in his writings typically connects Hamilton's theory with the Hymenoptera.[47]

It was not long before in the popular mind the haplodiploidy thesis became synonymous with Hamilton's inclusive fitness theory, which in turn became identified with kin selection. For Hamilton himself, one can imagine this created some discomfort. On one hand it was nice to be famous and have empirical backing for one's theories, but on the other, the coupling between his universal theory of inclusive fitness and this particular insect group was getting too tight. Hamilton, ever the practising scientist, knew that examples had sometimes been used because they appeared to illustrate a theory, but that they also sometimes had to be changed while the theory continued being valid. The Hymenoptera was only an example, at an early point of his inclusive fitness theory when there were few good examples available. On the other hand, he was not willing to totally dismiss haplodiploidy—he was convinced that there was an important truth to it, although it surely was not the only, or even the leading factor for eusociality. Ever the scientist, Hamilton was willing to consider new empirical evidence as it emerged.[48]

But at the same time, his belief that there was something truly important about genetic relatedness made him state this almost as a provocation—similar to Darwin's famous statement about a 'difficulty' that would undermine his whole theory. This was Bill Hamilton's stance in the mid-1990s:

> On whether *haplodiploidy helps shape in some degree* the pattern of sociality the case seems clear: the Neodarwinian arguments of Chapter 2 and 5 [his original 1964 paper on the evolution on social behaviour and his 1970 paper on spite] show that it has to shape that pattern; for this to be wrong the whole argument must be flawed and no one has shown that it is nor provided any alternative rationale for how automatic altruism can arise [italics added].[49]

This challenge was later to be picked up—or so it seemed—by a group of Harvard researchers, briefly causing major havoc in the international community of theoretical and practising sociobiologists in 2010.[50]

12

Priority Matters

Back from Brazil in May 1976, Bill Hamilton followed his usual routine, catching up on his journal reading. There were many interesting new developments. Young Robert Trivers, with whom he had been in correspondence, seemed to be doing very well, developing some Hamiltonian themes and getting well-deserved attention. He also followed the controversy that surfaced around EO Wilson's huge new book *Sociobiology, the New Synthesis*, published in May 1975. He would have to acquaint himself better with Wilson's *Sociobiology* book. *New Scientist*, his favourite popular science journal, had run a whole page review of that book back in August 1975. It was written by John Maynard Smith and cleverly entitled 'Survival through Suicide'.[1]

It was the middle part of the article that gave Hamilton a jolt. What was this? Maynard Smith was giving the readers of *New Scientist* a brief history of the biological explanation of 'altruism'—according to Wilson, the central problem of sociobiology:

> Darwin thought that the evolution of social insects could be explained by between-family selection, and he was right. The detailed working out of this idea, and its application, had to wait for a knowledge of genetics; oddly, it had to wait much longer. I first heard the idea in the now-demolished Orange Tree off the Euston Road; J. B. S. Haldane who had been calculating on the back of an envelope for some minutes, announced that he was prepared to lay down his life for eight cousins or two brothers. This remark contained the essence of an idea which W. D. Hamilton, a lecturer in zoology at Imperial College, London, was later to generalize. Unfortunately, Haldane, although he referred to the idea in an article in *Penguin New Biology*, did not follow it up, and may not have

appreciated its importance. The decisive step was taken in Hamilton's papers in 1963, in which he introduced the concept of 'inclusive fitness', and applied it to the evolution of the social insects.[2]

This was not the way Hamilton saw things at all, and so he sent off a letter to the editor of *New Scientist*. Bill began his letter explaining the reason for the delay: he had only just read Maynard Smith's review after spending almost a year abroad. The letter continued:

> I was astonished and rather dismayed by an historical anecdote in [the review] which I had not heard before. This concerns JBS Haldane's insight into the conditions for the selection of 'altruism'. According to Maynard Smith, after a calculation on the back of an envelope in a public house he announced that he was 'prepared to lay down his life for eight cousins or two brothers'. Haldane's phrase is closely similar to phrases which I chose in my first two papers on the same subject.
>
> For instance, my paper of 1964 had: '...in the world of our model organisms...everyone would sacrifice [his life] when he can thereby save more than two brothers, or four half brothers or eight first cousins...'
>
> The general impression given by this anecdote, by Maynard Smith's publications on kin selection in 1964 and on warning behaviour in 1965....is that ideas of kinship theory must have been in fairly general discussion at University College while I was a postgraduate student there. In fact, Haldane's former colleagues were wholly uninterested in my proposed research on 'altruism'. The only guidance to the literature that I received was various referrals to Haldane's book *The Causes of Evolution,* which gives no indication of the quantitative principle.[3]

Hamilton goes on to describe the generally indifferent or hostile attitudes he encountered at University College, until about the beginning of 1963, when his model was already written up as a manuscript submitted to the *Journal of Theoretical Biology*. Enter John Maynard Smith:

> As a referee for its publication Maynard Smith was one of the first to read the manuscript. Since earlier he had apparently no information to offer about Haldane's insight, it seems possible that a simplified version of my phrase later came to be associated in his memory with Haldane.[4]

Bill does not hold back. He has found a way of telling the readership of *New Scientist* that it was in fact Maynard Smith who was the referee for his paper.

Before that nobody, including Maynard Smith himself, had cited the pub story to him. Bill also elegantly criticizes the behaviour towards himself by Maynard Smith in his 1964 and 1965 articles. (In Maynard Smith's 1965 article on warning calls, Hamilton's name is not mentioned at all.) But the main point of Hamilton's letter is that he is here questioning the very existence of Haldane's pub quip. What is the real evidence that 'Haldane said it first'? Bill is putting Maynard Smith on the spot.[5]

At the time, many readers of *New Scientist* may have been surprised, or even embarrassed, by Hamilton's letter. What was Hamilton really protesting about? Wasn't Haldane's pub quip already almost 'common knowledge'? The joke seemed to be circulating around. And wasn't it natural that Haldane's closest student and 'heir', John Maynard Smith, would have personally heard the master making this statement? Well, Hamilton did not think so. Haldane's statement, as cited by Maynard Smith, that he 'was prepared to lay down his life for eight cousins or two brothers' was too similar to what he, Hamilton himself, had written.

In other words, Hamilton suggested that it was he, Bill Hamilton, and not JBS Haldane, who had originally coined that phrase about eight cousins or two brothers. (Actually, the pub quip as it has been floating around in the popular scientific literature typically tells it in a different order: 'two brothers or eight cousins', or 'more than two brothers or eight cousins'.)

For those readers who looked up the original offending document out of curiosity—Maynard Smith's review of *Sociobiology*—the pub anecdote may have seemed an insignificant detail. Taking into account the end of the cited paragraph, it would have seemed to them that Maynard Smith was being eminently fair. After all, he continued by saying that it was Bill Hamilton who had taken 'the decisive step'.[6]

The response from Maynard Smith comes almost a month later. He is 'distressed that Dr. Hamilton…believes that I have done him an injustice', but believes that his own memory is not at fault. He goes on to reiterate the point made by another correspondent in the magazine the previous week (Dr Maurice Dow from the Department of Zoology at Edinburgh): Haldane's *New Biology* article in 1955 contained a very similar formulation (Maynard Smith had hinted at the same in his review). Therefore, Maynard Smith argues:

> It is clear that Haldane understood the principle of kin selection before Hamilton. Nevertheless, I have no doubt that the credit for the idea should go to Hamilton. What matters in science is not merely to understand an idea, but to see its

relevance and to work out its consequences. Haldane did refer to ants and bees, but he did not develop the argument. It was left to Hamilton to show how the evolution of animal societies can be understood in terms of this idea. Hamilton, not Haldane, became the intellectual father of sociobiology. I attempted to make this clear in my review by saying that 'the decisive step was taken in Hamilton's papers'.[7]

As for not mentioning the pub joke to Hamilton in the early 1960s, Maynard Smith notes that he himself must have heard Haldane's remark in 1955, understood it, but not seen its importance—otherwise he would have included it in the first edition of his book, *The Theory of Evolution* (which came out in 1958). He continues:

> I was first persuaded of the importance of the idea when I refereed Hamilton's papers in 1963; it may be that I was able to understand Hamilton's argument because I had met part of it before. Since that time I have come to regard Hamilton as the most original and creative person working on evolution theory. I would be sorry if any words of mine were to delay a general recognition of his work.[8]

Both Maynard Smith and his Edinburgh colleague, then, were holding up a particular passage from Haldane's 1955 essay in *New Biology* as proof that Haldane said it first. What exactly did that passage say?

> What is more interesting, it is only in such small populations that natural selection would favour the spread of genes making for certain kinds of altruistic behaviour. Let us suppose that you carry a rare gene which affects your behaviour so that you jump into a flooded river and save a child, but you have one chance in ten of being drowned, while I do not possess the gene, and stand on the bank and watch the child drown. If the child is your own child or your brother or sister, there is an even chance that the child will also have this gene, so five such genes will be saved in children for one lost in an adult. If you save a grandchild or nephew the advantage is only two and a half to one. If you only save a first cousin, the effect is very slight. If you try to save your first cousin once removed the population is more likely to lose this valuable gene than to gain it. But on the two occasions when I have pulled possibly drowning people out of the water (at an infinitesimal risk to myself) I had no time to make such calculations, Paleolithic man did not make them. It is clear that the genes making for conduct of this kind would only have a chance of spreading in rather small

> populations where most of the children were fairly near relatives of the man who risked his life, it is not easy to see how, except in small populations, such genes could have been established. Of course the conditions are even better in a community such as a beehive or an ant's nest, whose members are all literally brothers and sisters.[9]

But before Maynard Smith's response appeared, in his answer to the letter from Edinburgh, Hamilton rejected the implied parallel between Haldane's pub joke and 1955 article. He explained that he knew the quoted passage well. It was not the same at all:

> I am aware that Haldane used the kinship principle in discussing altruism in 1955. His article on 'Population Genetics' in *New Biology* was cited in both of my first two papers. Haldane's wording there is quite different from that in Maynard Smith's anecdote: it is less explicit and suggests that he may have considered only the relatively simple case of a rare gene.
>
> What causes me to doubt the anecdote is not any improbability that Haldane would have talked about altruism but, apart from the similarity in phrasing, the back of the envelope (for something which cost me, starting wholly unskilled admittedly, reams of paper and two years of thought) and Maynard Smith's apparent unawareness of the insight, whether written or verbal, before 1963 and especially when I was introduced to him at University College.
>
> In my own approach I was not consciously helped by Haldane's comments although I must have read them since I used to read his articles in *New Biology* as they came out.
>
> Later re-reading the passage and a similar even briefer reference in a discussion of warning coloration by R. A. Fisher I felt disappointed that my idea had not been wholly original. Traces of a proprietary feeling appear to remain in me!
>
> I would be very interested to hear from any reader who remembers either Haldane in person talking about altruism or simply hearing (before, say, 1966) the story about him referring to the Orange Tree public house.[10]

It was obviously fairly easy for Bill to respond to this neutral-seeming colleague from Edinburgh. We also learn that he gets easily disappointed when he finds that he is not absolutely the first with an idea. (This feeling is probably common among scientists and inventors. Also, it is bound to happen more often to those who work in relative isolation from the mainstream, which was often the case with Bill.) But Bill did not give in an inch. If anything, he was extending his challenge. He was now inviting the whole readership of *New Scientist* to help provide proof of the existence of the Haldane pub joke.

What was the result of this gambit? Despite the query, no direct evidence appeared. What Hamilton got was the following response from Haldane's sister, Naomi Mitchison:

> Haldane (my brother) (see Letters, 1 July, p. 40, et seq) was constantly writing things on the backs of envelopes in the Orange Tree. I sometimes suggested a notebook: but no. He had probably been brooding over it at other times and getting the mathematics clear. But he had no brothers. He didn't apparently consider whether or not the person with the altruistic gene was likely (or not) to marry someone with the same gene.[11]

This did not seem to do much to settle the issue. Hamilton was asking for real evidence. But the purported pub joke may not have been the only thing that bothered him. The pub quip may in fact have become something of a symbol for Hamilton's long-standing uneasiness with Maynard Smith, a crystallization of his growing feeling that Maynard Smith was systematically trying to play down his discovery of kinship altruism, while appropriating for himself the recognition for this achievement.

It is this message that Bill is trying to convey to the readers of *New Scientist*. In his own way he is producing a piece of counterintelligence to the by now well-known pub joke. The reader learns among other things—somewhat gratuitously, it appears—that it was Maynard Smith that was the referee for Hamilton's important paper. Maynard Smith, though, is able to turn this around by suggesting that precisely because of his Haldane connection, he was in a unique position to be able to appreciate Bill's paper. It is an interchange between two skilled game theorists. Hamilton and Maynard Smith are operating in a public forum in a partly Aesopian language.[12]

But there may have been other formulations in Maynard Smith's original book review that upset Hamilton. Reading the text with super-sensitive 'Bill Hamilton glasses' on, let's see how his contribution is presented in context. Let's look at the paragraph following the 'decisive step' sentence:

> The decisive step came in Hamilton's papers in 1963, in which he introduced the concept of 'inclusive fitness', and applied it to the evolution of the social insects.

> Since that date, there has been a gradually growing appreciation, first, of the fact that apparently altruistic acts do call for an explanation in Darwinian terms, and second, that this explanation will have to be in terms of the genetic relationship between 'donor' and 'recipient'; that is, in terms of what I have called 'kin selection'. The significance of Wilson's book seems to me to be that it is the first sustained attempt at such an explanation, across the whole field of animal societies from slime molds to man.[13]

What is the overall message conveyed by this passage? The innocent reader may easily get the impression that Hamilton's inclusive fitness theory is really especially or uniquely applicable to social insects, while a more general theory is available in Maynard Smith's idea of 'kin selection'. Overall, then, in his review, Maynard Smith in fact succeeds in doing three things: 1) giving priority to Haldane (through his pub joke and 1955 article); 2) tying Hamilton's concept of inclusive fitness to the evolution of social insects; while 3) describing the general donor-recipient reasoning which in fact underlies Hamilton's kinship theory of altruism as his own idea of 'kin selection'. No wonder Hamilton was upset.

We can assume that for many reasons, on reading Maynard Smith's review of *Sociobiology*, Hamilton's pent up emotions exploded. Not only did he feel let down by Imperial College, but there was a nasty sense of déjà vu. Just as he had an unpleasant surprise last time as he came back from Brazil, now he saw Maynard Smith at it again, trying to usurp his theory. Last time, Maynard Smith had managed to publish his "'group selection and kin selection' paper in *Nature* just before the publication of Hamilton's paper in the *Journal of Theoretical Biology*. But this time Bill was determined that Maynard Smith was not going to get away with it—he decided to challenge the very core of Maynard Smith's claim.[14]

As we see, Bill does no less than suggest that Maynard Smith suffers from a scientific amnesia of sorts, an amnesia which would not only involve forgetting where he read something, but also misattributing authorship. He does not directly accuse Maynard Smith of plagiarism, although he comes close, since he refers to Maynard Smith picking up things when refereeing his manuscript. It is clear from Hamilton's *New Scientist* letter that he has at this point vividly in mind his own circumstances as a graduate student in the early 1960s as well as the frustration he felt at University College, including the failed meeting with Maynard Smith.

But in his response Maynard Smith offers proof: 'Although I understood Haldane's remark (which I would guess was made when he was writing the

Penguin New Biology article) I did not see its importance; if I had done I would have discussed it in the first edition of my *Theory of Evolution.*' This is a valid point. So let's see what happens in the second edition of the book (1966). Yes, there Maynard Smith does include a discussion of the evolution of altruism. But it is not clear that his recognition of Bill Hamilton is improving.[15]

The second edition includes a discussion of alarm calls. On page 157, the author proceeds in the following way:

> I have suggested the term 'kin selection' for the process whereby a characteristic is established because of its effects on the survival of the relatives of its possessor. Two other examples will be given. One concerns the differences in the length of postreproductive life of different species of moth, according to whether the species is cryptically coloured and palatable, or brightly colored and distasteful. Blest has pointed out that... [Here he goes on to explain how Blest uses kin selection to explain why cryptic species live shorter lives]
>
> Another example of kin selection is the evolution of sterile castes in the social insects. A worker bee is a sterile female who spends her life looking after her sisters, some of whom will be fertile queens. The difference between a worker and a queen is caused by the kind of nutrition given to the grubs. But the capacity to be sterilized by a particular diet is itself genetically determined, and its evolution must be explained by kin selection. *Hamilton has pointed out an entertaining twist to the story.* Social life has been evolved on at least four separate occasions by the social insects - by the ants, bees, wasps, and termites. The first three of these four groups belong to the same insect order, the hymenoptera. What features of the hymenoptera predisposed them to evolve societies? Hamilton suggests that the feature in question is their 'haplo-diploid' genetic mechanism... [italics added].[16]

Maynard Smith then goes on to explain how haplodiploidy affects social behaviour in Hymenoptera.

Hamilton's work as an 'entertaining twist' on Maynard Smith's theory of kin selection? And this entertaining twist having to do only with Hymenoptera? Reading this it is hard to avoid the conclusion that Maynard Smith *has* indeed had a consistent strategy in regard to his younger colleague. It appears that he exercised this strategy quite strongly at an early stage: by forgetting to mention Hamilton, by reducing his theory to a thesis about eusociality in Hymenoptera, and by implying that 'kin selection' was his independent theory. It is of course true that

Maynard Smith had coined the *term* 'kin selection', and this very fact made elegant formulations possible. When challenged, Maynard Smith could always step back and say that he took no credit for anything else than the *term* 'kin selection'.

It is not clear if Hamilton ever saw this particular 1966 formulation of Maynard Smith's, and he may well have been wrong about some details in what was to emerge as his life-long 'Maynard Smith paranoia'. But during the first decade of the life of his inclusive fitness theory Hamilton clearly felt that he had accumulated enough evidence of a wish on the part of Maynard Smith to usurp his theory. (Being an inductivist and modeller, Hamilton did not need a lot of data to produce a working hypothesis.)

As a result, for the rest of his life, Hamilton never quite relaxed in his attitude to Maynard Smith. He experienced recurrent emotional flare-ups whenever he was reminded of Maynard Smith's early actions. Maynard Smith over the years seems to have believed that things had gotten better, especially after he started making deliberate efforts to give Hamilton recognition. Bill, meanwhile, sometimes chose deliberately to avoid attending meetings where Maynard Smith was present, or, if the two found themselves at the same meeting, acted in a non-communicative or hostile way. Perhaps Bill experienced some relief as he shared his story with colleagues and friends (especially on long car trips, it appears). Still, rather than having a therapeutic effect, reliving his bad experience as he recounted it to others might just have ingrained his bitter feelings further.[17]

What made the situation with Maynard Smith especially difficult was a factor which had nothing to do with Maynard Smith. Bill had a pre-existing emotional schema for dealing with situations like these. His model for how to react to Maynard Smith was his father Archibald's reaction to General Bailey, the army engineer who tried to steal his bridge patent. As will be recalled, Bill's father responded by taking direct action, disputing Bailey's usurpation of his patent rights. He won and was awarded compensation, but continued to paint Bailey in the blackest terms for the rest of his life.[18]

So just as the Hamilton-Callender bridge was Archibald Hamilton's patent, so Bill Hamilton's kinship theory was his 'patent'. It was his investment in a brilliant idea which he hoped would bring in revenue in the indirect form of scientific recognition. And clearly he hoped that, at some point—sooner rather than later—his effort would translate into real money too.

***

But if Haldane in his 1955 article in fact provided the essence of Hamilton's kinship theory, why did that mathematically gifted population geneticist not proceed further? Various speculations exist. Here is an interesting psychological explanation provided by Robert Trivers:

> J.B.S. Haldane, the famous population geneticist, presented the essential features of kinship theory in 1955 in a popular essay on population genetics. To make matters dramatic, Haldane considered a gene that led an individual to try, at the risk of drowning, to save a drowning individual. He argued that you would have to save a cousin more than eight times as often as you drown yourself in the attempt in order for this gene to be positively favoured. Then he did a very curious thing: having stated the general principle, he at once severely restricted it in scope. He said that the two times he himself had jumped into the river to save a drowning individual, he had acted at once without considering the possible degree of relatedness. Thus he imagined that creatures in general would not tend to discriminate between possible recipients according to the relevant degrees of relatedness, but would instead have to react according to the average degree of relatedness they enjoyed to others in the area. If we travel in groups of 50 or more, as humans certainly do, then the average degree of relatedness must usually be fairly low ($r \ll 1/10$), so Haldane's restriction rendered kinship relevant only in special circumstances.[19]

Trivers believes that there were reasons why Haldane did not make more of his own argument:

> It is possible that a trace of egotism kept Haldane from seeing the importance of the general argument he was advancing, so quickly did he push forward his own two cases of altruism, but it seems more likely that he feared the political consequences of arguing, as we would, that selection will rapidly favour the ability to discriminate between intended recipients on the basis of degree of relatedness (so that average *r* is only relevant under certain circumstances). If this latter possibility is the reason, then it was certainly not the last time someone blinded himself for fear of the consequences of an idea. If Copernicus dethroned us from the center of the universe and Darwin from the centre of organic creation, then work on the evolution of altruism has dethroned us once again, making altruism more general than we had appreciated and more deeply self-serving. This has been a painful realization for some, generating minor spasms of resistance to this way of thinking.[20]

So, according to Trivers' interpretation, Haldane would have understood quite well the underlying principle (which Hamilton was later to develop), but he deliberately stopped himself from producing a full-fledged theory because of the political consequences he saw with the theory. A very interesting interpretation!

It seems to me that people are using a certain amount of hindsight bias when interpreting Haldane as 'on his way' to Hamilton's kinship theory, but 'not quite' getting there, although 'coming close'. The question is really whether Haldane when he thought about the evolutionary origin of altruism, ever thought in terms of the Hamiltonian type of inclusive fitness. My impression is that Haldane was largely concerned with a more limited topic: the role of small groups in the evolution and maintenance of altruism.

Time has come to examine what Haldane actually said.

It is a particular 24-line passage in Haldane's 1955 essay 'Population Genetics' in the semi-popular Penguin series *New Biology* that is taken by many as the smoking gun in regard to the argument that 'Haldane said it first', especially since it appears to relate to his pub joke, taken to be the real proof. (The passage was reproduced above as part of Dr Dow's letter to *New Scientist*.)

But let's look at the larger context of this passage (and better, read the whole Haldane essay). It turns out that the river-jumping story in this piece is really a pedagogical exercise to make a much bigger point. This small section is, after all, part of the longer essay 'Population Genetics', where Haldane defines basic terms and explains all kinds of basic principles. In the particular 24-line passage we are discussing, he is concerned with the advantages and disadvantages of small populations from a population genetic point of view. One clear advantage of small populations, according to Haldane, is the possibility of the spread of genes effecting altruistic behaviour.

It seems to me that he uses the river-jumping example as a vivid introduction to the topic that he really wants to talk about here: the so-called Sewall Wright effect. That is the phenomenon of random genetic drift that happens in small populations (small populations may by pure accident, 'sampling error', have a biased genetic structure, getting a higher concentration of some genes than the original population). Haldane is writing this essay at a point in time when the American population geneticist Sewall Wright and his idea of random drift have been severely criticized by his British colleagues RA Fisher and EB Ford. Haldane's real aim with this whole passage (perhaps even with the essay) is to defend Sewall Wright and the Sewall Wright effect.[21]

We only need to go to the part of Haldane's essay that immediately precedes the river-jumping story. That is a passage discussing Wright and supporting his idea:

> [Wright] thinks that if a species is split up into a number of nearby isolated populations they will diverge genetically through this chance process. Some, perhaps most, of the isolated populations will draw unfortunate gene combinations from the hat of chance... These will die out in competition with populations whose gene combination is more fortunate. Evolution will be much quicker than it would be in large mixed populations. Fisher and Ford have criticized this view sharply... But the Wright effect may have been important in human prehistory. It has been suggested that in Paleolithic times the human population of England was about 300 hunters. If these were divided into a number of tiny tribes or families, each of which killed trespassers into their hunting grounds, I think the Wright effect may well have worked.
>
> What is more interesting, it is only in small populations that natural selection would favour the spread of genes making for certain kinds of altruistic behavior. Let us suppose that you carry a rare gene... [and so on—this is the beginning of the 'river-jumping' passage].[22]

In other words, the main focus in this section of the essay is *small populations*, and the way in which Haldane treats altruism has to do with the particular possibility of establishing altruism in small populations. Haldane does not seem interested in developing a general rule for the evolution of altruism beyond the role of small populations. He is interested in supporting Wright, because the Wright effect appears to provide a convenient mechanism for generating groups with different genetic makeup, which can in turn provide a basis for group selection.

Remember that Haldane had earlier attempted a group selectionist derivation of altruism. In 1955 Haldane is still trying to solve his problem from his 1932 book *The Causes of Evolution*: to show that the evolution of altruism is possible and based on a group selectionist mechanism.[23] But now his colleague Sewall Wright has provided a great new tool: genetic drift in small groups. This natural process would seem to make it possible to have a concentration of altruistic genes in small populations. Is this kind of thing feasible among humans, Haldane wonders? Has it ever happened? Haldane looks backward to the Pleistocene. Yes, he coolly concludes, if there were a number of small tribes, each of which made a point of always killing any intruder, altruism might indeed have prevailed.

Let's take another look at those famous 24 lines. These lines do look similar to the pub joke at first. Haldane presents a situation where someone needs to decide whether or not to jump in to save another from drowning. He gives the person 1/10 chance of perishing in this attempt. Meanwhile he calculates the chance of the rescuer sharing genes with the person saved, first for closer relatives such as a child or brother/sister, then a grandchild or nephew, then a first cousin and finally a first cousin once removed, concluding that in the first two cases, altruistic genes are likely to be spreading as a result of the man's rescue mission, but in the later cases much more unlikely.

Pointing out that there is no time to think or calculate (Haldane says that at least he himself did not do so when *he* twice saved drowning people!), he suggests that one's spontaneous behaviour in such a situation is simply to act *as if* the people around you were close relatives—which they usually are in small populations. Haldane therefore concludes:

> It is clear that genes making for conduct of this kind would only have a chance of spreading in rather small populations where most of the children were fairly near relatives of the man who risked his life. It is not easy to see how, except in small populations, such genes could have been established.[24]

And that is exactly what Haldane wanted said: that it takes (or took) small populations for altruism to develop.

Then, after having discussed altruism as one of the potential advantages for small populations, Haldane considers instead the potential disadvantages that may result from breaking up a population into small fragments. (He appears to be thinking still of his earlier tribe splitting model.) He then continues his essay on 'Population Genetics' with brief popular overviews of other aspects of this new discipline.

Let's take a step backward. Back in 1932, in his *The Causes of Evolution*, what might have been Haldane's aim with his (ultimately fruitless) attempt to mathematically demonstrate in an Appendix that the evolution of altruism was possible based on a group selectionist mechanism?

The Causes of Evolution is a small book of essays, done in a typical Haldane style. Usually his essay collections are compilations of popular articles he has written for *The Daily Worker*. Also, in typical Haldane style, not much space is devoted to discussion of altruism. (Haldane typically writes short articles, and especially reading him in book form, one gets the impression that he is always impatient, flitting off to a new subject.) In his 1955 essay, he concludes his discussion of altruistic behaviour by saying: 'I doubt if man contains many genes making for altruism of a general kind, though we do probably possess an innate predisposition for family life'. He goes on to suggest that 'in so far as it makes for the survival of one's descendants and near relations, altruistic behaviour is a kind of Darwinian fitness, and may be expected to spread as the result of natural selection'. This is not much more than a reformulation of some of Darwin's insights, and no mechanism is mentioned.[25]

The answer may well be that Haldane in 1932 had a specific goal in mind. He was concerned to defend a left-wing alternative to the Darwinian 'struggle for existence': Prince Peter Kropotkin and his idea of Mutual Aid. During the early decades of the 20th century Kropotkin's book *Mutual Aid* had been much discussed in London, but thereafter largely ignored. Haldane may have felt that someone had to fight for this cause. Altruistic behaviour had to be shown to be compatible with natural selection. This was probably why Haldane, a man with left-wing sympathies, took up this cause, and even tried to demonstrate its feasibility with a mathematical model. Note that at this point Haldane was still inclined to socialism rather than being the card-carrying Communist that he would later become. And Kropotkin was the perfect example of small group socialism.[26]

***

I have in the preceding paragraphs tried to reconstruct the actual intent of Haldane's brief 1955 example, which I believe indicates that Haldane was not interested in developing a general theory of inclusive fitness *à la* Hamilton. But there is still the pub quip, which began this whole discussion. Assuming that Haldane actually made his remark as reported, what might have been the intent of that statement in the Orange Tree?

The joke is often told so that Haldane responds to the question: 'Shall I lay down my life for my brother?' with the response: 'No, not for one brother but for two brothers or eight cousins' (or 'more than two brothers'). What seems clear, however, is that in the pub, at least, what Haldane had mostly in mind was altruism in *humans*.

Although he realized that a trait for altruism can be passed on to offspring, and he mentioned social bees and ants, Haldane stayed very much at a general level, recognizing with Darwin that altruism can indeed be an evolutionary trait. He did not seem interested in finding a universal rule for how the particular relatedness between individual organisms would affect their behaviour toward one another, let alone seeing altruism as a phenomenon that would have to do with every aspect of nature, including plants. (Plants, meanwhile, were part of the inspiration behind Hamilton's early modelling efforts.)

The truly ironic thing is that it may not have been Haldane after all who 'said it first'. According to Fisher's student Anthony Edwards, the original idea of inclusive fitness came not from Haldane at all, but from RA Fisher, who said it as early as 1912 (and published it in 1914). According to Edwards, 'The idea of inclusive fitness goes back at least as far as Fisher's lecture to the Cambridge University Eugenics Society in 1912..., in which he used as his example the nephews that could replace 'genetically' a childless man killed in war'.[27]

Still, the answer to who said it first would seem crucially to depend on whether Fisher's 'it' is really the same as Haldane's 'it', and especially on how Hamilton's 'it' relates to these two. The debate continues.

13

When Leaving is Better than Staying

There was still one more unpleasant surprise waiting for Bill on his return from South America. He found himself still a junior lecturer at Imperial College. He had been passed over for promotion—again! Junior lecturer had been his lowly status back in 1964 when he got that first job, still without his PhD. Now 12 years had passed; he had received his PhD, he had published a number of papers, but in academic status he hadn't advanced a single bit.[1]

By the mid-1970s Hamilton had continued various lines of investigation and accumulated several important publications. In addition to his early altruism papers, he had produced a number of others, some of them in *Science* and *Nature*. His sex ratio paper (1967) was one of the few solid examples of predictability in evolutionary biology. In his *Annual Review* paper (1972) he had valiantly defended his original insights about kinship theory in the face of new evidence. His Price-inspired mathematical re-derivation of inclusive fitness from Haldane's earlier group selection theorizing had provided an important bridge between Fisher and Haldane, and he had extended his theory to the phenomenon of spite (1970). He had applied kinship theory and game theory to the human situation (1971). With the aid of Price's covariance formulae he had further explored the role of levels of selection in evolution (1975). His mathematical puzzle, Geometry for the Selfish Herd (1971), was becoming very well cited. Finally he had done naturalist work, including descriptions of fig wasp species and polymorphisms, and suggested the evolutionary origin of higher orders and taxa in a habitat of decaying wood (1978).[2] Admittedly, his teaching had not been a success. Bill, with his theoretical approach and gentle manner was probably not best-suited for teaching big undergraduate classes in a practically oriented place like Imperial College. As we saw, Bill himself even used to joke about the students' yearly delegation to the director, complaining about his teaching.[3]

In other words, it seemed that Bill himself realized that his teaching left something wanting. But here we have an example of the type of absolutist, normative reasoning that Bill would often use, and which would cause him difficulties with scientific and other authorities throughout his life. From Hamilton's point of view, teaching *should* not matter so much. In matters of promotion in science, the criteria that *ought to* be used, or at least be given greater weight, had to do with the quality of one's scientific contributions. Following his own criteria, Bill almost never put himself into the shoes of an administrator, editor, or journal referee, or asked himself what the prevailing standards actually *were*—he used his own absolutist standards derived from an ideal world of rules, which also assumed full appreciation of his scientific contributions. In this case of his non-promotion at Imperial College, Bill used his own assessment of the value of his science (perhaps combined with a view to seniority), took deep offence, and decided to leave.[4]

I asked Richard Southwood about this incident. It was he who had made the decision not to promote Bill. Southwood told me that he thought highly of Bill's papers, but at the time they were not that great in number, and remained somewhat controversial. In other words, Southwood did not yet see Bill's scientific production as academically fully validated. Moreover, as an administrator, Southwood also had to take into account teaching and service. Bill's teaching was poor, and he had not taken on any administrative duties. Meanwhile, there were a number of colleagues of Bill's who were excellent in all these fields: research, teaching, and administration. They deserved promotion first. Did this mean that Bill would have been up for promotion the following year, I asked? Yes, was the answer, and Bill was aware of this too.[5]

Hamilton may have thought he had published enough papers to merit promotion but from Southwood's administrator's perspective, the position may have been different. Although Southwood personally liked Bill and saw his promise, if one counted all Bill's peer reviewed journal articles from the time he came to Imperial College, that was a total of five (counting only the contributions after Bill obtained his PhD, the number would dwindle to three) and what is important in science is one's record of peer reviewed journal articles.[6]

Having taken over as the department chair and director of the Field Station after the famous OW Richards, Southwood must have felt himself under scrutiny as an academic administrator. With an eye to his own future career advancement, he may have wanted to err on the side of caution. Hamilton's fewer papers may well have weighed as much as the more numerous papers of his competitors for promotion, but their full scientific merit was not yet established, according to Southwood.[7]

It may also have mattered that Bill had had two almost year-long absences, the Mato Grosso expedition and the recent Brazil trip. All this had been arranged in an acceptable way with Bill being allowed to concentrate his teaching duties for a period before the trips so that free time was opened up for travel. Still, the results from those trips had not yet translated into obvious publications (for instance, Bill's work on fig wasps was not published until 1979, and then only as a book chapter).

But if Southwood believed that Bill, seemingly content at Silwood Park, would in any case not be going anywhere, while it would take promotion to retain his colleagues, he was badly mistaken. Bill almost immediately started looking for opportunities elsewhere, more specifically in the United States.[8]

It was clear that Hamilton was pretty angry at this time, and his anger gave him strength. This may have been a side of Bill's character that Southwood had not seen before. We already know of Hamilton's reaction to Maynard Smith's review of Wilson's *Sociobiology*: a surprising public challenge in the pages of the *New Scientist*. That challenge may well have been connected to Hamilton's keenly felt lack of scientific recognition in his own country at the time. He may have believed that one reason for this had something to do with Maynard Smith's appearance on the scene. Indeed, in Bill's view, this may have been part of the reason why his inclusive fitness theory was still not accepted, still 'controversial', as Southwood had said, after more than a decade.

It seemed to Hamilton that he was not appreciated enough in his home country. But why should he choose the United States? In fact, this was an obvious destination for him for several reasons. While catching up with the scientific literature after his return from Brazil, it had become clear to Bill that it was the United States, not the United Kingdom, where interesting research was being done relating to kinship theory. In the first place, Robert Trivers and others were actually working on it and extending it. In the second place, and perhaps more importantly, there was now this major controversy about Wilson's *Sociobiology*. From reports in journals everywhere, it was hard to miss the fact that the book was being much talked about, particularly at Harvard but rapidly spreading elsewhere.[9]

With the sociobiology controversy raging, the scientific basis of sociobiology was being examined, and here Bill Hamilton's contributions became relevant almost by default, and by general association, because on the very first page of his book, Wilson had declared altruism the central problem of sociobiology. (This did not mean, though, that he had quoted Bill Hamilton by name at that point. In fact, Hamilton did not appear in connection with altruism until chapter 5 of the book.)

'Sociobiology' was, for Wilson, a very general term designed to encompass all kinds of studies of social behaviour. The underlying assumption of the book was that just like morphological traits, behaviour, too, was a trait undergoing evolution. Wilson's basic aim was to extend the paradigm of the Modern Synthesis. It was clear to him that evolutionary biology, his field, needed to continue growing and asserting itself against the expansive attempts of molecular biologists.[10] However, the idea of behavioural evolution was news for most non-biologists, and bad news for all those (and there were many) who saw humans as purely cultural creatures, exempt from biological forces. What upset many was that Wilson in his last chapter had also brought our species into his sociobiological synthesis, painting a tentative picture of the evolutionary—read genetic—basis for a number of human behavioural traits.

It was the book's last chapter that angered a number of Wilson's left-wing biologist colleagues, too, including Harvard professors Richard Lewontin and Stephen J Gould. In autumn 1975 the so-called Sociobiology Study Group (later associated with Science for the People) published a critical manifesto against sociobiology. This pro-culturist attack presented sociobiology not as new science, but rather as a political ploy to promote 'genetic determinism' and in this way legitimize the idea of natural social inequality. The intellectual climate was to favour the critics for a long time to come.[11]

Rather than a single theory, Wilson's sociobiology represented a large and wide-ranging scientific programme. In addition 'sociobiology' had a more popular meaning as an umbrella term for all kinds of studies of social behaviour. And finally, as I have shown in *Defenders of the Truth*, Wilson's sociobiological programme can be seen as Wilson's broad moral/scientific agenda for taking charge of the future evolution of man. While Wilson's goal was not the narrow political one that the critics accused him of, they were right about one thing: Wilson's determination to break the prevailing 'culturist' taboo on biological explanation of human behaviour.[12]

Hamilton admired Wilson's 'brave' book, as he called it. Indeed, he compared his own graduate student situation trying to launch his ideas about altruism in the 1960s to the struggles of Wilson in the 1970s and '80s. He knew exactly where Wilson's current critics were coming from. In fact, Hamilton had ended his letter to the *New Scientist* (see chapter 12) by drawing clear parallels between himself and Wilson:

> The criticisms which used to be stated or implied about my research seem almost exactly those which E. O. Wilson now faces over the human implications of 'sociobiology', the main difference being the greater emphasis then on a supposed link between Fascist and Nazi ideas and more now on a supposed attempt to prop up unjust social differentials. On the whole epithets for sociobiology... 'tautology', 'pseudoscience', 'nonscience', 'ideological', have a very familiar ring. The ideas which a few biologists in the 50s felt to be neglected and important are shown in Wilson's great survey to have generated an unexpected wealth of new facets, predictions and evidence. In the face of this the sharp renewal of antagonism is disappointing.[13]

Bill was right: the critics of sociobiology were indeed operating largely within the post-war political paradigm, with special circumspection in regard to genetic attempts to explain race and group differences. To this had been added the concern about 'genetically determined' sex differences.[14] At the same time Wilson's *Sociobiology* and the controversy certainly drew renewed attention to the evolutionary problem of altruism, which meant that in popular overviews of sociobiology, Hamilton's name was getting to be routinely cited.[15]

But someone else was also being cited in the mid-1970s, and increasingly so, and that was Harvard's Robert Trivers. Trivers was moving forward with one important contribution after the other. His reciprocal altruism paper from 1971 had been soon followed by his theory of parental investment in 1972, and yet another paper on parent-offspring conflict in 1974. These were all high visibility papers, and written in a clear and popular style. Trivers was in fact in the process of extending Hamilton's inclusive fitness theory to human behaviour—the species that Trivers said he felt he knew something about. Trivers had written to Hamilton asking for comments on his papers and Bill was consistently encouraging about his work.[16] Meanwhile Trivers was teaching Hamilton's theories on Harvard undergraduate and graduate courses well before the sociobiology controversy. By default, it seemed, Trivers had become Hamilton's direct 'Harvard agent'.[17]

Hamilton was, of course, pleased, at Trivers' success. Trivers had followed exactly the same path as Hamilton himself—got some serious publications out before his PhD (and presumably making those count toward his PhD). But was this young man advancing a little too fast? For instance, here was Trivers in January 1976 with a 15-page lead article in *Science*. Trivers, in a paper co-authored with Hope Hare, had combined Hamilton's theoretical reasoning on kin selection and on the sex ratio. In his 1964 paper, Hamilton had called attention to the particularly close relationship between sisters as a way to get social behaviour started in the first place among social insects. But, argued Trivers and Hare, this was not enough. Later the female-biased sex ratio would need to be sustained. A systematic bias could be achieved as long as the sister-workers concentrated on initially raising their sisters or their sisters' sons (the sons of laying workers), rather than their brothers. They therefore suggested the following 'amendment' to Hamilton:

> The asymmetrical degrees of relatedness in haplodiploid species predispose daughters to the evolution of eusocial behavior, provided that they are able to capitalize on the asymmetries, either by producing more females than the queen would prefer, or by gaining partial or complete control of the genetics of male production.[18]

Trivers and Hare's study presented quantitative evidence for Hamilton's inclusive fitness theory (which they renamed 'kinship theory') and highlighted the special role of haplodiploidy in the evolution of social insects. Here is part of their conclusions section of their paper:

> The social insects provide a critical test of Hamilton's kinship theory. When such theory is combined with the sex ratio theory of Fisher, a body of consistent predictions emerges regarding the haplodiploid Hymenoptera. The evolution of female workers helping their mothers reproduce is more likely in the Hymenoptera than in diploid groups, provided that such workers lay some of the male-producing eggs or bias the ratio of investment towards reproductive females. Once eusocial colonies appear, certain biases by sex in these colonies are expected to evolve. In general, but especially in eusocial ants, the ratio of investment should be biased in favor of females, and in ants it is expected to equilibrate at 1:3 (males to female). We present evidence from 20 [ant] species that the ratio of investment in monogynous ants is, indeed, about 1:3, and we subject this discovery to a series of tests... Taken together, these data provide

> quantitative evidence in support of kinship theory, sex ratio theory, the assumption that the offspring is capable of acting counter to its parents' best interests, and the supposition that haplodiploidy has played a unique role in the evolution of social insects.[19]

It was quite by accident that Trivers had come upon the idea for this paper. One day he had simply tried to re-derive Hamilton's results in order to prepare for a class at Harvard. He was teaching Hamilton's kinship theory, and wanted to devote a follow-up lesson to the Hymenoptera and their special haplodiploid relationships, which he saw as the best evidence for Hamilton's kinship theory (at the time not much evidence existed yet for vertebrates). So, preparing for the lecture, Trivers set out to calculate the different degrees of relatedness to see that he understood this thing right. He then checked his result against Hamilton's original paper.[20]

Lo and behold, there was something wrong here. Trivers studied and studied Hamilton's derivation but did not understand it. He asked some colleagues and students around campus if they knew the answer, but the mystery remained unsolved. Trivers was puzzled that he came to such a different result when he so carefully tried to follow Hamilton's own logic in regard to the calculation of relatedness. Well, the obvious solution was to write to the Master himself. The response surprised him. Hamilton told him that as far as he knew, only one other person in the world except himself had detected the error (Ross Crozier in Australia), and that he was now correcting the error in an Appendix to his 1964 paper, which was being republished. This gave Trivers a tremendous boost of confidence—perhaps something similar to the boost of confidence that Hamilton had experienced when he originally found that Fisher's famous 1:1 rule for the sex ratio was not true under all circumstances. There is much to be said for finding an error in the Master's work.[21]

Incidentally, the refereeing of the Trivers and Hare paper was far from a smooth ride. Hamilton, one of the first two referees, generously recommended the paper for publication without revision. He pointed to the paper's general importance, while noting that some things could have been worked out in more detail. Hamilton's behaviour as a referee was clearly affected by the memory of his own time-consuming revision of his first paper and he wanted to speed up the review, but that was not to be, because of a negative review from Richard Alexander (who had just developed a different theory of his own. This now triggered a new round of refereeing, again with a split outcome. Finally it was up to a single last referee, George Williams. His verdict was positive, and that settled the matter. (Presumably, the original status of the paper as 'invited', suggested by Wilson, helped in the refereeing process.) So finally the Trivers and Hare article appeared, after some delay, but as the lead article in *Science*. Trivers had won the battle. And what was more, *Nature* went on to report on the article, giving it glowing praise. This was something that Hamilton could see with his own eyes, as he perused back issues of journals in the early summer of 1976.[22]

So, it was clear that matters relevant to kinship theory were going on in the United States, and reported on favourably in the scientific journals. And what was happening in the United Kingdom? Richard Dawkins book *The Selfish Gene* had appeared in 1976—with a Foreword by none other than Robert Trivers. *The Selfish Gene* popularized Hamilton's inclusive fitness theory (often using the helpful term 'kin altruism'), and also explained clearly just how to calculate the proportion of genes that an individual shares with various relatives. But it did much more. In this book Dawkins' heuristic device—'the gene's eye view'—which he used to explain Neo-Darwinian reasoning in a popular way, at the same time became the concept that unified new strands in evolutionary theory.[23]

'The gene's eye view', the idea that Hamilton had used in a number of his papers, was now transformed into the thread that ran through the new ideas of inclusive fitness, kin altruism, ESS, parental investment, parent-offspring conflict, male and female strategies, and more. This approach, in turn, was dovetailing nicely with George Williams' famous methodological 'doctrine', according to which one should not operate at a higher level than necessary when studying adaptation. The ideas that were presented in *The Selfish Gene* were largely those of Hamilton, Williams, Trivers, and Maynard Smith. A central feature of the emerging fields of behavioural ecology or 'functional ethology'—or what was later to be labelled 'sociobiology'—was their common game theoretical underpinnings.[24]

In other words, a new research field was slowly taking off. But the place to be would be United States, not Britain. At home, Hamilton did not feel appreciated enough. And this was not only his own impression. In *The Selfish Gene* Dawkins says about Hamilton:

> His two papers of 1964 are among the most important contributions to social ethology ever written, and I have never been able to understand why they have been so neglected by ethologists (his name does not even appear in the index of two major text-books of ethology, both published in 1970).[25]

Dawkins may have wondered, but there were at least two concrete answers at the time, and these had to do with resistance from other scientists. One group was indeed the traditional ethologists, working within a proud British tradition. They had adopted an outlook whereby animal behaviour should be explained in a multidimensional way, paying equal attention to each of 'Tinbergen's four questions'. Niko Tinbergen had pointed out that equally legitimate questions about behaviour could be asked at the evolutionary, developmental, causal, and physiological levels. In other words, for Tinbergen and for most other ethologists, evolutionary or ultimate questions were not necessarily the most important or interesting ones. However, for self-fashioned 'functional ethologists'—those who had decided to focus exclusively on evolutionary explanations—this was exactly the case.[26]

Many traditional ethologists were not well-versed in population genetics (which was not part of their curriculum), or may not even have approved of the whole Neo-Darwinian mathematical 'translation' of evolution as a change in gene frequencies. It is not surprising if Hamilton's papers remained obscure to them. Resistance to Hamilton on the part of individual ethologists may or may not have been also connected to a group selectionist conviction. That had been the case with his Cambridge tutors, and he appears to have regarded this as the general position still in the 1970s. In 1975 Hamilton wrote on 'the consensus of biologists...in believing that the generally significant selection is at the level of competing groups and species'.[27]

The second group who presented resistance and scepticism to Hamilton was—paradoxically, it would seem—the mathematically trained population geneticists. Here was a group that actually ought to have been able to appreciate Hamilton's contributions. The problem in this case was that this group was too 'pure' in their mathematical preferences. As we have seen, in his mathematics, Hamilton operated more like an engineer, making various types of approximations and building up toward universal principles. He was always carefully specifying the assumptions he was making and discussing under what conditions the result would be valid. The mathematically trained geneticists, however, wanted things to be universally true. This was one reason why they found Hamilton's formulae wanting. Moreover, they were inspired to make critical improvements. This is the way Alan Grafen describes the fate of a typical Hamilton paper—with a special nod to the mathematical critics, it seems:

> In review, it is panned by referees who demand shortenings and revisions. Immediately after publication, it attracts criticism for obscurity. Its significance

> slowly emerges through secondary works, further work is inspired, and one or more literatures develop around its themes. Later more mathematical work may even be rather patronizing about the paper, and emphasize discrepancies, while the primary finding is that the original idea is abundantly confirmed. The original paper is frequently, indeed often obligatorily, cited in papers in the new literatures, but is not read nearly as often as it deserves to be, since it retains a reputation for obscurity.[28]

But there existed indeed a more basic problem, and this had to do with Hamilton's ambition to connect himself to Fisher and the classical theory. Fisher himself had originally been greatly neglected by the population geneticists. In the early 1970s he was rediscovered, only to be examined, declared wrong, and dismissed. (This seems to have had to do with an unfortunate formulation on the part of Fisher, leading to a misinterpretation of what his Fundamental Theorem actually stated.) Now it is understandable that if Hamilton was seen as leaning on Fisher, and Fisher was declared wrong, obviously Hamilton had to be wrong too. This was why the point now became to show just how Hamilton was wrong. A burgeoning critical industry came into existence. Later, many population geneticists 'recanted', realizing that Hamilton had actually been right under his own assumptions.[29]

The stance of the population geneticists at the time was harmful to Hamilton. This was an academic group of high prestige, whose support Bill Hamilton 'should' have had, in order for his mathematics to be academically legitimized. It may have been partly the lack of support from this highly mathematized group that made Hamilton's theorizing seem academically 'controversial' to Richard Southwood.

So, there were two tendencies going on at the same time. Bill was undoubtedly slowly becoming better known—after all, we have to assume that it was on the strength of his reputation that he was able to pull off the *Nature* publication coup with Price in 1970—but at the same time, there were also sceptics in high places. In any case, Bill took his non-promotion at Imperial College as a clear sign that time had come to move. 'I almost immediately resigned', he tells us. One wonders what this meant in practice. This was May 1976. Academic job searches usually start in the summer or autumn and take some time, and new positions usually start in the autumn of the following academic year. Did this mean that as Bill was job searching he did not have a job? If so, that was a rather dramatic move, considering he had a family to support.[30]

In any case, Hamilton's reaction was to start immediately inquiring at some American universities which had earlier invited him for lectures, asking people there if they knew about any open positions. To his delight some of them answered that there was an opening at their own university. Soon Bill went off for interviews to the three places that he thought offered the best conditions. The result was that in the autumn semester of 1977 Hamilton held the position of Visiting Professor at University of Michigan, Ann Arbor. In the spring of 1978, again, he served as Honorary Visiting Agassiz Professor at Harvard University. Then in the autumn of 1978, he accepted a permanent position, a special Museum Professorship at the University of Michigan, arranged for him by the energetic efforts of Richard Alexander and Don Tinkle.[31]

***

In the mid-1970s, Bill appears to have been restless, with forces tempting him to leave England for greener pastures. First he took a year off to go to Brazil and later he started considering the United States. Let's take a look at Bill Hamilton's major intellectual pursuits during the early and mid-1970s. We see him pursuing two main themes: 'life under bark' and dispersal. (Life under bark, or life in closed spaces, was discussed in chapter 11).

These two themes were actually closely connected, as he had realized while poking around in rotten tree trunks and other closed spaces. Bill could not help paying attention to the interesting fact that males and females were sometimes winged and sometimes wingless. He started making mental notes, trying to see if he could find a clear pattern. (One thing he found was that winglessness in males and wings in females often seemed to correlate to brother-sister mating before dispersal).[32]

No general pattern emerged, however. There did not seem to be a clear system to being male or female and having or not having wings. Moreover, there were species where either the males or the females, or both, were polymorphic, that is, showed both winged and wingless forms. Hamilton concluded that there must be factors other than sex that influenced the outcome in these cases. Obviously the point of having wings was related to greater ability to disperse, something which in turn was subjected to its own considerations and pressures. Hamilton also reflected that it was not always obviously advantageous to have wings, because the development of wings was costly and cumbersome, possibly taking away the ability to lay more eggs. Also, wing development was typically irreversible. Interestingly, in some cases nature had found a compromise, allowing half-developed wings to be reabsorbed and the extra energy used for egg laying purposes.[33]

As Bill's own studies of fig wasps in Brazil had shown, there could be many different degrees of wing development and other modifications within a single species. (Think of those monstrous looking, wingless males in fig wasps.) Indeed, as already discussed, the dispersal polymorphism within a single species tended to give rise to such fantastic diversity of body forms that it was often believed that these were different species altogether, or even different genera.[34]

Bill now took a step further. He concluded that having wings was not really the crucial issue; there were other ways for organisms to disperse. Rather, the really important question had to do with *leaving or staying*. It seemed to him that organisms were largely preformed as either the stay-at-home or dispersal type (in the latter case, he speculated that dispersers may have some particular behavioural trait). Hamilton had no doubt that the problem he had addressed was one that applied more generally to all kinds of organisms—including humans. In plants, for instance, there were species with all winged and species with all wingless seeds (dandelions versus the common daisy) but there were also species with a mixture of both within the same flower head (some daisy species). And it was these that especially interested Hamilton. How was the proportion of the different types, leavers and stayers, in these species determined?[35]

Bill's scientific work had already for some time circled around the theme of leaving or staying. In turn this was a natural extension of his idea of kinship altruism. It was also a sort of correction of any overextension of that idea. In his studies of life under bark and other closed spaces, he had realized the premier conditions these offer for accelerated inbreeding among insects. In theory, inbreeding should serve to increase relatedness, and thus favour the evolution of sociality. But thinking about it more carefully, he realized that there was an important *ecological* factor involved: the availability of resources. Altruists, if they remained in or around the parental home, would be competing for resources with their own kin. The idea would be for them to compete with others and for that to take place, they would have to disperse.[36]

Bill was actually trying to find the optimum level of dispersion. Reasoning in the same manner as he did in the case of the sex ratio, he concluded that there must exist an ideal ratio, an intermediary optimum of an 'unbeatable' kind, between stay-at-home types and wanderers. But here he got stuck. Now what? Robert May to the rescue! May, a theoretical ecologist who often visited

Silwood from Princeton in the summer, was used to being asked to employ his mathematical gifts to clearing up theoretical problems for the Field Station ecologists. He cracked Hamilton's scientific nut over a weekend. May's trick was to use the notion of ESS, the idea of an evolutionarily stable strategy, instead of the more demanding concept of an 'unbeatable strategy' (from Hamilton's sex ratio paper) that Bill himself had tried to work with.[37]

Hamilton had to agree that the ESS idea was more manageable and useful—indeed 'elegantly simple'—and in practice often equivalent to the unbeatable strategy idea. He also realized that his own attempts had failed because he had wanted to do too much at once: look at all the frequencies at once at every stage of the game, just as he had been used to doing in his research on unbeatable sex ratios. ESS was, of course, Maynard Smith's concept (originally developed with Price but increasingly attributed to Maynard Smith), which could be described as a less demanding version of Hamilton's idea of unbeatable strategy. In his autobiographical notes on this case, Hamilton valiantly struggles to give the reader a fair presentation of the ESS concept and its applicability as a scientific tool. Still, he cannot keep himself from a final somewhat acid remark. He compares the burgeoning academic industry around ESS in the mid-1970s to an 'epidemic' similar to the deadly elm tree disease raging in England during that time, 'although fortunately the ESS disease was more benign'.[38]

Hamilton and May's paper, 'Dispersal in stable habitats' (published in *Nature*), presents the novel and counterintuitive idea that dispersal is advantageous even in totally stable and 'saturated' habitats (that is, the opposite of unstable and patchy environments, where dispersal would seem like an obvious idea).[39] And even more counterintuitively, it would seem that '[e]ven when migrant mortality is extremely high and the environment offers no vacant site for colonization it is still advantageous to commit slightly more than half of the offspring to migration'.[40]

How can this make sense? It seems like a high risk strategy. Hamilton and May explain that dispersal makes it possible for migrants to compete at new sites with individuals with unlike genotypes, and with a fair chance of winning the site. It is not evolutionarily stable to have a majority genotype that keeps all its offspring at its site, because an invading mutant genotype, prepared to send off a large number of its propagules to compete for new sites, will do better. In the end an optimum will be established, an evolutionarily stable probability of migration. Hamilton was pleased to note that this kind of reasoning about staying or leaving was universally applicable. For instance, the body of perennial plants or tubers could be seen as representing 'a plant's bid to retain the home site', while flowers and dispersing seeds, again, represented its investment in migration. And of course, it all also applied to those Victorian sons, one of whom stayed at home while the other went off to sea...[41]

Bill had found the answer to whether it made sense to leave or to stay. He had concluded, with May's help, that the balance was slightly more on the side of leaving than staying. Migration was worth the risk.

14

Encounters with Sociobiology

Moving to the United States was not easy. There were all kinds of matters to consider: new house, school for the children, and everyday, practical issues that come with adjusting to a new environment. On top of all this, of course, there was Bill Hamilton's adjustment to the University of Michigan and its faculty and students. He was going to stay at Ann Arbor only for the fall semester 1977, but already in the pipeline was a permanent appointment, starting in autumn 1978. In the space of a year Bill would be transformed from junior lecturer to a prestigious Museum Professor at 'the Harvard of the Mid West', the University of Michigan at Ann Arbor. The offer was finalized in the middle of autumn 1977.[1]

During the fall semester, the Hamiltons rented a house in the town of Ann Arbor itself. The university atmosphere was good, people were helpful, and they were involved in all kinds of activities. Bill's colleagues seemed keen on getting to know him, and the students were inquisitive and creative. This was a very different experience from Imperial College, and indeed, here Hamilton was teaching highly motivated graduate students, not indifferent undergraduates. His teaching style was typically of a more discursive and research-oriented kind, more like a philosophizing monologue than a persuasive oratory for a large audience. He was at his best in smaller groups. But this particular group of students did not just listen to what he had to say, they also challenged him. Michigan at the time was a place that attracted students interested in evolutionary theory, and many of these students later went on to become leaders in the field. It was a dynamic environment, with Donald Tinkle and Richard Alexander as motors of the action at the Museum of Zoology, arranging seminars and conferences, and inviting visitors (Tinkle was the director of the Museum and Alexander 'its guerilla commander and general bomb thrower').[2]

In the spring of 1978, the whole family moved to Harvard. Not only was Bill an Honorary Visiting Agassiz Professor in the Museum of Comparative Zoology, but he and his family also lived at the university's biological field station in Harvard, Massachusetts. (Later, too, university field stations provided accommodation for Bill and his family during trips.) The place was rather isolated, but new sports could be practised, especially cross country skiing. One of Hamilton's duties during the spring semester was teaching a course together with Bob Trivers. The course was called 'Social Theory Based on Natural Selection'. Had Hamilton's teaching improved at all? It seems that it had, to some extent, but he was certainly no match for Trivers' clearly pedagogical style. At one point, presumably after having got bogged down in some abstract reasoning or suddenly getting a new idea, Hamilton good-humouredly told the class that he would not be surprised if the class doubted if he understood himself what he was saying.[3]

That spring Hamilton was interviewed by the Museum of Comparative Zoology Newsletter, under the rubric: 'Professor William D Hamilton Visits the MCZ'. There he explained that the reason for his move to the United States was that so much more activity in the field of social behaviour was being done on this side of the Atlantic. This was true. In America Hamilton had attained the status of some kind of guru. People seemed to recognize him and invited him to give lectures at various universities. That year he was also made a Foreign Honorary member of the American Association for Arts and Sciences.[4]

***

But what was the relationship between Hamilton and the sociobiology controversy? After all, Hamilton's visiting professorships at University of Michigan and at Harvard coincided with the height of the conflict. By then the controversy was a national—if not international—matter, judging from the fact that on 1 August 1977, sociobiology and the controversy around it made the cover of *Time* Magazine. The cover featured two puppets attached to strings, a male and a female, presumably programmed by their genes for various behavioural strategies. The subtitle was: 'Why You Do What You Do'. Inside was a lengthy article covering general sociobiological ideas and an overview of the political criticism.[5]

In fact, Hamilton had been involved in the controversy to some degree when he was still in England, just before moving to America. Sherwood Washburn had not forgotten Hamilton's presentation 'Innate Social Aptitudes of Man' at the 1973 Oxford conference, especially since it had later reappeared as a chapter in Fox's conference volume, *Biosocial Anthropology*.[6]

In March 1977 Hamilton got an unexpected letter from Robin Fox. It contained a photocopy of the January *Anthropology Newsletter*. From the front page Bill learnt that a resolution condemning the field of sociobiology had been voted on at the American Anthropological Association Annual Meeting the previous autumn. It had received insufficient support and been defeated, but still, it had been a close call.[7] The topic had dominated the meeting—no less than five whole sessions had been devoted to sociobiology. But of particular interest to Bill was an attack on him in a letter from Sherwood Washburn, published in that newsletter.

Washburn seems to have regarded the general uproar about sociobiology as his opportunity to get back specifically at Hamilton. In his letter there is no mention at all of Wilson's *Sociobiology*, the usual target at the time. Instead Washburn brings up a particular paragraph that he regards as emblematic for the sociobiological kind of gene-focused reasoning. It is a passage in which Hamilton speculates about the possible historical consequences of genetic changes in human populations:

> The incursions of barbaric pastoralists seem to do civilizations less harm in the long run than one might expect. Indeed, two dark ages and renaissances in Europe suggest a recurring pattern in which a renaissance follows an incursion by about 800 years. It may even be suggested that certain genes or traditions of the pastoralists revitalize the conquered people with an ingredient of progress which tends to die out in a large panmictic population for reasons already discussed. I have in mind altruism itself or the part of altruism which is perhaps better described as self-sacrificial daring. By the time of the renaissance, it may be that the mixing of genes and cultures (or cultures alone, if these are the only vehicles, which I doubt) has continued long enough to bring the old mercantile thoughtfulness and the infused daring into conjunction in a few individuals who then find courage for all kinds of inventive innovation against the resistance of established thought and practice. Often, however, the cost in fitness of such altruism and sublimated pugnacity to the individuals concerned is by no means metaphorical, and the benefits to fitness, such as they are, go to a mass of

> individuals whose genetic correlation with the innovator must be slight indeed. Thus civilization probably slowly reduces its altruism of all kinds, including the kinds needed for cultural creativity.[8]

Washburn described this quote as 'reductionist, racist and ridiculous'. He suggested that sociobiologists should check their historical facts before they apply evolutionary theory to human behaviour and history, and concluded that 'when applied to mankind, Hamilton's genetic theory turns out to be no more than his political biases'.[9]

Fox found Washburn's letter quite ironic—according to him, Hamilton, in his presentation, had been just as speculative as Washburn himself, simply in a different way. Washburn, Fox mused, belonged to those physical anthropologists who believed that just a few bones and some primates gave anthropological licence to speculate about early humans. Fox had sent off a response to the Anthropology Newsletter. Once again he thanked Hamilton for his participation. He believed that they had made history. 'Up the revolution!' was Fox's final cheer.[10]

What was this about nomadic pastoralists? The offending passage was in fact not too representative of Hamilton's overall argument in that chapter, which was after all a longer discussion of the Price Equation and the levels of selection. It was at the explicit request of the conference organizer that Bill had been examining the implications of this kind of evolutionary analysis for humans and even the process of human history. But Washburn simply used this passage as an indicator of Hamilton's general political stance, which had by now been discussed in anthropological circles. It was generally seen as racist. Hamilton's Oxford colleague, biological anthropologist Vernon Reynolds, explained its racism as follows:

> The genetic aspect indicates that as human groups become spatially further apart, as their gene pools become less closely related, so they will be less co-operative and more competitive. Hamilton makes this perfectly clear in his chapter…He postulates an evolutionary model of small hominid/human groups that will inevitably, because of kin selection and reciprocal altruism, tend to select for co-operation between neighbouring groups and correspondingly to select for aggression and hostility to more distant, unrelated individuals and groups. Xenophobia and racial hostility come as no surprise on this hypothesis; indeed, the rigid application of sociobiological genetics shows that logically they must occur.[11]

And here is Washburn's interpretation (in a later piece on the same subject):

> In general, the further that two populations were apart at the time when races were forming, the greater the genetic difference; hence the less ethical responsibility people should have for members of the other group.[12]

'Should have'? This was not at all what Hamilton was saying. His was not a normative but a descriptive statement. He tried to give a realistic picture of what had probably been happening to human groups over time. That did not mean that he approved, he just presented his analysis as a naturalistic biologist. 'I was told to apply kin selection theory to culture for this conference, so I did,' he later told me somewhat defiantly.[13] Of course he was not particularly happy with the necessary conclusions he arrived at—he found it painful, just as he had cried at an earlier occasion when he thought about the truths he would have to tell about human nature. (Incidentally, these kinds of things were in fact what spurred him on to further theorizing in order to find an acceptable scientific solution.)

Another irritant for the critics was Hamilton's prediction that 'altruism slowly disappears' in modern civilizations. Not so, retorted Washburn, just the reverse, and the same applied to creativity! The critics were obviously taking Hamilton's various formulations at face value. Meanwhile, Hamilton himself may have been making more cryptic, partly self-referential points in this chapter (I will return to this later). It is not surprising if Bill's combination of new evolutionary principles with elliptic formulations remained utterly impenetrable to his anthropologist colleagues, and they in turn tried to make sense of them by following the mainstream critique of sociobiology.[14]

Where did Bill get the idea of the barbarian pastoralists from? Someone who had impressed him greatly with his application of evolutionary ideas to history was RA Fisher, whom he had read in Cambridge. Fisher had dealt with such things as altruism and self-sacrifice especially in the classical world, a period with which Bill, having studied classical history at Tonbridge as a member of the Athenian Club, was familiar. But the most important reference point for Bill was Charles Darwin himself. Darwin had discussed the phenomenon of self-sacrifice and courage in humans but left open the problem that while these traits would increase the fitness of groups whose members possessed these traits, they would be counter-selected within a group. This was now something that Bill felt he could take up exactly because of the new insights that he had obtained from working with the Price Equation.[15]

The Price Equation was perfectly applicable to Darwin's problem. So once again Bill found himself grappling with one of Darwin's unsolved matters. Those were the best puzzles around, and very stimulating for Bill's creative imagination. He knew of course that he was now treading on very speculative ground, but his friend Robin Fox had invited him to try out his new insights on humans, and knowing that the conference especially catered for biologically oriented anthropologists, he assumed that he would be among friends.[16]

Hamilton's pastoralist invasion argument has a number of aspects. In the first place it distinguishes between two types of intelligence, which he sees as prevalent in different types of societies: the mercantile type, characteristic of peaceful agricultural civilizations, and the self-sacrificial daring, characterizing more militant societies. Both involve abstract modelling, the first one of cumulative growth and organized military strategies, the second one for more risky and creative ventures.[17]

Interestingly, Hamilton makes a close connection between self-sacrificial daring and creativity. Perhaps he sees risk-taking as the crucial factor here. But of course, one would assume that a lot of creative effort also goes on in merchant societies. For some reason Hamilton is not interested in that kind of creativity—he is looking for big breakthroughs, artistic or intellectual renaissances. And here he makes the claim that it is the infusion of 'barbaric pastoralists' that have given rise to the creative geniuses in 'renaissances'. His two data points suggest that there may be a pattern of a lag of 800 years between the invasion and the Renaissance.[18]

Unfortunately, Hamilton doesn't specify what he has in mind, either in regard to the renaissances or the barbarian invasions. (Historians, of course have their own explanations for barbarian invasions and their consequences.) But probably more important is Bill's general vision in this part of his paper: a free-flowing romantic fantasy, connecting many of his favourite views with one another. And for much of what he says, he may in fact have had himself as a reference point.

Here Hamilton is defending the underdog: the barbarian pastoralists. He turns around the usual view of vandals destroying classical civilization and its subsequent slow recuperation—instead, it is the barbarians who bring in new genes for creativity, eventually sparking a renaissance when these genes combine in the right way with the existing genetic endowment of a few individuals. He also darkly hints at resistance to established thought and practice, and self-sacrifice, and that 'the cost in fitness' to such individuals is 'by no means metaphorical'. (This idea of rebellion against the establishment is one of Bill's favourite themes, which appears in various places in his writings. Here he may see himself as a risk taker, a daring rugged pastoralist, having something important to offer, but being misunderstood.)[19]

Hamilton also discusses reciprocity as a very important basis for all kinds of exchange relations underlying neighborly situations, trade, and the like—however, he refuses to call such reciprocal behaviour reciprocal *altruism*. Recall that, for Hamilton, in order for a behaviour to be classifiable as altruism, it would have to be ultimately possible to extrapolate it all the way to suicide. But this cannot take place with reciprocity, which is always based on the expectation of individual gain. This is what Bill at an early point tried to convince Trivers about, but with no success.[20]

In regard to the accusation about racism, Bill himself did not think that his statement was racist. He had in mind large panmictic populations (mixed populations with randomly mating individuals). What he was describing rather was the tendency for modern populations to be so mixed and carriers of similar genes spread so far apart that they would not find each other and the likelihood therefore was for genes to get lost. Obviously he was concerned about his particular kind of genes—the ones coding for altruistic behaviour—and was sad to have to report that 'altruism slowly disappears in modern populations'.[21]

The slow disappearance of altruism through the natural spreading of populations, however, stood in direct contrast to the clear enhancement of altruism taking place under inbreeding in closed quarters. That kind of enhancement he had already explored in his sex ratio research and his research on insects in rotten wood. For Bill, the enhancement of altruism that took place 'under bark' might have been a sort of counterbalance to the difficulty for altruism to evolve under the condition of nested hierarchies.[22]

With his letter and later publications in a similar vein, Washburn may have believed that he had succeeded in dragging Hamilton, too, into the sociobiology controversy, even accusing him of political interests and racism. Not much came of this effort, however. This was largely because Bill himself regarded his contribution to the Robin Fox volume as special rather than typical in regard to his central themes. He was not particularly interested in humans; his inclusive fitness principle was aimed at being universally applicable, but he had only extended his theorizing to humans because he had been asked to do so for the conference.[23]

Hamilton had in fact already entered the debate on his own initiative before coming to the United States, by responding to Richard Lewontin's fiercely negative *Nature* review of Richard Dawkins' *The Selfish Gene.* That review had infuriated Hamilton. Lewontin had called the book 'a caricature of Darwinism'. He had also pointed to 'errors' of Dawkins and other sociobiologists. One was 'the adaptationist program', another 'the confusion between materialism and reductionism'. Yet another was the belief that humans could be understood on the basis of their genes and societies on the basis of the properties of their members.[24]

In his letter to the editor of *Nature* Hamilton called Lewontin's whole review a 'disgrace'. Not only was its tone unpleasant, but it did not follow the game rules of science:

> It fails to meet any of the standards of informative value, objectivity and fairness to the views of others that are the part of the code of science... [A] reader unacquainted with the controversy which is its background may well be left with the impression that... the book itself probably is unsound and not worth reading. This is a great pity, since in fact the book is not only the best existing outsider's introduction to a new paradigm... but... itself a significant contribution to this field.
>
> For its intellectual worth, and seemingly in motivation as well, Lewontin's outline of the book is on a par with Bishop Wilberforce's notorious attack on Darwin and Huxley at the British Association meeting of 1860.[25]

What about those errors? Hamilton dismissed them as scientifically uninteresting. And now it was his turn to go on the offensive. Lewontin's review showed a total absence of the scientific spirit!

> Whether the literature of sociobiology reflects an ignorance of the difference between 'properties of sets and properties of their members' or confusion between materialism and reductionism are matters about which I feel little concerned. On the other hand, I feel a warmth very far from indifference when I encounter in much of the literature in question signs of a spirit which I share and which I have always assumed is the same as that which motivates scientific enquiry in all its branches. I can most simply express this spirit by calling it a desire to understand and communicate the nature of the world. I find it present in full measure in *The Selfish Gene* and totally absent in Lewontin's review.[26]

The contrast between the approaches of Hamilton and Lewontin is indeed great. Lewontin's review is of a critical, philosophical nature, questioning the scientific underpinnings of sociobiology. Hamilton is speaking as a positivist, naturalistic scientist, involved in modelling nature as best he can. This is a beautiful illustration of two totally different scientific styles, which at this time were colliding in the field of evolutionary biology.

Moreover, Lewontin's review in *Nature*, could probably be seen as a direct response to Hamilton's own positive review of *The Selfish Gene in Science* the year before. That review even included two poems.[27]

Still living in England, but with the Washburn attack and his own involvement with book reviews, Hamilton may well have felt part of the sociobiology controversy, at least in a small way. We saw how, in the *New Scientist*, he compared his own earlier experiences to the current political attack on EO Wilson. From a letter to his mother shortly before leaving for England for the United States we learn that he regarded himself as the source of the sociobiology controversy:

> Dear Mother,
> Here is some light reading for you—some items of the controversy that has sprung up around my work and its extensions by others. I tried to keep out of it personally for a long time but finally felt so incensed by Lewontin's review of The Selfish Gene that I felt I had to write something.[28]

In his 'light reading' package he enclosed the note about the vote to condemn sociobiology at the American Anthropological Association, Washburn's letter, Lewontin's review in *Nature* and his response to it, and his own review of *The Selfish Gene* in *Science*. This is typical of Bill, he was close to his family members and shared with them his successes and frustrations. Bettina, at the centre of the family information network, would in turn share news of the family with others, including their New Zealand relatives.[29]

In America, Hamilton could not avoid some minor run-ins with critics of sociobiology. During his brief stay there in autumn 1977 there was some friction at the University of Michigan—after all that was the home of the Ann Arbor Collective of Science for The People, a branch of the national organization whose Boston branch was leading the attack on Wilson.[30]

I have conflicting reports on the severity of the anti-sociobiological climate at Ann Arbor at the time. In interview, Richard Alexander played down the whole thing. Hamilton, however, told me of the betrayal he felt when an anthropology student in his seminar, with whom he had been on friendly terms, turned out to be a member of the enemy camp. And another former student of anthropology, whom I happened to meet at a conference many years later, described the then prevailing atmosphere in anthropology at the University of Michigan as simply unbearable for anyone who showed any interest in the evolutionary underpinnings of behaviour.[31]

At the time, social anthropologist Marshall Sahlins had just published his *The Use and Abuse of Sociobiology*, criticizing the whole idea of kinship altruism, and touring university campuses presenting his arguments. (Hamilton dismissed the book as 'religious' in nature and it was Dawkins who was to respond to Sahlins in a serious way in his '12 Misunderstandings of Kin Selection'.)[32]

In the spring of 1978, as Bill Hamilton started his visiting professorship at Harvard, one of the high points of the sociobiology debate occurred at the February meeting of the Association for the Advancement of Science in Washington, DC. The meeting featured a day-long sociobiology symposium, featuring among others EO Wilson, Stephen J Gould, Richard Dawkins, and George Williams. Bill Hamilton was visibly absent from this panel (probably because of his moving between countries and universities; he tended to turn down invitations around times of transition).

In general, too, Hamilton seems to have been able to escape too much negative attention in conjunction with sociobiology. One reason was that he was not perceived as someone particularly interested in human sociobiology; he was properly regarded as a more abstract theoretician. In the Harvard drama he was clearly a sideshow—the main focus was on the opposition between EO Wilson on one hand and Richard Lewontin and Stephen J Gould on the other. Also, during the early stages of the sociobiology debate, the bone of contention was the last chapter in Wilson's *Sociobiology*, which discussed our species.

At the AAAS conference, Science for the People (SftP) handed out (or rather sold) mimeographed copies of various critical treatises, such as a feminist critique of sociobiology. Sociobiology had by now become a political target for many groups of critics, seemingly representing all that was wrong with the United States—manipulation of the people by a power elite, sexism, racism, and you name it. In much of the critical literature sociobiology was presented at the same time as bad political ideology and bad science, and as leading to potentially terrible consequences. Moreover, the critics theorized that it was the bad political ideology that made for the bad science.[33]

But what people remember most about this symposium was the famous ice-water incident: a group of protesters rushed up to the podium from the audience and, pouring a conference pitcher of ice-water over Wilson's neck, shouted: 'Wilson, you are all wet!' This certainly attracted the media's attention.[34]

The year 1978 was also when EO Wilson published *On Human Nature*, an extension of the argument in his last chapter in *Sociobiology*, triggering hostile reviews, and Lewontin launched his critique of the Adaptationist Program, a criticism of evolutionary reasoning. Later this expanded into Gould and Lewontin's famous 'Spandrels of San Marco' paper which accused evolutionists of believing that the present state of affairs was 'the best of all worlds'.[35]

But being in the United States did not isolate Hamilton from potential frustrations. There was still the John Maynard Smith issue, of which he was continually reminded. In autumn 1977 a letter had arrived from Maynard Smith, inviting him to a conference to be arranged in 1978. This was a perfectly normal-seeming request, but it made Hamilton see red. The old anger flared up again.[36]

But hadn't the problem with Haldane been resolved already? For many readers of the *New Scientist*, Haldane's sister's mention of her brother scribbling away in pubs may have provided enough of a basis for supposing that Haldane had indeed come up with that formulation about two brothers or eight cousins, and let the issue rest. Not so Bill Hamilton. His challenge to the *New Scientist* readership had not met with a satisfactory answer. As far as he was concerned, the issue was wide open.

Was it the distance from home or was it the knowledge that he would soon be offered a permanent position in the United States that prompted Hamilton to finally let Maynard Smith know what he thought of him? Here is his response:

19 October 1977

Dear Professor Maynard Smith

Thank you for your letter about the meeting in London in 1978. It sounds as though it should be a very worthwhile symposium. Unfortunately I have to write that I do not wish to attend it myself. This is for two reasons.

One is that I can't think what I could talk about that would not be just a reworking of old ideas. With the prospect of moving home, new teaching duties, etc., in the coming year it is difficult to imagine much time for ideas that could possibly prove presentable by the end of 1978.

The other reason is rather painful to have to write but, since it concerns a view which it is also rather painful to hold and especially so when I know that I have never made the view really plain to you, I had better write it. I have actually begun letters to you on the same subject before but did not finish them; this accounts for my failure to answer one or two letters from you in the past year or two, for which I apologize. I do not believe your anecdote about what J. B. S. Haldane supposedly said in a pub about the kinship principle. This means that while I continue to have considerable respect for your versatility as a scientist and for your contribution in making our common field of interest advance as rapidly as it has, I am unable to respect you as a person. Consequently I would be very ill at ease and a poor performer at a symposium that you are organizing and attending.

Either you are some kind of amnesiac capable of unconciously [sic] fabricating an anecdote harmful to the reputation or else you are a person capable of fabricating such an anecdote conciously [sic] as part of an attempt to avoid the discomfort of admitting intellectual indebtedness to a younger man. The first supposition is the best I can think of you. But your response to my letter to New Scientist protesting about the Haldane anecdote gave no hint of agreement on the possibility of this view.

I am very much aware, of course, that in your letter to N. S. as in other places, at the same time that you attribute more than a true share of originality in kinship theory to Haldane, you have praised my work in publicising the theory extremely highly—more highly than it deserves. It seems in practice, as an expression of inward accounting that surprises even me, that the one thing does not at all balance the other. At the end of three or so very miserable years trying to find what a kinship factor in evolution could possibly amount to I was extremely proud to have been able to prove (at least to my own standard of allowable approximation) that particular principle and to be able to state such a simple but non-obvious truth in that form which, if you are to be believed, was almost exactly prestated by Haldane. I felt that in my early papers I had mentioned all my debts to predecessors as accurately as I possibly could and that the assessment of my originality could therefore be fairly based on those papers, with the modification of course by any historical resurrections of authors that I missed (so far I am only aware of having given slightly less than due credit to Darwin and, on the group-selection aspect, to Sewall Wright, whose papers at that time I found almost incomprehensible and therefore read less than I should have). Your anecdote radically changes the situation and the principle in its general form is now attributed to Haldane—as shown, for example, by the citation in the recent Time Magazine article on sociobiology. Since I feel I have good reason to doubt the authenticity of the anecdote, I continue to resent it extremely, almost like the theft of a child.

My disbelief in it has four main grounds:

i. The similarity of wording in the anecdote to the way I expressed the principle in my first two papers makes a most improbable coincidence,

ii. Had such a coincidence really occurred you would have mentioned it soon after 1963 if not at once, assuming that you intended to mention it at all. In fact, no one seems to have heard the story until about 1974.

iii. It seems to me inconsistent with your position as a university teacher as well as with your usually open and argumentative character that you would have told me nothing of what you knew of Haldane's recent interest and unpublished comments had you really known of these (even of his brief comments in New Biology, which no one at U.C.—or anyone else—ever referred me to) at that time when I was introduced to you at the Galton Laboratory. Cedric Smith, with his usual unambiguous clarity told you that I was working on 'altruism' and I had the impression that he was glad of the opportunity to introduce me and that he hoped that you would talk to me and give me some guidance. I think he always found me a rather awkward student to advise, not having himself much enthusiasm for my evolutionary field. You showed absolutely no interest in or knowledge of the subject of biological altruism, so that I at once wrote you off as yet another failed hope of encouragement in a research ambition which even I at the time was half persuaded must be due to a mental aberration peculiar to myself.

iv. Had Haldane really taken the trouble to derive the principle in its general form, as opposed to working out the easy 'rare gene' case which he mentioned in the New Biology paper, his interest in it must have been strong and in that case it would be very puzzling that he never published his derivation. Further to this point, I still find it wholly incredible that he could have worked out the entire basis for what you claim he said on an envelope. It is quite conceivable (as his sister pointed out in a letter to N. S. subsequent to ours) that he could have been just finishing an algebraic argument that he had begun elsewhere but in that case I still remain puzzled as to why he did not publish such an obviously striking idea. In contrast his note in New Biology could easily have been based on a few calculations on a scrap of paper—indeed his failure to mention a factor of 8 connected with his case of cousins even rather suggests this.

Of course if you could provide corroborating evidence from someone else who remembers the incident in the pub as you described it, or else produce notes in Haldane's handwriting showing that he did really go much further in genetical kinship theory than his published work indicates, then I should feel extremely [underlined] apologetic about the kind of misappreciation of your scientific accuracy that is shown in this letter, although I would still find it puzzling and, I think, not quite gentlemanly in you not to have told me (or, seemingly, anyone else) about the coincidence at an earlier time, it being evident from events that you did not intend to keep it unknown indefinitely. I wonder if you know of anyone who might be able to provide such corroborating evidence—might Dr. Spurway for example? I hope you will believe me that it would be a load off my mind if I could put aside altogether the kind of miserable conjecturing that is evidenced in this letter. But this is not going to happen easily and not without some sort of hard evidence. Suspicion about your scientific honesty (or reliability as a historical source, should a case of amnesia be involved) extends long before I learned about your anecdote about Haldane, in fact to the time when I found you had been so quick to publish papers on 'altruism' after you had seen my long paper in 1963, and at the same time, in yours, gave my papers very scant recognition. I realize that this is consistent with a claim, whether it is based on a mis-memory or not, that you were familiar with my arguments and examples from what you had been told by Haldane and consequently felt no obligation to cite my works. Since later you changed to giving me what I felt to be more than a due share of recognition at least as a publicist of the kinship principle I came to feel that I had been unjust to you about that, but the suspicions came back very sharply when I read the Haldane anecdote.

My suspicion and low estimate of you runs counter to that of many colleagues and I think most regard it as odd of me to be resentful towards someone who promotes my work and reputation as energetically as you do. But equally people reading news accounts of legal and illegal struggles over possession of children can be amazed at the apparently destructive irrationality of the behaviour involved, whereas people actually involved in such struggles themselves understand this behaviour very easily. I make this comparison to try to make myself more understandable, not to justify my being so resentful. I know you have claimed that in science the main thing is to advance human understanding as fast as possible, in comparison to which end scientists' individual reputations are very secondary. I am not absolutely sure about your sincerity if you imply that this is why you don't bother much about mention of the sources of the ideas that you use, but basically I agree with your claim and hold that ideal too; I am ashamed that in practice I find myself much more selfish than my ideal.

I am afraid this letter must be as unpleasant for you to read as it has been for me to write. I hope it makes plain, why, while I wish your symposium every possible success, I can not accept your invitation to attend it.

Yours

W. D. Hamilton (written signature)

W. D. Hamilton[37]

John Maynard Smith's response (in a hand written letter):

October 27, Sussex

Dear Bill,

I am very distressed to receive your letter. Although I have known (since George Price told me) that you have some such feelings. I have not known how deep they were. All the same, I am very grateful to you for writing with such honesty. I will try to do the same. As you will see, I do have some feelings of guilt about the way I have treated you, but I do not accept all the criticisms you make of me.

The thing about which I feel worst is that I did not help you when you were a graduate student. I ask you to believe that this was stupidity, not malice, on my part. I simply do not remember the occasion on which Cedric Smith introduced you to me. Obviously, I failed on that occasion to recognise the importance of what you were doing. As an explanation, not an excuse, I suspect that you may not have been very articulate, that although I like Cedric I have no great respect for him as a biologist, and most important, that at that time I did not [underlined] see the evolution of altruism as an important problem. I could have helped and encouraged you, but failed to do so. In the long run, I have probably suffered more because of my failure than you have. I can only ask you to forgive me. But I must emphasize that my failure was that I did not recognise a promising student in need of help—it was not that I recognised your promise and decided not to talk.

I also think that I may have been insufficiently generous in quoting you. I did quote your paper (as well as Haldane's Penguin article) in my 1964 paper. As far as I was concerned at that time, the point of my paper was to argue against Wynne-Edward's group selection thesis. My only reason for mentioning your ideas was to distinguish them from his. You may feel that I was unjust in coupling my reference to you with one to Haldane—but at least there is nothing in my 1964 paper to imply that I was trying to take the credit for your ideas. I think you could criticize me because I was still underestimating the importance of your 'inclusive fitness', but not for trying to steal the idea. More serious, I should, but didn't, have quoted you in my paper on Alarm Calls. Incidentally, I not only did not quote you, I didn't use $k > 1/r$ either!

Since that time, I believe I have been pretty generous in giving you credit, I know it looks a bit as if I have given you credit only when I had to, because the rest of the world was doing so, but not in the early days when it would have helped. If so, it is because, like others, I was slow to see the full significance of what you were doing. I must add that since George Price discussed the matter with me, I have consciously tried to be just. However, I do not think my behaviour has been as bad as all that. I was glad to be told by a colleague who was an undergraduate in Cambridge in 1965 that when I spoke there in that year I spoke about your work in very glowing terms.

There remains the question about Haldane. If I invented his remark, then there is no defence. I can only say that I am quite certain I did not. Further, I do not think I need any evidence in support of my memory beyond Haldane's New Biology article. Although not identical, the two things are so similar it really makes no odds.

Taking your points in turn:

i. The similarity in wording between my anecdote and your first papers exists also between Haldane's article and your papers.
ii. I didn't quote the anecdote in '63, but I did quote Haldane's article, in 1964, which as far as I am concerned is equivalent.
iii. I did not tell you of Haldane's idea when you were a student, because I failed to understand what you were up to—stupid of me, but not wicked.
iv. I have no reason to think Haldane worked out the idea in general form. Both the New Biology article, and my anecdote, are explicable if he understood the easy 'rare gene' case, or, more likely, that he had not seen the snags which arise if the gene is not rare. As to what he was calculating, I have no idea. My guess is that he was calculating r_{ij} for first cousins. You cannot even do the rare gene case unless you know the chance that a relative has a copy of your genes.

However that may be, I am quite clear that he did not see the significance of what he had got hold of—this becomes clear if you read on in the same article. I was perfectly sincere in the N. S. when I said that the credit for the idea is yours. People give Darwin the credit for the idea of natural selection even though the idea was hinted at before him. There is not the faintest doubt that you will get the credit for inclusive fitness. And so you should.

There is one other matter between us which you do not mention, but which has been on my mind. That is the origin of the idea of ESS. When George and I published the idea, we did not quote your 1967 use of an 'unbeatable strategy', although I had read your paper and the idea is basically the same. Since I became aware of this, I have tried to put this right (e.g. in the American Scientist last year). There are really two points here. One is that I must have been influenced by your paper, but was not conscious of it at the time. Although I am not a complete amnesiac, I don't always know where my ideas come from. The other is that I think I did a good deal more with the idea than you did, and feel I deserve the credit for seeing its generality.

One last point. I would be a hypocrite if I pretended that priorities do not matter to me. We would not be human if we did not mind about these things. I certainly mind, although I try, unsuccessfully, not to be unreasonable about it. I fully understand and sympathise with your feelings of protectiveness about inclusive fitness. I can understand your resentment towards myself, although I think you are being in some ways unjust to me. I cannot think of anything in my scientific career which would give me more pleasure than if you and I could somehow learn to discuss science without any feeling of distrust. I do not ask you to think that I have always behaved well—only that I am not more dishonourable than most men.

Yours,

John Maynard Smith (signature)[38]

What was going on here? Clearly Hamilton had not been able to forgive Maynard Smith. The old anger had flared up again. This time the reason for Bill's emotional upheaval may well have been the recent article on sociobiology in *Time* Magazine (which Bill himself mentions in his letter to Maynard Smith).

Let's take a look inside that 1 August 1977 issue of *Time*. We find a lengthy article covering general ideas about sociobiology and the controversy around it. Once again, Haldane figures prominently with his pub joke, whereby he is said to have 'anticipated the gene-based view of sociobiology'. The *Time* article gives considerable space to Trivers and Wilson, while Hamilton's contribution is mentioned very briefly. Also, it would seem that his theory is specially related to social insects:

> British biologist William Hamilton in 1964 explained how altruism could help an individual spread his genes; he argued that the principle explained the social life of insects.[39]

Where was *Time* getting its information? Of course, this had been a time of transition for Bill, and he had been hard to reach. Once again, because of his travels, he had been away from the scene of the action and others had been speaking for him. This was certainly not the way Bill Hamilton would have liked his contribution to be known. By now he had also come to realize that although he shared some general goals with EO Wilson, their sense of 'sociobiology' was rather different. Moreover, it may have seemed to Bill that the Haldane pub joke, repeated time and again, was taking root in articles and books—including *Sociobiology*. And here we get to one more reason for Bill's general frustration at the time: EO Wilson's treatment of him in his book.

We can get straight to the point by looking at Hamilton's own review of *Sociobiology*. He was one of the very last to review this book; his review in *Journal of Animal Ecology* was not published until 1977.[40]

For Hamilton, 'the book does indeed make a strong and satisfying new synthesis of fact and theory'. But he has some objections to the way in which Wilson goes about this and especially how he defines kin and group selection. According to Hamilton, Wilson's kin selection as a term 'is rather ill-defined, as if he were undecided whether actually to try to define "a kin" as a unit at somewhere about the level of an extended family, or to endorse the more traditional view that it can concern interactions between *any* related individuals' [italics added]. He also notes that Wilson chooses to restrict the use of 'kin' to related by at least the degree of third cousin but does not justify his choice.[41]
Moreover group selection for Wilson is restricted to lineage groups. This means that he excludes 'groups which get their genetic correlation other than by kinship, although these too, to the extent that they have the correlation, can evolve altruism.'[42]

Hamilton therefore suggests the following categorization:

> In my opinion the present confused situation could be best tidied by using the term 'kin-group selection' for the area of overlap of the two common terms, covering all these cases where there are identifiable groups on the ground whose members are kin. Equivalently this specifies as well all those groups where members are genetically correlated because of restricted migration/exogamy: it is often not realized that genetic drift implies and reflects a raised level of kinship among group members. Then, alongside the term 'kin-group selection', a term such as 'nepotism' could be used for the remaining part of the present kin selection concept, covering those cases where interactions are too ephemeral or too

> interwoven for an epithet of 'group' to be really appropriate. Finally, for the remaining cases where kinship is *not* involved, we would have 'reciprocant selection', 'assortative selection' (or equivalent more felicitous terms if these could be produced) and perhaps others as well.[43]

But he goes on to say:

> It cannot be too strongly emphasized, however, that *all such terminology is just differentiating natural selection of replicating molecules according to the various kinds of assemblage into which they are formed or form themselves.* In my opinion (doubtless very biased) the best general guide to which way the cat of adaptation will jump is provided by the concept of *inclusive fitness*... [italics added].[44]

Inclusive fitness is Hamilton's 'eldest child', and he is particularly protective of this concept. At this point in the review, he cannot contain himself any longer. Wilson is not doing justice to inclusive fitness!

> In my opinion Wilson is clearly unfair to the concept in treating it as only a rough heuristic guide to the phenomena which he subsumes under 'kin selection', while at the same time giving more space to some rather special group selection models of the 'kin-group' kind. As if to justify this bias, he says of my approach that 'The conventional parameters of population genetics, allele frequencies, mutation rates, epistasis, migration, group size, and so forth, are mostly omitted from the equations'. Well, I don't see at all how he can say this for allele frequency, which has always been the dependant [sic] variable of my arguments. Moreover, but for the 'mostly' I would equally contradict the claim that the approach has failed to consider migration, mutation and group size. And as for epistasis, I am not aware that this is given any greater prominence in the special kin-group models that Wilson treats in more detail.[45]

Hamilton is particularly puzzled about Wilson's wish to make 'kin' a unit of selection (or a level in a hierarchy based on relatedness). He says he believes that this is related to Wilson's reliance on Darwin's view that in the social insects, selection operates at the level of the family. According to Wilson, Darwin was the first to think of the idea of kin selection. 'This is roughly true', Hamilton agrees, but adds that Darwin's vision was limited. There was a necessary *quantitative* aspect missing in his logic (something that Darwin failed to notice). In his logic, Darwin was relying on the practices of animal breeders, who were breeding from the close relatives of animals and plants (which had been killed in the very process of assessing their excellence!). 'Darwin realized that the sibs and parent would differ slightly in heredity from the slaughtered animal but he seems not to have noticed that *how much* they differ sets a requirement in rigour for the breeding programme' [italics added].[46]

So Darwin just proceeded to draw a direct parallel between the practices of the breeders and the self-sacrifice of a worker among the social insects.

But, Hamilton notes, 'his argument has to be confined to the problem of how evolution is *carried on*, in an insect that is already perfectly eusocial; as regards the *origin* of a social insect it glosses a difficulty.'[47] Darwin's parallel probably held very well for the careful and controlled breeding that he knew about, but it did not hold up when it came to the origin of insect sociality. 'The situation of an incipient social insect could be better likened to culling from half-wild herds, and with this practice it would be very far from certain the selection would make the cattle fatter, as the owners might desire, rather than thinner.'[48]

In other words, according to Hamilton, Darwin did *not* in fact solve the problem of the origin of eusociality of social insects. That problem still required an explanation.

Was there an explanation? By the time Hamilton wrote the review, Trivers and Hare had, with their 1976 paper, provided empirical support for the idea that haplodiploidy was a key component for the evolution of eusociality in social insects. Their contribution had been published as a sizable lead article in *Science* and it had been positively commented on in *Nature*. Trivers had been speaking to large audiences across the United States. In other words, it looked as if Hamilton's early suggestion in 1964 about the possible connection between haplodiploidy and the origin of eusociality was being supported.[49]

But Hamilton was not so sure. We know from his 1972 *Annual Review* paper, and his subsequent under-bark and fig wasp papers that he still regarded the origin of eusociality as an unsolved and complex problem—with an important part of the answer to be found elsewhere—particularly underground.

15
The Parasite Paradigm

By autumn 1978 Bill had taken up his position as Museum Professor of Evolutionary Biology at the University of Michigan, Ann Arbor and the family had settled down in a nice house which looked out over River Huron. The house that they had bought was some distance away from the university, but it could be reached fairly easily by various means—Bill tried walking, biking, skiing, and even skating to work. He was delighted with all the new flora and fauna that he could observe and be inspired by on the way. His trips to and back from the university became veritable adventures in comparative zoology and botany. But just as the unexpected singing of a red cardinal in the middle of the winter snow triggered in Bill research ideas of a positive kind, so the various hospitals that he regularly passed started him on a more pessimistic train of thought about health and illness—what he called 'the dark side of my walk'.[1]

The Hamilton house was one of only three houses on that side of the river, next to the house of Peter and Rosemary Grant, the famous researchers of the Galapagos finches and Bill's university colleagues. That arrangement was ideal, and the Grants and Hamiltons became very close. Just like the Hamiltons, the Grants also had two daughters, somewhat older than Helen and Ruth. The river was a special treat and Christine taught the children to sail in the summer. There were schools close by for her daughters to attend (Ruth went to a Montessori school). Bettina and Bill's sisters Mary and Janet came to visit.[2]

Bill enjoyed the new intellectual climate at Ann Arbor—it really was a hotbed of evolutionary discussion and a place where interesting people were regularly invited to spend a semester or give seminars. For instance, George Williams came there for a term and he and Bill gave a team-taught seminar on sexual selection. And the students, typically bird specialists, many of whom had spent time with the Grants on the Galapagos Islands, were inquisitive, challenging and resourceful. 'I found them boldly and almost aggressively interested in what I could teach', Bill observed. It was a stimulating atmosphere, and the Museum was also used as the venue for parties. Sometimes it seemed to Bill as if the evolutionists had taken on the role of 'high priests of Ultimate Explanation'.[3]

Bill's teaching experience was also quite different: here he was dealing with research students. Many felt that he treated them as fellow researchers. According to his colleague Peter Grant, Hamilton was a great teacher and handled his seminars well. There are two surprising elements surrounding his stay at Ann Arbor. It seems that Hamilton was good at teaching his students the Price Equation, enabling them to effectively use it in their own work, and at least two students later published in this vein. Another surprise was that Hamilton rarely discussed scientific matters with Dick Alexander, his 'patron'.[4]

Bill also travelled a lot. He was a popular lecturer on American university campuses. By the end of the 1970s kinship theory, or kin selection (a term that Hamilton had started using more and more after the mid-1970s) was a very popular topic. But Hamilton wanted to pursue another new topic, and that was the question of the origin of sexual reproduction. So, ironically, his pursuit of Darwin's second big problem, the origin of sexual reproduction, was being obstructed by his many invitations to speak about his solution to Darwin's first big problem, altruism. He usually solved such conflicts by quickly establishing a connection between the two competing topics, and then concentrating on his new interest. What was the connection here? As Hamilton put it, 'kin selection lay in the shadow of sex'. And just as his kinship theory had been earlier, this new topic would soon become an obsession with him (as he stated himself).[5]

***

Sex had emerged as an intellectual problem for Hamilton well after his 1960s' research on inclusive fitness (or kinship theory) and sex ratio. In fact, his interest in sexual reproduction followed partly as a natural consequence of his sex ratio research. As he explained it, once he realized how easily he could justify the existence of uneven sex ratios, that is, the overproduction of females, sometimes at the ratio of 50:1, the real problem became why males existed at all! Why would a species ever put half of its effort into producing males, he asked himself? Why was humankind, for instance, not a parthenogenetic species? Looking, say, at the extreme efficiency of ever-increasing armies of aphids attacking rosebushes, females seemed alarmingly self-sufficient. There was clearly a battle to be fought. Bill now embarked on nothing less than a universal defence of males, and their right to existence. Males must be good for something, and he would prove scientifically what that was.[6]

As he put it himself—in typical Bill fashion—he came to a point where he felt as if all those females in the experimental vials from his old sex ratio research were now shouting at him in unison to go and do something useful! (He was familiar with this type of female shout.) Well, he would show them just how useful males could be. But he knew he would have a lot of explaining to do:

> Why all this silly rigmarole of sex? Why this gavotte of chromosomes? Why all these useless males, this striving and wasteful bloodshed, these grotesque horns, colours . . . and why, in the end, novels . . . about love?[7]

There was another influence of a more theoretical kind. In the 1960s James Crow and Motoo Kimura had reopened the whole topic of sex, pointing out that Fisher and other individually oriented theorists had relied on surprisingly 'groupish-sounding' arguments as soon as it came to sex. These had presented sexual reproduction as important for the species rather than explaining it in terms of short-term individual benefits. Hamilton realized that his own awe of Fisher must have blinded him to this fact.[8]

In 1971 John Maynard Smith had usefully characterized the problem when it came to sexual versus asexual reproduction as 'the twofold cost of sex'. (This refers to the fact that in meiosis half of the genetic material is lost, whereas in asexual reproduction it is kept intact.)[9] In other words, sexual reproduction was costly. Why then was it maintained? In the early 1970s, Hamilton—ever the puzzle solver—had already begun systematically addressing this question, partly by mining library collections and partly by tentative modeling (his usual exploratory strategies).[10]

The concrete stimulus for Hamilton came in 1975 in the form of the request for a double book review of Michael Ghiselin's and George Williams' books, both dealing with the evolution of sexuality. Hamilton himself likened reading these books to hearing 'Beethoven's four muffled chords of fate'—they ruthlessly lay bare 'biology's most outstanding problem', opening it up to close scrutiny. Both books insisted that sex must be shown to be beneficial at the level of the individual, not just the population or species. They also both saw sexual reproduction as connected to some type of environmental uncertainty.[11]

Hamilton agreed with this general view, but not with the pessimistic sounding conclusion of Williams. Williams declared: 'Maynard Smith's...analysis convinces me of the unlikelihood of anyone ever finding a sufficiently powerful advantage in sexual reproduction with broadly applicable models that use only such general properties as mutation rates, population sizes, selection rates, coefficients, etc.'[12] This was obviously a challenge that had to be met. The road ahead was clear. The task was to find a universal explanation for the evolution (and maintenance!) of sex. Williams in his book had provided a set of special case scenarios (such as 'the strawberry-coral', the 'elm-oyster', the 'cod-starfish', and the 'aphid-rotifer') but Bill was sure that a general rule could be found. After all, he had done it before with inclusive fitness after working through a lot of special cases himself.[13]

The request for the review came just before the beginning of what Hamilton describes as the 'sex wave' in evolutionary biology. It seemed that soon everyone was getting in on the problem of sex. Many have pointed to Graham Bell and his magisterial overview, *The Masterpiece of Nature*, as the most important stimulant for the new trend in the early 1980s and the many international conferences that followed.[14]

But Bill felt an increasing dissatisfaction with what he was reading. People didn't seem to understand just what a formidable problem sex was! Losing half of the biomass to produce males seemed like such a waste from an efficiency point of view. Or, as he formulated it, now taking the gene's point of view:

> Clearly, if you are a gene with some say in what the genome does, then to allow your bearer to waste half her descendant biomass in each generation on a seemingly pointless production of maleness certainly isn't itself a good idea; in fact you lose out and disappear extremely fast compared with a gene preventing maleness. But if being around with males, however idle they may be, or having your daughters be around with them, mysteriously brings into your stock some splendid and at least twofold vigour... the case might be very different.[15]

Using a kind of reverse engineering type of thinking, Hamilton reasoned that the production of new combinations—the only advantage that sexual reproduction has over asexual reproduction—must be of central evolutionary importance. When would such a solution be typically called for? The answer was: when the environment was changing rapidly and the changes would require recombination of existing material rather than the production of novel material. Both Ghiselin and Williams had seen sexual reproduction as connected to environmental uncertainty. What exactly might this mean? This had to be examined more closely.[16]

The natural next step for Hamilton was to model various scenarios. He regarded the computer as a 'magic carpet' that could be used when insufficient data were available. Through simulations, it would be possible to use reasonable assumptions to create different scenarios and in this way find out how the system hypothetically behaved under various conditions. Hamilton teamed up with Peter Henderson (his computer specialist from his Imperial College time) and Nancy Moran, a doctoral student at the University of Michigan, in investigating the role of various types of environmental changes. This resulted in a joint paper, but the puzzle was far from solved.[17]

Hamilton continued working on his models by himself: 'I mathematized them as far as I could and then used simulation on the Michigan main computer', typically accessing it from the basement of the Herbarium building. There he experienced all the well-known frustrations of computer programmers until he finally found the right way to achieve what he wanted. The key question was: Under what conditions would sexual reproduction have an advantage over asexual reproduction? The paper with Henderson and Moran had explored the role of random physical changes in the environment, including cyclical patterns. But these factors were not good enough—they were too unpredictable and too easily led to destruction of valuable variation. What was needed was some kind of responsiveness in the environment. Predators seemed like one possibility, but there was typically not a one-to-one relationship between prey and predator. Finally, the idea of parasites occurred to Hamilton almost by a process of elimination. Parasites were interactive and responsive to host genotype frequencies.[18]

He had a clear idea what he wanted his models to accomplish:

> Their aim was to simulate a coevolutionary process in which a population of hosts was being affected by a population of parasites and the parasites by the hosts. At least in principle the hosts always and the parasites sometimes were arranged to be active in the models in two alternative versions: in one they reproduced sexually while in the other they were parthenogenetic. I was trying to see what environments and what interactions would make life easy for the sexual variants and hard for the parthenogenetic.[19]

What ensued was a sequence of model generation and selection whereby 'model grew from model under a kind of secondary Darwinian selection'. He continued developing the ideas that contributed to the success of sexual over asexual reproduction and abandoned those that didn't. Moreover, any candidate model had to be able to work under low fecundity, and follow what he himself judged to be a 'credible fitness pattern'.[20]

The hallmark of sexual reproduction is genetic recombination, the result of meiosis (the process which divides the double set of chromosomes that has resulted from fertilization, mixing their genes along the way). It was clear to Hamilton that sexual reproduction had to be a universal principle:

> Sex of the kind that was interesting me most...seemed to me clearly an evolutionary unity with all its procedures and genetic results much the same whether one looked at it in plants, animals, or unicells. This unity and the implied age for sexual phenomena—back to the earliest single-celled organisms in the Precambrian era—made for me an awesome phenomenon, a spectacle I viewed as I might a cathedral.[21]

But if sex was universal, that meant that it would have to apply equally to haploid and diploid organisms. (Examples of haploid organisms are male bees, wasps, ants, and mosses.) That is why it was irritating to Hamilton to see that some recent models published in *Nature* had all been assuming diploidy. He thought this very assumption made them unconvincing: 'For me the papers were nowhere near to showing practical ways to pay for the costliness of sex'. Still, *Nature* had published those models.[22]

Moreover, it also seemed that the models getting published were simple one locus models. From Bill's point of view even two-locus models were 'absurdly limited compared with the real multi-locus stabilization [of sex] that occurs in nature'. 'I would not even have thought of sending such a model to, say, the journal *Nature*', he observed. For him, two-locus models were at most acceptable as parables. What then was the proper way to go about model building? Well, his own approach was to start off with a haploid model in a simple two-locus situation. If all went well, he would be able to continue from there and capture reality in all its complexity:

> If support could be found in this then surely I could get more support, possibly very much more, when numerous interlocus interactions were brought in. Eventually I'd try out the whole necklace of all the chromosomes all at once.[23]

But there was a question that needed to be answered. Recombination, the essence of sexual reproduction, did not only mean that new associations between genes were formed. It also implied that existing precious combinations were being destroyed. How could the decay of associations over generations ever be a good thing, Hamilton asked himself? There was one clear condition: the destruction of existing combinations could be good if the selective advantage had changed. And under what condition was this most likely to happen? Answer: co-adapting parasites, because parasites uniquely evolve to quickly exploit that which has become common in a particular environment—in this case particular gene connections.[24]

But the process of recombination meant that it would take many generations to destroy old combinations. Wasn't this process too slow? Hamilton had an immediate answer: slowness was in fact a virtue. Those earlier, well-tested combinations might be needed again. A slow process would be able to retain precious gene combinations, while a process that was too fast might result in their extinction. And it was important not to lose them:

> Yes, it is slow but in the long run too that can turn out an advantage. In a world full of cycles perhaps the host may soon again need the very combination its recombination is currently destroying, so it can be best not to destroy it too fast and too completely.[25]

In other words, the point is not to lose genetic material that might come in handy when the environment changes again.

However, this kind of model is very difficult to handle mathematically. Hamilton was keenly aware of this: 'Even two-locus population genetics under constant selection is difficult mathematically if one demands complete algebraic accounts. Three-locus genetics is worse; here even excellent mathematicians soon admit to too much terra incognita.'[26]

But Bill, ever looking for approximations and simplifying assumptions, found a solution. He decided that in his model the genetics of the parasites themselves could be omitted, since the parasites' countermoves were so big and predictable. He was very pleased with the result: 'the neatest and most suggestively open-ended model for sex I had yet devised'. The model showed that 'under sufficiently strong selection a frequency-dependent response will stabilize sex against the twofold cost and can do this even when two loci only are varying'. Moreover, it could come about quite easily. '[T]he process needn't involve any dramatic changes in gene frequency: statistical changes of gene associations (LD) are enough' (LD is linkage disequilibrium, a measure for the strength of gene association). In other words, Bill had been able to show to his own satisfaction that 'very simple conditions might set in motion a process that was highly beneficial to sex and yet at once give absolutely no danger of extinction to any allele'.[27]

It was by an interesting coincidence—the need to solve a purely technical problem with computer printouts—that Hamilton had an epiphany, which brought to him a key aspect of his parasite paradigm. Computer problems were nothing new for Bill. Invariably, he was having problems not only with the programming but also with the people in charge of main frame university computers. He also had his own ways of getting around them.

At Ann Arbor, Bill continued to be a threat to the system. His host-parasite cycling simulations often resulted in curves that oscillated so much that they messed up the ink plotter and tore the wet computer paper. Angry calls followed from the computer station, Bill promised to take action, but it happened again. In the end he realized that he had to do something concrete, because the problem was not going to go away. Well, he would simply have to find a way to keep the curves on, rather than off, the printout! How was that to be achieved? One answer was: make the cycles slower and give the ink some time to dry.

> The most sure way to produce the slow cycles was to introduce a lag. This meant that instead of having the selection determined just by the last generation's genotype frequencies I would make it respond to what they had been several generations back. In nature this might happen if it had taken time for a parasite or predator to complete a dormancy... Much longer delays could also be speculated...[28]

So, through introducing minor lags, Hamilton was able to get his model to perform permanent and slow cycling. At the same time, he was modelling a process that he knew could be found in nature. But as he compared the numbers indicating the long-term average fitness for sexual and asexual reproduction for each simulation run, he found that the cycling motion was usually not forceful enough to beat asexual reproduction. Also, the swings were too wide, which meant that genes might get lost, and that was something that he did not want to happen. Bill went on tinkering with these models, feeling increasing disappointment that he could not quite achieve a situation where sexual reproduction was demonstrably capable of winning over asexual reproduction in the long term and beating its two-fold advantage.[29]

It was through almost physical interaction with his own diagrams that Hamilton got his crucial insight. He did nothing less than step into his own model—in his mind's eye, that is. Once there, he imagined himself to be one of the animals in the host cycle, moving about in three-dimensional space:

> Gloomily I traced the back and forth of this neatly drawn curve…noticing where the line in true 3-D space must pierce and pierce again the…plane which implies zero association of the genes…If I were one animal in the population…could I know whereabouts in the cycle we—I, my fellows, and our parasites—were, I wondered idly?…Where in the cycle would I feel at my best? Obviously the answer to the last would depend on my genotype. Suddenly I stared at the coloured lines snaking my page with renewed interest. Feeling good, being fit; what did this imply to be happening to biological fitness of the various genotypes as this cycle goes around. Of course I could tell if I was fit and others could tell too. So, apart from how I might feel about myself, which would be the best genotype at my point for me to be with?…When, if ever, would fitnesses of parent and offspring be negatively heritable—like I was now realizing that they had to have been all the time in my two-point cycle—and when instead would they be positive? This last plainly held right here in the diagram and it wasn't a failure after all.[30]

At this point Hamilton felt that he had hit upon a 'minor grail' when it came to population genetics. His model showed that the genotype that was unfit in this generation was bound to be very fit in the next. There was a clear cycling of genotypes due to pressure from parasites. He had found a rationale for the permanent heritability of fitness! The genotypes changed, but fitness remained.[31]

How long did this epiphany last? Hamilton himself describes it as a half-hour episode of revelation, during which lots of ideas raced through his head. On a more sombre note, he admitted that it was probably 'more like several weeks of deepening conviction'.[32]

This was how Hamilton's new addiction (his own term) started. He plunged into the bird libraries:

> My first target was of course the cardinal, this most local example and so abundant in Ann Arbor. I became obsessed with cardinals…[they] sang, puffed brilliant feathers for me on snowy trees…Books, bones, and birds of many kinds swayed and swooped around me. Hoopoes it was first and then blue jays and Canadian jays ('whiskey jacks') that soon eclipsed the cardinals and the waxwing as my focal birds.[33]

It soon became clear to everyone that it really *was* an addiction. The Bird Library curators took Bill's enthusiasm in their stride, but among Bill's students, asking a lot of critical questions, it was finally only Marlene Zuk who supported him.[34] Together they embarked on an interesting intellectual journey. Was it possible to reinterpret Fisher's idea of 'runaway selection' so that the trait causing the runaway was not arbitrary (as Fisher had argued), but had some specific significance. From an evolutionary point of view, what did females want in their mates? Hamilton and Marlene Zuk argued that what females wanted was healthy males. Birds might in fact have developed their costly ornaments, colourful plumes, and complicated songs as honest signals of health. The prediction was therefore that colourful species would be particularly parasite-prone.

The task would be to select a suitable sample of birds, grade them on their showiness, and correlate them with their proneness to parasites. Marlene was to take care of the rating of the birds (in this case North American birds on a scale from 1 to 6), while Bill was scouting around in libraries for any possible information about bird parasites. The predicted relationship was shown to hold up statistically in Hamilton and Zuk's parasite theory of sexual selection, 'the Hamilton-Zuk hypothesis'. It was published in *Science* in 1982, fuelling wide enthusiasm for an exciting new framework for research.[35]

***

It was an invitation to Uppsala in 1980 and the subsequent publication of his paper by a 'friendly editor' in the English language Scandinavian journal *Oikos* that made it possible for Hamilton to quickly disseminate his new ideas about host-parasite co-evolution to a larger audience. Indeed, many quote the *Oikos* paper as their first inspiration to a new way of thinking. What they don't know is that there was a good chance that Hamilton might not have gone to Uppsala at all. Here is the story.

In June 1979 Bill responded to the invitation from Staffan Ulfstrand from the department of Zoology at Uppsala University to a symposium to be held there the next year. As the topic for his contribution Bill argued for 'The puzzle of sexuality' instead of the one suggested on sex ratios, since this was more in line with his current interests. This was typical Bill. He almost always succeeded in turning a conference topic into something that he was currently working on and wished to talk about. His next request was also typical:

> As regards preparing a written and publishable manuscript, I would like not to be *expected* to produce one. It is just possible that my models and ideas on maintenance of sexuality may have reached a stage where I could have them written up in time, but I do not wish to commit myself to producing a manuscript at this stage.[36]

But the last paragraph revealed Bill in paranoid top form:

> I regret now to have to write about one misgiving which I have about the symposium. I do not wish to come if Professor Maynard Smith is also to be invited. I note that he is not on your list of speakers at present. If he should be added I would wish to drop out. I have respect for his work in general but dislike him as a man on account of what seemed to me an attempt to attribute more credit to Haldane and less to me for the origins of genetical kinship theory, this being done through a barely-believable anecdote which he has put about in recent years concerning insights said to have been revealed during his personal contacts with Haldane. I am afraid I should not enjoy the meeting and would perform poorly as a discussant if he were present. Of course I realize that if he were to decide to come on his own initiative near to the time of the meeting it might upset your plans too much if I were to drop out and in that case I would have to take things as they came. I am sorry to have to mention this 'incompatibility'; apart from this I am enthusiastic about participating.[37]

Judging from Hamilton's response to Ulfstrand, it appears that Maynard Smith's previous letter had had little effect.

As it happened, Maynard Smith did not come to the conference and Hamilton attended as planned. Still, as far as the development of the parasite theory was concerned, it was a close call. Had Bill not gone to this Uppsala meeting, he would not have made his mark there with the talk that would later become his 1980 *Oikos* paper: 'Sex versus Non-Sex versus Parasites'. More importantly, he would not have encountered a friendly and helpful editor, who made it possible for Bill to quickly get his paper published and out into the scientific consciousness. Because for many younger scientists, it was that paper which opened up new vistas and a new paradigm of research. In fact, it set the basis for a whole new research industry. If the name of the game was to avoid parasites, what methods might organisms use to achieve this? This opened up new avenues of creative speculation followed by rigorous hypothesis testing.[38]

Uppsala itself was dark and cold when Hamilton attended the February meeting called 'Theories on Population and Community Ecology'. The aim of the meeting was largely to acquaint Swedish doctoral students with international luminaries, which meant that the setting was fairly informal. This was exactly the kind of arrangement that Bill liked and had become used to at Ann Arbor. He enjoyed discussions with students—in fact, he often felt more comfortable with students than scientific colleagues. He was also eager to present the result of his 'epiphany'. Although he earlier had told Ulfstrand that he didn't want to promise a paper, by the time of the meeting he had undergone a complete change of heart.[39]

The Uppsala meeting was a success. In general he felt comfortable with the friendly Swedes and amazed at their English language ability. Bill was usually quite sensitive to the atmosphere of conferences and his varying performance as a speaker at least partly depended on that. Ulfstrand, with his keen eye for scientific excellence, was an early supporter of Hamilton, and would later become a major positive force in Bill's career.

Although the Uppsala meeting had set Bill nicely off on his new project connecting the maintenance of sexual reproduction with the need to avoid pathogens ('parasites' as he called them), he was still involved with his earlier inclusive fitness theory. Sometimes he felt as if he were in a rearguard battle: the population geneticists were catching up on his theories and seemed to delight in finding faults with them. A whole critical industry had developed around Hamilton's models, and mathematically inclined purists were fixing now this, now that perceived error of Bill's, arriving at different coefficients of relatedness. The field looked more like a jungle.[40]

Of course, Bill knew that he had made various kinds of approximations in his modelling of inclusive fitness, but he also knew under what conditions they were approximately right. In addition, he had explicitly stated these conditions. Sometimes it seemed to him as if population geneticists just ignored what he was trying to do in order to be able to do what *they* wanted to do.

Bill had nothing against population geneticists—in fact, he admired their mathematical skills—but for any collaboration to work, his partner would have to be friendly. One great example of a mathematically sharp, while friendly, specimen of this sort was Rick Michod at the University of Arizona, Tucson. So when Hamilton had finally agreed to visit his university to give a talk in November 1979, Michod remembers how Hamilton and he spent a whole night sitting in a tepee in his Tucson garden engaged in intense discussion. It is not clear what role the tepee played, but the result was their 1980 paper, 'Coefficients of relatedness in sociobiology'. Going through the different recent publications about coefficients they concluded that it would be important to try to sort out the tangle of alternative suggestions and find a way to 'translate' between them. Michod's mathematical skills were combined with Hamilton's biological insight. The result was a tour de force. Their joint paper showed, in a manner satisfactory even to the most mathematically exigent, that all the supposedly alternative derivations of coefficients of relatedness that had been proposed over the past few years, were in fact saying the same thing.[41]

Fast forward to October 1980. Bill is writing a letter to John Maynard Smith from the University of Michigan. He is responding to the latter's request that he be an external PhD examiner for Mike Orlove (his former student at Silwood Park), saying that, unfortunately, he has already promised to read too many manuscripts and will have to decline. The letter begins 'Dear John', which is an improvement over the last letter from 1977, where he addressed his correspondent 'Dear Professor Maynard Smith'. But the real surprise comes in the middle of the letter:

> Changing to another topic, it seems that I owe you an apology for having doubted your anecdote about Haldane. Very recently H. Eysenck has written in a book review that he heard substantially the same statement by Haldane when he was a student under him. It can't have been on the same occasion because the wording is slightly different (3 and 9 instead of 2 and 8 and no back of envelope) and in fact Eysenck goes out of his way to add that, as with some other sayings of his that he knew to be important, Haldane repeated it to different groups of students in various pubs. That all this is stated rather 'out of the blue' in a review where it seems hardly relevant and without any further detail of the kinds of context in which Haldane's remarks arose, still seems odd to me and I still wish that the verification could have come from someone who did not have a fairly obvious incentive for wishful thinking on this theme—which incentive, incidentally, the review as a whole does seem to reveal. However, your version and his now fit together in a reasonable scenario which, not having known Haldane personally, it would be wrong for me to think improbable. So I am very sorry that I doubted your veracity on this.[42]

Well! He hadn't only doubted it but had gone so far as to ask the readers of the *New Scientist* to find someone who had actually witnessed Haldane making his famous joke in some pub.

Hamilton now goes on, expressing in more explicit language what he tried to hint at more cryptically in his *New Scientist* 'challenge letter' of 1976:

> I am afraid that I still feel rather resentful about the matter in a general way. I would have hoped that the purpose of Haldane's aphorism was precisely to encourage young men to take up the study of what he considered an important topic. Yet at U.C., surrounded presumably by a number of scientists who had

been primed in this way, all my own attempts to tell people that a principle of this kind was needed in biology and this was what I was working on were met by the sourest rebuffs—words to the effect that I was wasting my time, there was no such thing as genetic altruism, or at best total disinterest. No one repeated Haldane's sayings to me, no one even referred me to his New Biology article. I can remember persons telling me that everything worth doing on altruism had been done by Fisher and Haldane but when questioned as to where the work could be read I never found that they knew of anything other than what was in the books of 1930 or 1932, which of course I had read long ago.

I suppose it is *understandable* that a disciple without a master must expect a hard time and that statements that seem significant or humorous according to the hearer's mood when uttered by a famous scientist seem to be heretical and ridiculous pretension when uttered by a detached graduate student, but I can't help feeling that it is wrong that this should be so in a university. But I also have in mind that I was a rather retiring kind of student who preferred to withdraw from people who found me ridiculous rather than argue back. This may be the reason why I never got to hear much about Haldane. (I needed, I suppose, a streak of Mike Orlove's persistence towards hostile or indifferent listeners, for which most people do thank him in the end!)

...

Yours sincerely,
Bill (written in pen)
W.D. Hamilton
Professor of Evolutionary Biology[43]

John Maynard Smith responded a couple of weeks later with a handwritten letter:

Dear Bill,
November 14, 1980

Thanks for the letter. I'm sorry you can't manage Mike's thesis, but I quite understand why. I agree that he has merits. I have written to Michod to ask whether he will do it.

On the Haldane thing—I'm glad about what you say. I think you have every right to feel resentful in a more general way. You needed and deserved support for your ideas, but did not get it—from me or anyone else. There is one aspect of the situation which helps to explain the situation, even if it doesn't excuse it. I really don't think anyone was very interested in the evolution of social behaviour. Haldane may have repeated his aphorism several times, but there is no reason to think he thought the idea more than entertaining. He didn't so far as I know, even mention it in a lecture. He published it as a throw-away paragraph in a popular article. He never followed it up. My own interests at the time were in the evolution of ageing.

I suppose that if you had been a more pushing kind of person, you might have got our attention. As it is, I'm aware that I failed in my job as a university teacher. However, in self defence I must point out that I was in a different department, and I imagine we cannot often have met. Surely the people who really ought to have helped were the people in your own department, and in particular your supervisor (?I believe that was Cedric Smith?).

The two things that made me interested in the evolution of social behaviour were first your 1964 paper (which, as you know, I saw earlier—? In 1962 or 1963), and in particular your application of the idea to social insects. I do remember feeling that it was stupid not to have thought of that. The other was Wynne-Edwards' book, or, more precisely, the fact that most of my students and colleagues swallowed the group selection argument.

I suppose all this has a moral. But the people one should really feel sorry for are those who have an important idea and never got the credit—at least not until too late. Mendel is the classic case.... There must be many others. Did you know that Castle thought about the Hardy-Weinberg ratio before they did?

...

Best wishes
John[44]

So with this information one would assume that the Haldane pub story was now settled once and for all. This was also Bill's chance to personally convey to Maynard Smith how he actually felt as a student at University College, and Maynard Smith's chance to explain himself, and (once again) reassure Hamilton that Haldane did not follow up his pub joke in any serious way.[45]

***

Hamilton's whole parasite theory sometimes came into direct conflict with the 'official' take on host and pathogen among bona fide parasitologists. This is something that both he and Marlene Zuk observed.[46] But there were exceptions when it came to people who knew his work well. In 1982 Bill was invited to an important parasite conference at Dahlem by Robert May, with the expectation that his paper would be published in the resulting volume. May's instructions had been to come up with 'a readable, speculative, and potentially controversial paper to start discussion'. Hamilton devoted his paper to a discussion about various ways in which large multicellular organisms can resist pathogens, with special attention to the pathogens' potential for molecular mimicry. Among the organisms' various defences he pointed to the many ancient variants of resistance genes that all seemed to co-exist (he likened them to old tools in a garden shed, just sitting there waiting to be used again). He also presented his most recent model for the maintenance of sex as a defence against parasites.[47]

But Bill wondered if he had upset the conference participants with his theory, especially his own group discussion leader, Dr AC (Tony) Allison, the famous discoverer of the important phenomenon of heterozygote resistance to malaria (the resistance is conferred by a gene for sickle-cell anemia if it is exists in only one copy in an individual; the same gene is severely handicapping when existing in two copies in homozygotes). This was seen as a major step of progress in medicine, and was generally believed to be a possible model for other diseases as well. Bill realized that if his own parasite-and-illness theory was correct, this went in exact opposition to the malaria paradigm. The notion of heterozygote resistance presupposed a stable equilibrium in which the heterozygote was reliably recreated. Bill had not designed his theory to run up against the malaria idea. Rather, it was supposed to be an alternative to other theories of sex, but this was the practical outcome. (Meanwhile he did not see the malaria case as evidence that his own parasite/pathogen theory was incorrect—rather, he regarded the malaria heterozygosity case as an exception.)[48]

But there was still work to be done to convince sceptics in regard to the parasite avoidance theory of sex that had come to Hamilton in that moment of epiphany. There were some obvious reasons why it was hard to believe that positive heritability of fitness could come out of co-evolutionary cycles between parasites and hosts, something that sceptics did not fail to point out.[49]

In cases when a genotype became common, fitness would quickly decrease, and the cycle would swing rapidly, altering the optimal type. The result would be negative fitness heritablitiy. On the other hand, if the cycles were to swing so slowly that the gene frequencies would go near to fixation for most of the time, that would bring the mean heritability to near zero. Was there any hope for a situation that would be conducive to positive heritability of fitness? For Hamilton, the answer had been 'intermediate cycling', a cycle length that would be optimal for 'sosigonic' mate choice (Hamilton's term for a health based selection criterion). The big question was how to model this mathematically in a convincing way.[50]

Hamilton decided to enlist his Israeli colleague and friend Ilan Eshel to help generate the necessary models. The real advantage when it came to working with Eshel was that Hamilton believed the two thought—and what was more, felt—in the same way about Nature.[51] For example, in their 1984 paper they argue that it is indeed possible to have a selection process that does not exhaust the additive variance that it feeds on. Also, their paper shows that variance in fitness can indeed be maintained at a steady value as long as there can be periodic reversals of selection at many different loci, with the timing of reversals uncorrelated.[52]

An interesting generalization (and some backtracking?) also takes place in this paper. The authors point out that the case of parasite co-evolution was an illustration of a more general principle. It need not be parasites that cause the cycling, it could in principle be various features of the environment. But it is necessary to maintain a steady parent-offspring correlation, and this requires many co-occurring cycles in order to prevent drastic changes. And the numerous host-specific parasites that exist in birds or mammals are indeed a good source of such cycles.[53]

The paper with Eshel represented an intermediate stage in the development of Hamilton's thinking about the evolution of sex and sexual selection.

A distinction made by Hamilton around this time was between what he called classical Red Queen and neo Red Queen manoeuvres. The classical Red Queen has to do with the species-in-community level, while the neo Red Queen has to do with the dynamics of alleles and genotypes in a sexual species. He later reflected that the term Red Queen was in fact 'kidnapped' from Leigh Van Valen, but commented that '[a] notion of appellations controlées for our concepts have seemingly not yet developed in biology, so I am very happy to go along with the preferences of others who would call my present theme a "parasite Red Queen paper" (or as I now often abbreviate it, a PRQ)'.[54]

There were more questions that Hamilton was able to ask and answer in regard to his parasite paradigm and its working in nature. It seemed to have more general application. He realized that diversity of genotypes could be a strategy for avoiding parasites for non-sexually reproducing species as well. It turned out that populations of parthenogenic species, clones, in fact were not homogenous, but consisted of mixtures of many genotypes and these showed dramatic changes in frequencies over time.[55] Why was this? The answer was to make it harder to be found by parasites and more costly for them to change their system in order to attack a host. Each clone is in fact hiding among others that are genotypically different. In other words, when it comes to avoiding parasites, there is an advantage to being rare. Hamilton mused that this might be the explanation for biodiversity, and also how biodiversity could cause its own increase—and why the tropical forest was not totally eaten up by herbivorous insects, despite the fact that the richness of species in a tropical forest is orders of magnitude higher than in a temperate one.[56]

And there were additional possible strategies. Speciation was a good way to avoid parasites. Here we might have a possible explanation for 'the superabundance of speciation events that have occurred on our planet', Hamilton reflected.[57] Indeed, figuring out possible strategies for parasite avoidance and testing these empirically was to become the basis for a burgeoning new research enterprise for quite some time to come.

16

Cooperation without Kinship

Bill enjoyed himself intellectually and socially at Ann Arbor. There was always something going on, some symposium or some interesting visitor giving a seminar. Bill and his neighbours Peter and Rosemary Grant numbered among the few faculty members living relatively close to the university, which meant that Bill and Christine ended up doing a fair amount of entertaining. A great moment for Bill was having Sewall Wright to come and stay with him and his family. The stamina of this almost 90-year-old man was formidable—after giving a talk at the university, he held forth at a reception given in his honour at the Hamilton home, staying on his feet for hours.[1]

But there was a new source of delight in store for Bill at the university. He was to discover a group of people with whom he could feel close intellectual fellowship, and who seemed to accept him as one of them, too. And this group had nothing to do with the biology department. This was the interdisciplinary BACH group. BACH was an acronym that stood for the first letters of its members' last names. Actually BACH was the most abbreviated version of the group name, which depending on who was around at the time could take on names such as HBACHS or even HHBACHS. This was a group interested in modelling and simulation that met informally over lunch and occasionally collaborated on projects. During Bill's time, B stood for Arthur Burks, A for Robert Axelrod, C for Michael Cohen, H for John Holland and S for Carl Simon. Hamilton was added as the extra H in front. A graduate student named Rich Riolo was also an active member of the group.[2]

What united the group was its members' curiosity, interdisciplinary orientation, and deep wish to make scientific sense of the world. Mathematically gifted, but realizing that many real world problems were intractable to traditional mathematics, they were interested in creatively using the power of the computer instead. Bill, with his early interest in programming and simulation, fitted right in.[3]

It was Bob Axelrod who introduced Bill to the BACHS group. And who introduced Bill to Bob Axelrod? Axelrod had asked Richard Dawkins for the name of a biologist with whom he might discuss the evolutionary implications of various game theoretical models. Dawkins immediately suggested Bill Hamilton, and told Bob Axelrod that he just so happened to have Bill Hamilton in his own backyard! Dawkins had been an entrant in Axelrod's famous game theoretical experiment at the time. Axelrod had been running computer tournaments between various iterated Prisoners Dilemma games suggested by his entrants, looking for the strategy that would hold up best in the long run. But how universal were the conclusions? He had started to feel the need for evolutionary expertise. And here was an answer—Bill Hamilton.[4]

This is Bill's description of his first contact with Axelrod:

> Now on the phone to me was someone out of political science who seemed to have just the sort of idea I needed. A live game theorist was here on my own campus! Nervously, and rather the way a naturalist might hope to see his first mountain lion in the woods, I had long yearned for and dreaded an encounter with a game theorist. How did they think? What were their dens full of?[5]

To his surprise, even during that first telephone call, it was apparent to Bill that Axelrod's thinking was 'more than a bit' biological.[6] At a subsequent lunch it was Bill who suggested to Axelrod that they collaborate on a piece for *Science*: Axelrod's work with game theory and computer tournaments was of clear biological interest. In their joint paper Axelrod would describe his idea and results and Bill would give it 'a natural scientist's style' and provide some biological illustrations. Axelrod, who knew of Bill's 1964 paper, was delighted to learn that Bill had already worked on the problem of the Prisoner's Dilemma.[7]

Axelrod soon realized that Bill was absolutely crucial when it came to finding persuasive biological examples:

> Bill's naturalist's style included having at his fingertips an astonishing knowledge of species from bacteria to primates. His knowledge would be equivalent to

> an economist knowing much of what there is to know about hundreds, if not thousands, of companies of every type from GM and Microsoft to a self-employed sidewalk vendor.[8]

As Bill later observed, the naturalist examples given in the paper were later replaced with other, better ones. In fact, he seems to have thought that his examples were not very good at the time. But the real value was not in the specific examples, it was in providing a model which gave the possibility of making testable predictions over a wide range of species. Time was ripe for this idea, it seemed. The paper was accepted in *Science* with considerable ease and later awarded the AAAS' Newcomb-Cleveland prize for the best paper published in 1981.[9]

The basic point of 'The Evolution of Cooperation' is to show how game theory can be used to formalize the various potential strategies for social actors in real life and identify the conditions under which cooperation can come about. To simplify issues the authors focus on interactions between pairs of individuals:

> The Prisoner's Dilemma game is an elegant embodiment of the problem of achieving mutual co-operation... In the Prisoner's Dilemma game, two individuals can either co-operate or defect. The payoff to a player is in terms of the effect on its fitness (survival and fecundity). No matter what the other does, the selfish choice of defection yields a higher payoff than co-operation. But if both defect, both do worse than if both had co-operated [hence the dilemma].[10]

The Axelrod and Hamilton paper demonstrates how game-theoretical insights from an *iterated* (repeated) Prisoner's Dilemma framework can in fact illuminate the conditions under which cooperation between unrelated individuals can in principle evolve. What is needed is a high probability that individuals will meet (and 'play') again.[11]

The paper does indeed suggest that a game theoretical approach is applicable to quite a range of biological reality. For instance, since employing a strategy does not require a brain, the Prisoner's Dilemma game can be played even by bacteria. Bacteria are highly responsive to their environment (especially its chemical aspects) and can develop conditional strategies of behaviour depending on what other organisms around them are doing, and these strategies can be inherited. At the same time, bacteria and their surrounding organisms both affect each other's behaviour. And when it comes to organisms with higher intelligence, there are possibilities for richer game playing, especially as the capability for individual discrimination now allows for rewarding cooperation and punishing defection in others.[12]

With this collaborative paper, then, Bill Hamilton had now further expanded his original insight about the conditions for evolution of other-directed behaviour. Originally he had looked at the probability of sharing genes as a condition for the evolution of altruism. The framework had been one of clear kinship nepotism, but also including the more mysterious 'green beard effect', that is, some phenotypic marker whereby altruists might be able to recognize one another and cooperate.

Such kin selected (or 'phenotype matching') altruism had its inbuilt system for overcoming the situation otherwise plaguing social life: the inevitability of Prisoner's Dilemma situations (which when generalized can lead to such larger social problems as depletion of resources). In 1969 at the Man and Beast conference he had felt depressed by the thought of the Prisoner's Dilemma, being unable to see a way out.

Now through collaboration with Bob Axelrod a solution had been found. Individuals did not necessarily have to be related for cooperation to work. What it took was for individuals to *have disincentives to act selfishly, or to have incentives to act cooperatively.* One such condition was simply not to be able to get away with acting selfishly. This meant that individuals would be counted on to meet again, and that they would be recognizable.[13]

Another way of putting this was to introduce the idea of an Evolutionarily Stable Strategy in the context of the Prisoner's Dilemma. 'Deductions from the model and the results of a computer tournament show how cooperation based on reciprocity can get started in an asocial world, can thrive while interacting with a wide range of other strategies, and can resist invasion once fully established.[14]

The strategy that attained the highest score in Axelrod's computer tournaments was 'tit-for-tat'. The strategy is simple: it is based on cooperation on the first move, and then doing whatever the other player does.[15] The robustness of this strategy was shown to be dependent on three features: the tit-for-tat player was nice (never the first to defect), it could be provoked into retaliation by a defection of the other player, and it was 'forgiving' after just one act of retaliation.[16]

But tit-for-tat is not the only evolutionarily stable strategy. A strategy called ALL D [all defect] is especially evolutionarily stable, and it is not 'nice'. The question is: with such a strategy around, how can cooperative behaviour have a chance of starting at all? Axelrod and Hamilton had the answer:

> ALL D [all defect] is the primeval state and is evolutionarily stable.... But cooperation based on reciprocity can gain a foothold through two different mechanisms. First, there can be *kinship* between mutant strategies, giving the genes of the mutants some stake in each other's success, thereby altering the effective payoff matrix of the interaction when viewed from the perspective of the gene rather than the individual.
>
> A second mechanism to overcome total defection is for the mutant strategies to arrive in a *cluster* so that they provide a non-trivial proportion of the interactions each has ...
>
> The basis idea is that an individual *must not get away with defecting* without the other individuals being able to retaliate... Higher organisms avoid this problem by their well-developed ability to recognize many different individuals... The other important requirement to make retaliation effective is that *the probability... of the same two individuals meeting again must be sufficiently high* [italics added].[17]

This kind of situation can be realized in several ways: by maintaining continuous contact (inter-species mutualism), by employing a fixed location (as done by cleaner fish), territoriality (eg in birds), ability to recognize faces (humans), or cues for the likelihood of continuing reciprocation.[18]

The authors summarized their conclusions as follows:

> TIT FOR TAT is an extremely robust [strategy]. It does well in a variety of circumstances and gradually displaces all other strategies in a simulation of a great variety of more or less sophisticated decision rules. And if the probability that interaction between two individuals will continue is great enough, then TIT FOR TAT is evolutionarily stable.[19]

Together then, Axelrod and Hamilton in their 1981 paper had been able to demonstrate that it was possible for cooperation to originate and evolve: '[C]ooperation based on reciprocity can get started in a predominantly non-cooperative world, can thrive in a variegated environment, and can defend itself once fully established.'[20]

Incidentally Hamilton and Axelrod had slightly different views when it came to the significance of kinship. Hamilton believed that this was typically the way cooperation got started, while Axelrod theorized that cooperation might come about if only individuals meet often enough. Hamilton agreed with this in theory, but could not quite imagine how it could happen in practice: 'The only way to avoid such kin-based clustering is to allow phases of population where the clustering ends and all organisms mix.... In humans, for example, they could sort themselves by their love or hatred of vintage cars, or living in hilly country... Bob's line was: well, why not? And mine: but how unlikely and where was even a single plausible example?'[21]

For Bill, all this had actually been a distraction from his main business at the time: the parasite paradigm. Indeed, he had thrown himself into this new area of the evolution of sex with the deliberate wish to get away from his earlier research on kinship theory. But Bill was used to handling parallel projects, and this one was particularly fun and yielding. It was also unusually satisfactory emotionally and intellectually to work with someone who actually thought alike.

Bob Axelrod was pleased with their collaboration:

> I was delighted to accept Bill's invitation to collaborate. Despite coming from different disciplines, Bill and I shared not only mathematical training, but a love of formal modeling. Bill had even published one paper using the Prisoner's Dilemma, although he was hoping to get away from that when I dragged him back....
>
> Bill's proposed division of labor turned out to be a good description of how the collaboration developed. I gradually realized, however, just how much was included by Bill's modest formulation of adding 'a natural scientist's style and some biological illustrations.'... His experience as a naturalist often gave him the capacity to check out the plausibility of an idea with pertinent examples right off the top of his head. It also helped him to generate surprising new ideas.[22]

If Bob appreciated Bill, so Bill appreciated Bob. Here is how Bill saw the two of them working together:

> That brilliant cartoonist of the journal *American Scientist*, Sidney Harris, has a picture where a mathematician covers the blackboard with an outpouring of his formal demonstration.... [I]t starts top left on the blackboard and ends bottom right with a triumphant 'QED'. Halfway down, though, one sees a gap in the stream where is written in plain English: 'Then a miracle occurs', after which the mathematical argument goes on. Chalk still in his hand, the author of this *quod est demonstrandum* now stands back and watches with a cold dislike an elderly mathematician who peers at the words in the gap and says: 'But I think you need to be a bit more explicit—here in step two.' I easily imagine myself to be that enthusiast with the chalk and I also think of many castings for the elderly critic.

> Yet how easy it is to imagine a third figure—Bob—in the background of the picture, saying cheerfully: 'But maybe he has something all the same, maybe that piece can be fixed up. What if . . .[23]

Bill would never have imagined that he would be working with a political scientist:

> I would have thought it a leg-pull at the time if someone had told me of a future when I would find it more rewarding to talk 'patterns' to political scientists rather than to fellow biologists.[24]

But Axelrod noted:

> Had Bill known of my long-standing interest in evolutionary theory, he might not have been quite so surprised that my thinking was more than a bit biological. For example, in high school I wrote a computer simulation to study hypothetical life forms and environments. This early interest in evolution was nurtured during college by a summer at the University of Chicago's Committee on Mathematical Biology.[25]

He shared Bill's surprise at how well they worked together. Of course it did take some getting used to:

> [W]hen I asked Bill a question he sometimes thought long and hard before saying a word. His face took on a blank look, his gaze was in the distance, and I could almost hear the wheels spinning inside his head for the longest time. I learned to be patient.[26]

What was the reason for their collaboration working so well, they both wondered—and as curious scientists started investigating right away. They found that they shared certain convictions and taste preferences. 'Perhaps the most important thing we shared was our aesthetic sense', Axelrod mused.[27]

Bill made the following analysis of their similarities:

> [A]n intuitive understanding between us was immediate. Both of us always liked to be always understanding new things and to be listening more than talking; both of us had little inclination for the social maneuvering, all the 'who should-bow-lowest' stuff, which so often wastes time and adrenalin as new social intercourse starts. Bob is the more logical, but beyond this what we certainly share strongly is a sense for a hard-to-define aesthetic grace that may lurk in a proposition, that which makes one want to believe it before any proof and in the midst a

> confusion and even antagonism of details. Such grace in an idea seems often to mean that it is right. Rather as I have a quasi-professional artist as my maternal grandmother, Bob has one closer to him-his father. Such forebears perhaps give to both of us the streak that judges claims not in isolation but rather by the shapes that may come to be formed from their interlock, rather as brush strokes in a painting, shapeless or even misplaced considered individually, are overlooked as they join to create a whole...[28]

Axelrod, too, considered the fact that they both had artists in the family:

> I see a further connection between art and modeling. My father painted to express how he saw the world that day, highlighting what was important to him by leaving out what was not. Likewise, I see my modeling as an expression of how I see some social or biological dynamic, highlighting what I regard as important, and leaving out everything else.[29]

In general, Axelrod was impressed by Bill's knowledge and sheer intuition in regard to game theory and the many connections and associations he made across different fields. There was no doubt that Bill was a most welcome addition to the BACH club:

> Bill's disciplinary training as an evolutionary biologist and a naturalist proved essential to making our theoretical work compelling to biologists. He was adept at identifying pertinent biological examples so that biologists could see what we were talking about. While not all of his proposed applications have been borne out, he was able to demonstrate the potential relevance of computer tournaments for the major biological puzzle of why individuals cooperate with unrelated others. He was also able to explain what our contribution added to what was already understood about evolution. Specifically, he showed how our modeling work provides a solid foundation for many of the insights about altruism formulated years earlier by Robert Trivers (1971). Bill was also able to show how our model could be used by other evolutionary biologists to formulate and test new hypotheses about animal behavior, as well as explore dozens of variants of the simple iterated Prisoner's Dilemma.[30]

***

In 1981 Trivers received a reprint of the cooperation paper from Hamilton. It came with the following inscription: 'To Bob, Tit-for-Tat or Hamilton's Revenge, Bill.' What was this all about? What was the revenge supposedly for? Trivers' first thought was that this was a friendly 'retaliation', 'the tit for the tat of Trivers and

Hare (1976)... That is, I had stolen a portion of Hamilton's thunder on the haplodiploid Hymenoptera and he was now returning the favor on reciprocal altruism.'[31]

What Trivers had in mind here was that the Trivers and Hare paper had actually been able to demonstrate that Bill's calculations about Hymenoptera were in fact correct—this was the needed empirical test of Bill's inclusive fitness model, and made his whole theory more believable.[32]

But there was more to it, Trivers' wife Lorna insisted. Bill had a subtle mind. Trivers soon arrived at the following more complicated interpretation of the meaning of Hamilton's Revenge:

> Earlier I had on the one hand prevented Bill from publishing his first formulation of game theory applied to reciprocal altruism (when he asked permission to include it in his 'Man and Beast' paper), and then had turned around and stolen it back (to be sure while citing Hamilton), and this was my tat for that tit as well.[33]

But what was it that Bill himself meant by 'Bill's Revenge'? Here is his own version:

> In a letter I sent to him [Bob Trivers] along with a copy when our paper came out, I told him that the theory I had helped with was to be considered my revenge—my 'tit' for a 'tat' of his from long ago. In that he had shown how a 'patents-pending' sex ratio idea of mine put together with a clearly 'patents-granted' kin-selection idea could yield, greatly to my chagrin, something completely new and unexpected about the social insects—about, in fact, various 'patents-uncertain' as well as erroneous examples I had used to illustrate kin selection. Just as I'd kicked myself for not having noticed the 'Trivers and Hare' applications in that case, so the present paper showed something I thought Bob T would be sorry to have missed—a way to quantify in effect the poetic 'long shadow' he had plainly, in its essentials, recognized.[34]

With the 'long shadow', Bill referred to an expression used by Axelrod which he liked very much because of its poetic appeal: 'the shadow of the future'. It is the shadow of the future in iterated Prisoner's Dilemma games (the fact that they will meet again and be subject to the consequences of their actions for a long time) that keeps the players in check and makes them choose more cooperative alternatives than in a 'one-shot' Prisoner's Dilemma situation.[35]

In any case, Trivers wrote back to Hamilton that his 'heart soared'. As he describes it in his autobiographical notes, 'For one wild moment, I kidded him, I actually believed there was progress in science!'[36]

There was one thing, though, in Axelrod and Hamilton's 1981 paper that Bob Trivers absolutely refused to go along with. That was a piece that he was sure that Bill had written. Trivers recognized the situation—it seemed he was being lured by Bill to go way beyond other people's comfort zone.

> I remember the sensation vividly as if it were yesterday. It was as if Hamilton and I were exploring unknown territory together. I was, once again, following Hamilton through some kind of tangled undergrowth, and he was confidently hacking his way through with a machete and assuring me that he knew just what lay ahead. But the terrain was becoming steeper and more dangerous as we descended, Bill still confidently telling me that the water hole, or the river, or whatever it was that we were looking for, was straight ahead. But finally I was too frightened to continue and called out: 'This far, Bill, and not *one* step further!' Many, many people have had that sensation far earlier in their explorations with Bill and have turned back, to their own disadvantage. But that is the only time I can remember, in reading any of Bill's works, where I drew back and said, in effect, 'I am not going there with you, Bill, I am turning back.'[37]

The situation that Trivers was referring to was the final part of the Axelrod and Hamilton tit-for-tat paper, where there is a suggestion that the Down's syndrome, trisomy 21, produced by an extra copy of chromosome 21, may come about through a game theoretical strategy on the part of one of the paired chromosomes in a germ cell. Trivers was sure this was Hamilton's idea.[38]

This was the relevant passage:

> Our model (with symmetry of the two parties) could also be tentatively applied to the increase with maternal age of chromosomal nondisjunction during ovum formation (oogenesis). This effect leads to various conditions of severely handicapped offspring, Down's syndrome (caused by an extra copy of chromosome 21) being the most familiar example. It depends almost entirely on failure of the normal separation of the paired chromosome in the mother, and this suggests the possible connection with our story. Cell divisions of oogenesis, but not normally of spermatogenesis, are characteristically asymmetrical, with rejection (as a so-called polar body) of chromosomes that go to the unlucky pole of the cell. It seems possible that, while homologous chromosomes generally stand to gain by steadily co-operating in a diploid organism, the situation in oogenesis is a Prisoner's Dilemma: a chromosome which can be 'first to defect' can get itself into the cell nucleus rather than the polar body. We may hypothesize that such

> an action triggers similar attempts by the homologue in subsequent meioses, and when both members of a homologous pair try it at once, an extra chromosome in the offspring could be the occasional result... For the model to work, an incident of 'defection' in one developing egg would have to be perceptible by others still waiting. That this would occur is pure speculation, as is the feasibility of self-promoting behaviour by chromosomes duing a gametic cell division. But the effects do not seem inconceivable: a bacterium, after all, with its single chromosome, can do complex conditional things.[39]

Bill provided a justification for his reasoning by calmly calculating that from each individual chromosome's point of view, Down's syndrome fitness was certainly more than zero fitness, which is what would have happened to each of them if they had gotten into the polar body and been eliminated. In other words, this was a typical Prisoner's Dilemma situation, with each chromosome taking the safer way out. Trivers had a particularly hard time believing that a chromosome in a neighbouring egg would be able to 'learn' about the behavior of the selfish chromosome. But Bill saw a parallel between oocytes and unicellular organisms, which can certainly process various types of information (usually of a chemical kind).[40]

The Axelrod and Hamilton paper had a major impact in its time—it soon became textbook knowledge applicable to anything from international relations to marital strife. Tit-for-tat sounded simple and effective. But of course, this was just the beginning of model development, and the model later faced challenges. For instance, it can be invaded by 'foolishly nice' strategies (meaning they are never the first to defect, but at the same time won't retaliate), and become vulnerable to take-over by 'nasty' strategies. One of the rivals for tit-for-tat is arguably Pavlov, a strategy which plays an 'asocial' game, responding only according to whether it has been 'fed' or not (that is, achieved a 'good' result). Other models are being developed all the time and tested for robustness, stability, and initial viability, as evolutionary game theorists continue their simulations and experiments. Tit-for-tat, though, appears to be holding up rather well.'[41] The evolution of cooperation was a starting point. As Hamilton dramatically put it, 'as a chapter in the history of a theory of Good versus Evil, our paper records a battle won but not an end to the war.'[42]

17

The Oxford Move

Bill Hamilton was obviously doing well at the University of Michigan, having received the AAAS award for the most important paper for 1981, and having found Robert Axelrod and the BACHS group to work with. He had also bought a small piece of land a short drive away from Ann Arbor. The idea was to use it for excursions and close study of nature. The landscape was quite varied and there was even a little brook running through it. This piece of land meant a lot to Hamilton, and he would proudly take his family and friends there for nature walks.[1] Meanwhile Hamilton and Marlene Zuk had been continuing their work on mate choice and health. Zuk's critical fellow students may not have been convinced that showiness in birds was really correlated with proneness to parasite infection, but some journal referees thought differently. Hamilton and Zuk had been able to produce good statistical evidence and their short paper in *Science* seemed to inspire both fellow researchers and journalists. A longer article in *Scientific American* featured Bill's parasite paradigm and the Hamilton-Zuk hypothesis, and there were several reports in more popular journals, too. The ideas were exciting and seemingly broadly applicable. A new scientific subfield was on the verge of being established.[2]

But there were those who thought that the time had come for Hamilton to return home to England. And one such person was Bill's former head from Silwood Park, Richard Southwood, now the Chair of the Zoology Department at Oxford. In 1976, having been denied a promotion at Imperial College, Hamilton had been looking for greener pastures—and had found them overseas. Obviously, he had now proven himself, and it seemed only natural to Southwood that the prodigal son should return.

Southwood had identified the ideal position for Hamilton—a Royal Society Research Professorship. That position involved no teaching duties (save for one public lecture a year), while providing generous support for travel and research. Southwood could not see how Hamilton would be able to resist such an offer. A minor obstacle—the fact that this professorship could only go to Fellows of the Royal Society—was easily overcome by making Hamilton a member of that illustrious body in 1980. Southwood took charge of this minor problem, supported by another Fellow of the Royal Society: John Maynard Smith.[3]

The initial offer for this career change came in 1981, and Hamilton travelled to England to discuss the matter in person. But as Hamilton suggests, at that point he was perhaps 'too contented' in his work.[4] Also, the Hamilton family was not yet ready to move. One important consideration was to the children's education.[5]

The Oxford move was postponed to 1984, at which point the Hamilton family—which now consisted of three daughters, Rowena (Rowie) having been born in 1979—moved back to the United Kingdom. There it was arranged for them to rent the old Groom's House adjacent to the only pub in Wytham village near Oxford and close to Wytham Woods, a nature reserve frequented by Oxford University naturalists. Bill, who had sold his American house at a loss, realized with dismay that he could not afford to buy a house in Oxford.[6]

How did Bill feel about moving to Oxford? He explained that he had been getting tired of political correctness and worried for his daughters about increasing violence in the United States. With regard to political correctness, there was a particular incident that irked him. He thought he had written what he considered to be his best letter of recommendation ever for an outstanding female student of his, commenting on her mathematical skills as unusual for a woman. He had believed that this would make the letter stronger sounding. Instead it was regarded as sexist by the people processing recommendation letters. He took this reaction extremely seriously, reflecting that it would obviously be 'difficult to be myself if I remained there'.[7]

In Oxford, one of the first things Hamilton did was to try and finish his 'Big Novel', *The Dark of the Stars,* something he had been working on for about six years. To speed up the process he hired a special typist to type for him. He did finish the manuscript, but did not succeed in getting it published. *The Dark of the Stars* was an adventure novel with many dramatic features and a protagonist called Robert (just like his brother). The novel was described as 'a swashbuckling tale of intrigue, romance, diamond smuggling, gun-running, and exotic flora and fauna'.[8]

Also high on Bill's agenda when moving to Oxford (and also connected to the world of novels) was a bicycle ride up to a point where he assumed Thomas Hardy had been standing as he reflected on the events taking place in his imaginary Christminster. Thomas Hardy was one of Bill's favourite authors.[9]

Hamilton initially felt somewhat awkward in the new Oxford environment. He had come some weeks ahead of his family; his older daughters needed to finish the school year. In the United States he had had a difficult time adjusting to the American academic style—he had noticed that academics there constantly seemed to want to test people's intelligence by engaging in very fast and witty verbal interchange. Now he had entered another academic culture:

> If I was leaving, in Ann Arbor, one world I found hard to understand, the one I was entering turned out not so transparent to me either. Meeting Oxford academics I felt a little like Gulliver just departed from the friendly giants of Brobdingnag amongst whom he feared to be stepped upon and arriving in the land of Laputo where he found himself politely but harmlessly ignored.... [A]rriving in Oxford was like landing on another planet.[10]

But measures had been taken to help Hamilton integrate into the Oxford culture. Richard Dawkins had arranged for Hamilton to be a Fellow of New College, of which Dawkins was a member (incidentally, this had also been JBS Haldane's college). Although Bill regularly lunched there he kept much to himself. On the other hand, there were often interesting people from other academic fields that could be met at the luncheon table. Bill actually got some useful pieces of information from colleagues this way—for instance he was able to derive the term 'sosigonic' to refer to his new health-related criterion for mate choice (although he soon found that the term was not to catch on).[11]

Unlike many other professors, perhaps deliberately breaking an unspoken rule of academic life, Hamilton felt free to talk to graduate and undergraduate students. In general he was resistant to adopting any particular type of academic style that seemed to be required. In the United States he had disliked the whole political correctness thing. What was good, though, was that Americans were quite direct. Here at Oxford the game was different and much more opaque. Still, Bill had already succeeded in discovering some social rules:

> Here in Oxford all such PC issues—the supposed unisex brains, the smoking issue, etc seemed a lot more complicated than they had in Ann Arbor and they also had to be talked about in more oblique ways: you might believe in PCness or you might not, but you didn't discuss your view directly and, above all, not at lunchtime.[12]

There was also the problem of when and how to laugh at jokes. Still, he was hopeful that he would get the knack of this new culture, just as he had done after a while among American academics.[13]

Together with his New College membership, Hamilton had been given rooms where he could work and even sleep, if needed. He also had a spacious office in the Zoology Department on South Parks Road, and that even came with a paid assistant. As usual, he needed professional advice with computer modelling, especially since he had become increasingly reliant on simulation. He usually ended up becoming good friends and writing joint papers with his computer experts, especially Peter Henderson and Brian Sumida. Brian Sumida was Hamilton's student and first assistant, helping him translate programmes into UNIX and later working with him on the development of his host-parasite models.[14]

There are lots of stories about Bill at Oxford. One of the more famous ones tells of Bill, an avid cyclist, flying through the rear window of a car. According to Richard Dawkins' brief version, Bill is supposed to have told the driver: please take me to the hospital! Is this story true? Well, this is Brian Sumida's report on Bill's 'famous cycling episode':

> Bill and I had arranged to meet at his office in the Zoology department to work on a computer simulation. Due to the accident, Bill arrived at his office late, as was not uncommon, appearing with multiple lacerations in his face where the glass shards had imbedded themselves. He was very irritated because the doctor who extracted the glass fragments had made an appointment for Bill to have his skull X-rayed for possible fractures. This irritated Bill immensely since he was completely unconcerned about his injuries. What *did* concern Bill was the fact that he had to leave his bicycle on the pavement of Woodstock Road. Bill asked me to fetch the bike with urgency since it was quite precious to him. He said that the bicycle had belonged to his brother who had been killed in a climbing accident. Bill said that whilst riding on the bicycle he would often think of his brother.[15]

Bill was a daredevil cyclist who as a daily sport used to compete with the bus while riding to Oxford from Wytham village. Add to this that he did it without a helmet. Of course an explanation could have been that it was not possible to find a helmet large enough for his head, but that excuse was no good after he turned down someone's kind offer to have a helmet specially made for him.[16]

Being at Oxford did not end Hamilton's travelling—if anything, it gave it new wings. In the mid-1980s he was invited to a number of symposia abroad. He cleverly combined these trips with side trips. In this way he got to see Israel in 1985, and Japan in 1986 with Taiwan as an extra bonus. He was also invited to yet another Dahlem conference in 1986, this time on sex and parasites to which, interestingly enough, he did not contribute a paper. In late 1987 he was a participant in the so-called Nobel Symposium at the Gustavus Adolphus College in Minnesota. This is a yearly gathering of outstanding scientists discussing their own work and addressing important issues in terms understandable to laymen—in this case undergraduate students. (There is no direct connection between the Nobel Prize and the symposium, except the timing.)[17]

The papers presented at these conferences were later published in edited volumes, and in addition, Hamilton had been asked to contribute chapters to other volumes. In other words, in the mid- to late 1980s Hamilton was being asked to do a number of overviews. These were opportunities for him to reflect on his work in multiple areas, as well as the way in which his various underlying themes related to one another. Hamilton used to call book chapters 'his wild oats' in contrast to research papers contributed to scientific journals, and seemed somewhat embarrassed about this (presumably because of an easier acceptance by interested editors in such cases). But he did, in fact, produce overviews that were very useful, both for himself and others. They forced him to update his knowledge in different areas, to develop some themes further, to make connections, and to continue reflecting on some long-standing concerns of his, both intellectual and ethical. Having an interested audience and friendly editors may also have encouraged bolder conjectures than usual. And so we get a number of papers with different titles and emphases, but which all in some way address the common themes of social relatedness, the evolution of eusociality, the nature of clonal colonies, and the role played by parasites and pathogens in sustaining diversity. He also took on big and difficult issues such as the future of the human race and the meaning of telling the truth in science and in life.[18]

The paper that Hamilton presented in Israel took his host-parasite coevolution model one step further. He was now experimenting with using ecological models for understanding host-pathogen interaction, a type of 'translation'. In fact, this paper is one of his few ecological explorations. But how ecological was it really—Hamilton admitted that he was doing the 'translation' merely in order to be able to later translate his ecological insights back into genotypical processes! He also argued that one could actually regard pathogens and parasites as 'environment' too (although he did not usually act on this). He was aware of the importance of the role of both ecological factors and epidemiological factors, though his models soon started concentrating on host-parasite interactions.[19]

In typical Hamilton style, Bill also addressed some questions not connected with the parasite paradigm—or maybe they could be connected? Travelling in the beautiful landscape with his friendly host Eviatar (Eibi) Nevo, Bill was reminded of the chalklands that he knew so well from Kent. He was thinking about the past and future of humankind and the importance of sustaining diversity. This was the Fertile Crescent where the first crops had been found and slowly domesticated, laying the foundation for agriculture. But the wild varieties always needed to be sustained. He thought of the diversity in human health and the fact that different populations were typically susceptible to different diseases as a result of their history. What if a rational international system could be organized for potential marriage partners to be able to improve the health of their offspring? Bill was ever the rational policy maker, at least in theory.[20]

On the Japan trip in 1986 Bill's fellow travellers were Naomi Pierce, Andrew Berry, and Mary Jane West Eberhard. Bill was mesmerized by the country and its people (later on he modelled his collected works *Narrow Roads of Gene Land* on what he gleaned from this visit). But when it came to biology in Japan at the time, the situation was rather delicate. The academic world had been dominated by Professor Kenji Imanishi, who advocated a theory of social cooperation and harmony. Conflict was simply not dealt with, because it didn't exist. His observations of animals, be they macaques or freshwater organisms, supported the social cooperation paradigm, and that was also what was being published (and becoming widely accepted). This inbuilt bias set back the field considerably. So in the mid-1980s Hamilton's biggest contribution to the development of evolutionary science in Japan may have been simply to appear in person.[21]

The biological field was slowly trying to pull back from the Imanishi paradigm, but it was still shy about sociobiology. Was that because sociobiology substituted Imanishi's holism for an unfamiliar, reductionist approach, or was it because this alternative approach to evolution recognized (or even seemingly promoted?) selfishness? In any case, the transformation took its time. It seemed to me when I was in Japan in 1989, talking to biologists and giving a presentation about the sociobiology controversy to Professor Hidaka's graduate seminar in Kyoto that the students were unfamiliar with the American controversy about sociobiology. I had to explain the hostile reception that sociobiology had received in the United States, and why. In Japan, in contrast, sociobiology was hot! According to Hidaka, Wilson's book *Sociobiology* was read in a clandestine way, very hush hush. I also learnt from Shigeyuki Aoki that it was still difficult for younger biologists to give sociobiology-based presentations at conferences, because of the dominance of the field by adherents to the old school.[22]

The central question that Hamilton addressed at the Kyoto Symposium in 1986 was the following: why do closely related individuals cooperate less than one would assume, and less related individual cooperate more? This was at the same time a puzzle and an undeniable fact because by now it had been possible to assess relationships with the help of biochemistry. For instance, in beehives sisters turned out to be only 3/8 related to one another, not 3/4 which would have been expected. The task now was to explain this phenomenon. This was also one more reason why the haplodiploidy hypothesis in its strong form did not hold, obviously exceptional relatedness was not required:

> There is in fact a growing opinion that the high relatedness of 3/4 that arises in full sisterships in Hymenoptera due to their male haplodiploidy has not been the most critical factor for the common evolution of eusociality in the Hymenoptera, even though theorists have on the whole upheld the idea that the exceptional relatedness, if present, can strongly favour eusociality.[23]

But why was the relatedness lower than expected? Hamilton had the answer: parasites! The issue had to do with the health of the colony. It was in the interest of the hive as a whole not to have too close relationships between its members, because in that case they would become more easily infected by parasites (pathogens). There were examples of wild colonies destroyed by disease. In other studies the rate of disease appeared correlated with the rate of unification of the genotype.[24] In other words, the point would be to *increase diversity of genotypes*, in order to make it harder for parasites to catch up. One solution would be polyandry, which is particularly pronounced in large colonies of social insects:

> In your life as a larva with half-sibs pressing around you sired by the dozen or so fathers and partitioned away from you only by the delicate wax walls, it is much less likely that any single strain of a deadly bacterium could make a quick culture medium of the whole brood comb.[25]

Therefore, the idea was that of mixed genotypes, and that had actually already been realized in the field of animal husbandry and plant pathology, Hamilton noted, referring to such things as 'multiline' planting of crops. (Interestingly, there almost arose a question about the priority for the idea of polyandry as a way to avoid parasites. Hamilton's Kyoto symposium paper was shortly afterwards followed by a paper by three Cornell biologists making the same point. According to Hamilton, the Cornell trio did a better survey of the evidence, and therefore should get more credit. 'First discoverers' often get too much credit, Hamilton mused—and one can guess what he is referring to here.)[26]

Another possible explanation for the question of how weakly related organisms could be so cooperative, again, was *reciprocity*. Hamilton recognized this as an important factor, at least as a subsequent development. He also pointed to accumulating evidence that cooperation within separated clones served the colony's need for *defence*. In general, the importance of defence in the evolution of social insects was being increasingly recognized, he noted.[27]

What about the evolution of human society? The development of reciprocation would require an appreciation of distant future possibilities, in other words what we call intelligence, Hamilton reasoned. Intelligence, again, opened further ways to social progress, because we could now develop such things as laws and religions to help cooperation and avoid cheating.[28] Laws and religions? This does not sound much like Bill Hamilton. Well, he found a way out:

> The laws and religions are themselves being generated, of course, by memetic rather than by genetic processes of evolution.... But it is well to remember that they are so on the basis of a pattern of human impressionability that was established by ordinary genetical natural selection as it acted on a social primate living in the savannahs of Africa...[29]

But Hamilton did not like what he called 'the memetic revolution' in regard to belief in inevitable medical progress. There was a current enthusiasm for various cures, doctors, and hospitals and a faith that the creation of ever new and better cures would be able to keep pace with any degeneration of the human genome due to mutation. But was this faith justified? 'It is possible to imagine the current trend leading to an increasingly delicate balance, a metastability', he warned.[30] He went on to envision how a sudden crisis would start a chain reaction:

> With the degree of interdependence assumed above, civilizations would have become *superorganisms*...they would begin to suffer from sudden death like metazoans. Correspondingly they would probably have to find ways to compete with each other and reproduce like metazoans and *Astegopteryx* aphid colonies [italics added].[31]

As usual, when Hamilton reasons about the future of humankind he uses a model applicable to some aspect of the general historical process of evolution. For him, this comparison is unproblematic: he uses biological processes as thinking tools. The message Hamilton delivered to the Kyoto symposium was the same as that he would deliver a year later at the Nobel Symposium and elsewhere: Questions like these need consideration! There is a problem with the genome as a whole! We should not allow the pool of deleterious genes to increase. '*Do we want to become mere cogs in the wheel, lose our individuality? Is it inevitably to this end that the human pattern, like the metazoan and the social insect patterns before, had to evolve?*' Hamilton asked, and added: '[I]f we should not want to evolve this way, are we still able to halt the trend that is already in progress, and if so, how? [italics added]'.[32]

It was in the face of these kinds of alarmist projections that Hamilton's colleagues and friends tried to nudge him a little. Bill's former student and good friend Steven Frank wrote in a letter to him that the idea of superorganism was controversial at the time and that Bill would probably be misunderstood and criticized.[33] In another letter he cautioned Bill that the issue of deleterious genes had already been debated and resolved, and that Bill might look out of date in regard to the actual ratio of spread of deleterious genes, and end up being criticized.[34] (It was probably these kinds of reactions Hamilton referred to a little later when he wrote a note to himself that Maynard Smith and Dawkins disapproved of his 'social line').[35]

And there were more requests for contributions to volumes. Hamilton was extremely pleased and honoured to be asked to write a chapter in a book on kin recognition edited by Fletcher and Michener. (Bill had been addressing Michener as early on as his 1964 paper while discussing the origin of eusociality. He considered Michener an exciting writer and possibly the best expert on social Hymenoptera and co-inspirer of his first Brazil trip).[36]

Hamilton was certainly familiar with Michener's strong conviction that it was sufficient for bees to share burrows for eusociality to start building. (He had already commented on this in his 1964 and 1972 papers.) This was now Hamilton's chance to present his alternative vision of how the building of eusociality might be taking place. His focus was on *'evolution mediated by genes of small effects not obligately expressed in their bearers'*.[37] According to Hamilton, under slow evolution of these adaptations, complex traits of altruism could be built up by accumulation of genes. A good example would be for instance the set of traits involved in the adaptive suicide of the worker honeybee:

> Some traits are structural (e.g. barbs...) and some are behavioural (e.g. ferocity). The queen whose safety the acts of suicide serve is carrying the same gene unexpressed, and it is through her breeding success that the genes are maintained... *In situations approaching eusociality, long before caste differences are apparent, there are numerous asymmetries that make conditional expression the natural course, and in my opinion, genes with small effects are likely to be by far the most important*... Conditionality, although mentioned, was insufficiently mentioned in my work [italics added].[38]

Genes with small effects were also likely to be responsible for kin recognition:

> It should be pointed out that although the discrimination has to be a positive adaptation if it exists at all, it need be nothing mysterious and its arrival in a series of small steps is easy to imagine.[39]

Also in this paper Hamilton was dealing with both insect eusociality and human societies. Again treating the hive as a superorganism, this time he addressed the question of *the division of labour*:

> If it makes sense at all to talk about the efficiency of a community—that is, if the community is integrated enough to have 'functions' that can be measured as more or less efficient—then the ideal road to division of labour is probably that taken by the cells of the embryo of the worker honeybees of a hive. *Here every individual carries the code both for all the varied preliminary paths of ontogeny, and for the detailed acts of living*. All these are to combine together for the wellbeing of the superorganism [italics added].[40]

For humans he offered a different model:

> I postulate a mechanism basically analogous to the way interspecies mutualism is expected to arise out of the goalless meandering of interacting species in an ecosystem.[41]

This was not only a time for kin recognition but also for general recognition of the work of Bill Hamilton. In 1987 he became an honorary member of the Uppsala Academy of Science, cited especially for his parasite theory of sexual selection. (Staffan Ulfstrand, not surprisingly, was instrumental in his being awarded this honour).[42] In 1988, Hamilton's original 1964 kin selection paper became a *Current Contents* Classic.[43] That same year he was also elected president of the newly established Ann Arbor-based Human Behavior and Evolution Society. These were all encouraging developments. But all this time Bill Hamilton had something else on his mind that was even more important. That was his quest for a really convincing model for his parasite avoidance theory of sex.

18

Defending the Queen

The big background project for Bill during his early years as an Oxford don was the Parasite Red Queen hypothesis. The move to Oxford had not signalled an end to his work on this, quite the opposite. This hypothesis had to be pinned down for good. There had to be ways in which to make it so compelling that it would meet all kinds of possible criticisms. He had been invited to present his hypothesis at the 1982 Dahlem parasite conference and contributed a chapter in the book of conference papers edited by Robert May. It looked as if people were increasingly starting to consider the role of parasites, both for health and for the evolution of sex. But Hamilton knew that there were still many obstacles to overcome.[1]

One of the major things he wanted to do was to make the model more realistic. So far his model could be shown to work for two hosts and two parasites, and involved two loci. Clearly, in nature the situation was much more complex, involving many hosts and many parasites and many genetic loci. The problem with introducing any realistic-seeming numbers in his model populations, though, was that mathematically, the result easily got out of hand. With all the possible combinations, there simply was not enough computer power to simulate the resulting population genetics. This is why, from the very beginning, Hamilton had felt the need to resort to various visual and even physical means to capture and demonstrate how the host-parasite dynamics worked.[2]

Bill had a wonderful visual imagination and could truly imagine himself into the curves generated by his computer printouts. However, those curves had the limitation of being two-dimensional, hence Bill later resorted to three-dimensional perspex models of host-parasite interaction, which he used to bring with him to illustrate his talks on sex and parasites. His audiences often did not have Bill's visual imagination, and had a hard time seeing exactly what was going on in the curves that Bill himself described as 'yarn messed by a kitten' and others just called 'spaghetti'.[3]

By building mechanical models Bill also made things clearer for himself. Once he baffled his student Marlene Zuk by asking her to play with a physical model to better understand host-parasite interaction. The model he had built consisted of a board with two sets of long screws—for hosts and parasites, respectively—some long rubber bands, and a couple of bicycle spokes! She also had to admit that she did not understand what he meant when he tried to make her visualize another dynamic situation as 'water sloshing in a bath'. It was only when she worked out the dynamics mathematically that she could see his point. (We can assume, too, that she didn't appreciate the long spikes sticking up from underneath Hamilton's 'fakir' version of the bath.) Later Bill was to invoke something called 'the floating ball' model, another three-dimensional (almost four-dimensional) visualization.[4]

At Oxford in the mid-1980s Bill was having a difficult time convincing colleagues that the Parasite Red Queen theory could in fact account for the twofold cost of sexual reproduction. At the same time he felt stuck with his two locus model for host parasite interaction. What to do? Bill presented the problem to Bob Axelrod, with whom he kept in touch and used to meet when he visited Michigan. For Bill, Axelrod was a wizard who came up with solutions to seemingly impossible problems. And Axelrod again waved his magic wand. This is how he recalled the situation:

> The problem was that the equations that described the process were totally intractable when the genetic markers had more than two or three loci. Yet, the whole idea relies on there being many loci so that it would not be trivial for the parasites to match them. When I heard this I responded to Bill with something like. 'No problem, I know a method to simulate the evolution of populations with a lot of genetic markers. The method is called The Genetic Algorithm, and I've already used it to simulate a population of individuals each of whom has seventy genes'.
>
> I explained to Bill that a computer scientist, John Holland, had been inspired by the success of biological evolution in finding 'solutions' to difficult problems by means of competition among an evolving population of agents. Based on the evolutionary analogue, including the possibility for sexual reproduction, Holland developed the Genetic Algorithm as an artificial intelligence technique.

> I could simply turn this technique around and help Bill simulate biological evolution, with or without sex. Since Bill was used to thinking in terms of heterogeneous populations of autonomous individuals, he readily grasped the idea of agent-based modeling.[5]

What was still needed was a computer specialist, and Michigan graduate student Reiko Tanese was invited to be part of the team. The result was an agent-based model with two co-evolving populations: long-lived hosts and short-lived parasites. The parasite population was tracking the constantly changing host population in which hosts with matching marker genes were killed off by parasites, while hosts with different genes survived. Meanwhile, as the previously rare genes became the more common ones, the process would repeat itself.[6]

Agent-based modelling, then, rather than mathematics, was the tool for out-of-equilibrium models. For Bill Hamilton, the Genetic Algorithm was a revelation. (Later, working with other computer specialists, especially Brian Sumida, he was to use the Genetic Algorithm for other projects.)

In the spring of 1986 Bill was working with Bob Axelrod on 'valley crossing'. The trip to the Kyoto conference got in the way of the collaboration, but Bill caught up with the project again when he attended a meeting in Santa Fe in early August on his way back from Kyoto. This marked the start of the model that Bill was to call HAMAX, after Hamilton and Axelrod.[7]

But soon there was trouble. Bill's note from the period after Santa Fe up to the end of October says: 'Busy with HAMAX but it wasn't successful'. On 31 October we learn: 'Bob writes [he] can't find much support for my theory'. The next entry is for 10 November: 'I suggest "soft sel"'. (Soft selection is selection which affects only the most vulnerable genotype and then stops. This is why it is also called 'truncating' selection. Under soft selection, 'the weakest genotype' means weakest in relation to the other genotypes present, that is, it does not have an absolute value. Soft selection typically gives rise to polymorphisms.) Towards the end of November: 'Working on putting in "soft sel"'. 10 December: 'I am much more keen on HAMAX than on earlier valley crossing'. 19 December: 'Bob pretty gloomy. Even soft sel not working'. 22 December: 'I suggest coming over'. And on 9 April 1987 we see an entry with an arrow to 'AA' (Ann Arbor) and the note: '14, 16, 21, 23 HBACHS'. On 11 May there is a note by Bill: 'Bob writes glad I came'.[8]

It was proving difficult to conduct this collaboration across the Atlantic, using the computer at University of Michigan to do all kinds of runs, executed by the computer student Reiko Tanese under the supervision of Bob Axelrod.

The sheer physical distance between the researchers created problems of communication. Bill was typically emailing his Ann Arbor collaborators to do particular computer runs, based on which he would then suggest modifications in the model. At one point they misunderstood one of his emails. He was telling them to 'undo recent changes' and try something else. By 'recent' Axelrod and Tanese thought that Bill meant undoing last *week's* change, while he meant the last *day's* change. This confusion was not detected until a month later.[9]

For these reasons, Bill spent a couple of weeks in the United States in April that year. He returned again for a week in January 1988, and in April and July for a few days each time. By then the work had become more hectic, involving nightly runs and at one point even Reiko's fiancé and his computer were put to work on the project. But Bill was happy. For him, it was fascinating to see the HAMAX model working:

> "On my visits I used to sit in Bob's office watching displays... I could see the Queen's toolkit... gene variation being impoverished... unable to resist.[10]

Bill was extremely pleased with the result.

> Our model had achieved results that others had stated impossible with the tools we were allowing ourselves. Many of the dragons that had oppressed individual models in the past seemed to us to be slain. Part by part, perhaps, not much in our work was new although our explicit modelling of a large number of loci in a Red Queen situation certainly was and the increase of stability of sex that came with the growth of numbers of loci made the most dramatic feature in our results. Simulation in itself admittedly isn't understanding, and various previous papers, including some of my own..., had already drawn attention to the possibilities we were now testing. The simpler analytical discussions and models, however, including again my own, all had severe snags and none showed any chance to be general. Besides testing many loci and many parasites at once—obviously much closer to the real situation...—we had brought in a variable life history that I consider to be much more realistic than is typical in most evolutionary modelling.[11]

Writing up the paper was largely left to Bill, since according to Axelrod, 'it was his idea'.[12]

Having the final product in hand was liberating for Bill. He had worked on the parasite paradigm with the same intensity and obsession as he demonstrated in regard to inclusive fitness. Now it was over. For Bill the effort and the result were comparable in significance. He regarded the HAMAX paper as 'among the most important . . . perhaps my second most important ever'.[13]

In 1988 he also felt that the tide had turned in his favor. It seemed that his fellow scientists had become more willing to seriously consider the parasite avoidance theory of sex. Even a couple of those who had earlier resisted had come round to his way of thinking. Equally important: new empirical research supported the parasite avoidance theory of sex. Now considered a major player in the international discussion about the evolution of sex, Hamilton had been invited to contribute to two books. He was especially pleased with the chapter on host-parasite cycling he had written together with Jon Seger. It was so clearly written that people told him that they finally 'got' the basic idea (Hamilton gives Seger credit for this). And as mentioned earlier, in 1987 Bill had become a member of the Uppsala Academy of Science, which had especially cited his work on sexual selection. He felt appreciated and vindicated. Clearly, his struggles had been worthwhile. But the best thing was that he had finally finished the paper with Axelrod and Tanese. A decade's work of model building was finally over.

It is a long way down from the top of the world to criticism and rejection. The contrast between Bill's sense of accomplishment and triumph in 1988 and his sense of being attacked one year later could hardly have been starker.

In 1988 both Hamilton and Axelrod were convinced about the value of their achievement. As Bill put it, they were both well-known scientists, not known for making mistakes or for being frivolous. After all, hadn't they jointly been awarded the Newcomb-Cleveland prize for the best paper in *Science* in 1981? Bill was certain that their present paper had achieved as much as was scientifically possible at the time:[14]

> From Erasmus Darwin to the present time, sex had repeatedly been saluted as one of biology's supreme problems, perhaps its very greatest. Hence Bob Axelrod and I at first believed that our model, with its realism and its dramatic success under conditions others had deemed impossible for it, was virtually sure to be acceptable to one of the major general scientific journals such as *Science* or *Nature*.[15]

It therefore came as quite a shock when the referees of the paper did not appreciate it.

The paper was first submitted to *Nature*, where it was rejected. Bill immediately proceeded to submit it to *Science*, where it met the same fate. An inquiry into the possibility of submitting to one of the Royal Society journals met with lack of interest.[16]

But this paper was Hamilton's baby! It was the result of a decade of work and hundreds of models. He was determined to make the scientific community appreciate its value. He fought tooth and nail, resubmitting it to the journals with what he believed were satisfactory responses to the referees' queries. Axelrod let Hamilton handle these resubmissions, just as he had left the writing up to him, since he regarded this as largely Hamilton's paper.[17]

In the second volume of *Narrow Roads of Gene Land* Bill devotes a long chapter to this rejection and more commentary can be found in an Appendix. It is clear that this sequence of rejections was a devastating blow. Once again he had taken a chance on *Nature* and been turned down. He could not help but remember earlier rejections he had suffered by that prestigious journal (he had had triumphs too, of course). More puzzling was *Science*, where he had contributed a number of papers, and which 'should' have been happy to publish his and his collaborator's comprehensive host-parasite model. Bill also felt guilty—by now, Reiko Tanese had left science altogether. He hoped it was not because of the fate of their paper.[18]

What was it that the referees did not understand? Hamilton obsesses over this in his autobiographical notes and flatly refuses to make any concessions. It was a good and careful paper, and the referees should have engaged with it in an intelligent way! Instead it seemed as if they did not understand the points that the authors were making:

> One of the puzzles about the dislike, even contempt, of the work...seemed to arouse in my evolutionary peers is that it was as if we had been unable to explain what we were thinking...I believe that the work in the present paper is unusually well explained. In so far as it is true, it is due to Bob Axelrod who has always been a restraining influence on my flights of both prose and fancy. He is insistent that we should say just what would be helpful to our reader on our theme and no more.[19]

This may well have been true—Axelrod's style is indeed crystal clear—but as we know he did not have much involvement in the writing of the paper. What did Axelrod have to say about all this? He was surprised at the stark differences in the way his collaborations with Hamilton had been received. Their work on cooperation in biological systems had been easily accepted, while their simulation of Hamilton's idea that sex could be an adaptation to resist parasites was proving almost impossible to get published.[20]

Axelrod agreed with Bill that the complaints of *Nature*'s referees were mainly about the robustness of their results. The obvious response to such a criticism was to demonstrate that the result in fact held up under a wide range of conditions. This Axelrod and Hamilton did by doing a great number of new runs before resubmitting their paper. Were the referees satisfied? Not at all. As Bill observed:

> [W]e found all our new points left uncommented and the manuscript rejected by the referees even more curtly than before. Two of them indeed dug out new objections they hadn't thought of [the] first time and claimed to see no substantial changes in the rest.[21]

Bill had a ready explanation for the referees' reaction. He guessed that the main reason for their distaste was that while doing new runs, he and Bob did not abandon their initial focus on the Homo-like life history of their model organisms. Axelrod agreed that using human-like organisms had been part of the problem. He also agreed with Bill's Freudian style explanation of the reviewers. (Bill had suggested that the model's human-like organisms were too similar to the reviewers themselves, who resented having their own sexuality explained as an adaptation to avoid parasites!)[22]

However, we learn that although he deferred to Hamilton on this point, Axelrod had not been happy with the inclusion of the humanoid organisms. He had have wanted to make the models as simple and understandable as possible:

> Since we wanted to demonstrate that his theory could explain sexual reproduction in humans, Bill thought that it was important that we include salient characteristics of human reproduction. For instance, he wanted to include the fact that humans are not fertile for the first dozen or so years of their life. I, however, wanted our model to be as simple as possible to make it easier to understand and appreciate. This was the only significant disagreement we ever had. Since it was Bill's theory and Bill's audience, I deferred to his preferences in this regard. So one reason our model might have been so hard to sell is that it included some realistic details that may have obscured the logic of the simulation.[23]

Here we seem to encounter one of Bill Hamilton's recurring problems: his models are too ambitious. He tries to do too many things. There is too much detail and some important point may get lost. So Axelrod's natural deference to Hamilton in this case, suggesting he write the paper because it was his idea, may not have been the obviously best solution. Others presenting Hamilton's ideas have often improved their accessibility. Take for instance the case where he worked with Seger, which resulted in a paper that made his point clear to people—not to mention Dawkins' lucid exposition in *The Selfish Gene*, or Matt Ridley's in *The Red Queen*.

Bill apparently didn't appreciate the tremendous requirements the HAMAX paper put on referees. He demanded that the referees fully appreciate what he and his colleagues were trying to accomplish. To do so, though, they would also have to share Bill's criteria for a successful model, which would presumably lead them to see the advantages of the HAMAX simulation model over a traditional one.

It appears that in this case Bill, as in many others, could not overcome his natural tendency to over-optimize, to do it all at the same time, including even outsmarting the referees as to potential objections (this is why Bill wanted to bring in human-like organisms in the first place). Bill wanted to deliver his absolute best, but his taste was surely for complexity rather than simplicity.[24]

Meanwhile the real problem with the referees may have lain exactly in their particular expectations of 'models'. While Bill was clearly aware of this problem too, his would-be Freudian explanation for the referees' resistance to the HAMAX model suggested either a defence, or that he had a hard time putting himself in the shoes of referees unfamiliar with (or hostile to) agent-based modelling. Axelrod, though, believed he knew what might be going on:

> There are at least two factors, however, which make it harder to sell an agent-based model than a model that can be analyzed mathematically. The first problem is that most reviewers (and potential readers) of theoretical work are familiar with the logic of deductive mathematics, but not with the logic of agent-based modeling. Indeed, they often demand that the results of an agent-based model must be as general as the results of an analytic model... A mathematically inclined reader is likely to want to know how robust the results are, and agent-based modeling may not be able to provide a definitive answer to that question... An agent-based model typically needs to assume specific values for certain parameters in order for the simulation to run... Demands to check new variants of the model as well as new parametric values in the original model can make the review process seem almost endless. What is worse is that a reviewer with a not-so-legitimate problem with the submission can always use 'insufficient' checks for robustness as a cover for a negative review.[25]

But Hamilton and Axelrod kept at it, and tried to take on board the reviewers' criticisms. The paper was rejected twice from *Nature* and once from *Science*. What more could have been done? As Axelrod later reflected:

> We might have tried another tactic. We could have first introduced a minimal version of the model to highlight the essential mechanism to demonstrate that Bill's theory could, in principle, explain how sexual reproduction could overcome its two-fold cost. We could then have provided a more realistic and detailed simulation to show the theory also applied to situations characteristic of human life spans. Unfortunately, the journals we aimed for had such strict page limits that we were not able to write our paper this way.[26]

The resubmissions and corrections had taken over two years. The version that was finally successful and published in the *Proceedings of the National Academy of Sciences* (PNAS) was their fifth. It sailed through quite easily. (It could not have hurt, either, that Bob Axelrod was a member of the National Academy of Sciences.) Finally they had found two reviewers who saw their point. One was even enthusiastic.[27]

But let's look at the larger picture, too. Is Hamilton, perhaps, taking the whole thing too personally? One factor at work was the frustration Bill had personally experienced with his modelling efforts over several years, and the work he had put into writing the paper (it took a year, he said). Quite another was how the HAMAX paper impressed the reviewers at that particular time. Bill was not operating in a vacuum. The response of the referees may well have been 'frequency-dependent', that is, influenced by what other evolutionists were doing at the time. Let's quote Bill himself on this (in a different context, however, than his discussion of the reception of the HAMAX paper). Here's his own overview of the evolutionary interest in sexual reproduction:

> Begun in the 1960s and put into perspective, along with a summary of a lot of the relevant evidence, in 1982 by Graham Bell's massive review, the debate seemed to reach a high point. In 1987 and 1988 two multiauthor books came out on the subject...After that the activity declined a little. The fever to organize 'sex' meetings likewise began to abate.[28]

Now, as we learnt already from Hamilton, 1988 was a year particularly favourable to the host-parasite hypothesis:

> In 1988 we were...at about a high point of the favour that would be given to the host-parasite idea of 'sex itself' by theoretical geneticists, or indeed by any evolutionists, for years to come. Graham Bell of McGill and John Maynard Smith of Sussex had swung across in this direction in 1987 from espousal of other possibilities.[29]

In addition, by now field observations and experimental findings were coming in in support of the parasite hypothesis, and it was encouraging to see these reports published in *Nature* (in 1987 and 1988).[30]

Still, Bill notes, in the end the situation was left unresolved. He describes the various scientific players having 'chewed and pulled the matter about in all directions like a litter of fox cubs gnawing a dead goose, but even then we could hardly decide what to leave and what to swallow'. The abatement of interest was, he believed, attributable to the circumstance that 'we evolutionists were committing ourselves increasingly to particular theories and becoming less inclined to listen to others . . . at least until there was much more evidence.[31]

This was, then, the scene around 1988 for the general group of evolutionists interested in sex. For his own part, although he recognized that there were 'several other reasonable theories', Bill stuck to his parasite paradigm.[32]

What does this tell us? It seems we had a situation with an emerging body of research and various contending theories, but insufficient evidence to persuade evolutionists in one direction or the other. It was a jungle out there—Hamilton was not alone. The field was getting saturated. With so many contending views around, it would seem that a contribution to the evolution of sex towards the end of 1988 would have been especially critically scrutinized. In other words, when Hamilton finally got his act together with his big HAMAX model in October 1988, the editorial interest of *Nature* may well have been to shut down discussion rather than continue it.

The submission in 1988 was especially unfortunate for yet another reason. 1988 was exactly the year that *Nature* published Aleksei Kondrashov's big Deleterious Mutation Hypothesis (DMH), the would-be rival to the Parasite Red Queen one. (According to this theory the reason for sexual reproduction is to eliminate deleterious mutations that tend to accumulate in a population.) That theory in many cases led to the same predictions as the Parasite Red Queen (as Bill later pointed out himself), and both theories were interwoven in many respects. Although Bill thought that an important difference was that the Parasite Red Queen theory could singlehandedly sustain sexuality while the DMH could not, this was not necessarily the general view. Appearing first in print, DMH, a mathematically based model, may have formatted the expectations for what a model ought to look like. HAMAX, on the other hand, was an agent-based simulation model of a quasi chaos type. In order to buy into HAMAX, one would have to understand and appreciate agent-based modelling as a useful substitute for intractable mathematics, and one would have to accept the idea and necessity of non-equilibrium dynamics when it came to host-parasite co-evolution. Both ideas were outside mainstream thinking at the time.[33]

Hamilton devotes much space in the second volume of *Narrow Roads* to the whole HAMAX case, and it is clear that the rejection of the paper was a source of major frustration for him. It irks him particularly that other, much simpler, and unrealistic, models were published in *Nature,* while his was not. He never complains about Kondrashov, whom he treats with great respect, but he finds faults with the authors of two other papers, whom he chooses to call 'X and W', and 'Y and Z', respectively, who did get their models published in *Nature* in 1987 and 1989.

Well, who *should* have been published, asked Bill? Who was most worthy of being published in *Nature* at the time? Let's measure the value in terms of citations received for each paper, he suggested. Checking the Science Citation Index numbers for each article over 10 years, 1988–98, Bill found that the papers that had attained the most interest were HAMAX and Kondrahsov with 17.6 and 15.0 average citations respectively per year, while the two contender pairs, X and W, and Y and Z, had got only 2.9 and 3.8 each. In other words, HAMAX '*should*' have been published in *Nature,* and *Nature* made a mistake by rejecting it. (Hamilton later told his collaborator Brian Sumida, that the Hamilton, Axelrod, and Tanese paper actually *would* have been published in *Nature,* had 'a former colleague' not written a negative review.) Meanwhile he consoled himself with the fact that he obviously could still trigger hostile reactions from referees. That was always something![34]

Who were these mysterious X and W and Y and Z? And why don't we get their names? A little searching in the references in the chapters of the second volume of *Narrow Roads of Gene Land* brings up the only matching candidates for *Nature* for these years. (I came up with Graham Bell and John Maynard Smith for the 1987 paper, and M Kirkpatrick and CD Jenkins for the 1989 one).[35]

The first model by X and W dealt with the crucial aspect of host-parasite co-evolution: the need for gene frequency change over generations. It was artificial, but showed how this idea worked. But this looks a little like déjà vu. As Hamilton was struggling to finish the complex HAMAX there was already another simpler parasite model being submitted.[36]

What happened in 1989, however, was not only the unexpected difficulties with getting the HAMAX model published. Hamilton also experienced what he perceived as a hostile attack from members of his own Oxford Zoology department, something akin to friendly fire. The methodology of his and

Zuk's 1982 paper on sexual selection was unexpectedly being criticized by two of his colleagues, and in *Nature* of all journals! There was among other things a seeming suggestion that the scorer of birds had not been objective when rating their showiness, and may have been aware of the parasite load of different species. A restudy had been arranged, in which the judgement of a single expert was now replaced by a panel of experts consulting bird books, and that study did not produce a positive correlation between showiness and proneness to parasites. Marlene Zuk responded by providing more details about their own previous study and questioning, among other things, the ability of a panel using bird books to assess such important matters as showiness of birds in flight.[37]

This could of course have been regarded as a methodological dispute. Indeed, one of the critics had just co-authored a book on such matters, and the famous Hamilton-Zuk hypothesis was a natural target. But Hamilton was upset. This was not the way scientists ought to behave! Why not settle the matter in person within the departmental walls? Why publicize it in *Nature*? Moreover, this time it was not only Hamilton himself who was being attacked, he was worried about the career of Marlene Zuk, his young collaborator.

Hamilton's frustration with the overall situation that year manifested itself in a number of ways. In 1989 he embarked on a series of worldwide travels, more intense and complicated than before. A large part of 1989 and 1990 was spent travelling, it seems. In 1989, in addition to visiting the Amazon, he visited America a couple of times (once as the President of the Human Behavior and Evolution Society in August), in September and October attended meetings in Rome and Pavia (afterwards walking in the Ligurian Alps), and in November an international conference in Bangalore, India (with field trips and talks in other places). In March 1990 he attended the Kyoto Evolution of Life conference, where he met Allan Wilson and Motoo Kimura, and also visited Nagoya. Soon after he went on an extended expedition to the Amazon and to see Warwick Kerr in Uberlandia, Minas Gerais (where Kerr had now moved), and in July and August travelled to Moscow, Sinferopol, Karachi, Bombay, Brunei, Singapore, and Tokyo, ending up spending some days at Mount Aso (a Buddhist monastery) with Jon Seger and David Cohen in early September. From Brunei he had also been able to fit in a brief visit to see his brother Robert and his wife Tiang, who lived in Borneo.[38] But the most significant thing that happened during this period was that Hamilton seriously considered moving to Princeton. He went there with his whole family for a visit, discussing with his old friend

Peter Grant a position which would involve minimal teaching and then only to graduate students. In the end, the family chose to remain in England, but as Peter Grant said, Princeton came close.[39]

***

Hamilton was acutely aware that his Parasite Red Queen hypothesis, however compelling, was not the only game in town. The competing explanation for the evolution of sexual reproduction was Kondrashov's mutation elimination hypothesis, or as Hamilton called it, the Deleterious Mutation Hypothesis (DMH), or even 'the Black Queen'. That idea had, after all, been published in *Nature*, and so could be regarded as having the scientific establishment's imprimateur. Kondrashov basically argued—and with quantitative demonstration of his reasoning—that the real reason for sexual reproduction is to eliminate deleterious mutations that tend to accumulate in a population.[40] An accumulation of deleterious genes would result in a deterioration of the human genome. Moreover, Kondrashov had suggested something called the Kondrashov threshold, on which his hypothesis depended. The DMH requires a mutation rate of at least one mutation per individual per generation.[41] The problem is that for many organisms, the rate seems to be just around this K limit.

So how did the Parasite Red Queen compare to Kondrashov's Black Queen? Hamilton clearly had a hard time with this question, spilling a lot of ink discussing the merit of each hypothesis. The issue is clearly on his mind in *Narrow Roads of Gene Land*, and he devotes one chapter and one very detailed Appendix to this comparison. The tension is obvious between his subjective preference for his 'baby', the Red Queen, and his objective scientific self which acknowledges that Kondrashov does have a point (in fact, many).

While Hamilton's Appendix provides a comparative list that seems to give an edge to the Parasite Red Queen hypothesis, in the text he presents the hypotheses as 'far from mutually exclusive'.[42] He even suggests that his hypothesis may be paving way for DMH[43] and that '[i]t is obvious that the PRQ and DMH processes cannot avoid interweaving'.[44] Is there, then, some obvious way to determine which of the two theories are operative in a particular case? Here's the problem. The conditions under which the hypotheses work are often very similar. For instance, the mutation clearance idea also works best under soft, truncating selection. Also, the predictions about behaviour are often similar in DMH and Parasite Red Queen, for instance in regard to mate choice.[45] Also, under both the Black Queen and the Red Queen regimes the genotypes are degrading—in the first case because of the accumulation of bad mutations and in the second because of parasites attacking common genes.[46]

There is an important difference, however. In the case of the Parasite Red Queen genes are not eliminated, rather, they are stored for later usage. There is a reserve of old alleles that have already been tested for usefulness in a particular environment. Under this hypothesis, an individual with 'bad' genes can actually be described as the 'guardian of alleles'. Under DMH, however, what is expected is a series of medical interventions.[47]

The idea of gene storage of ancient alleles is important, Hamilton points out, and something that population geneticists have not paid attention to:

> We present a quite new way in which ancient alleles—that is, preserved in polymorphisms for longer than the lives of the species they inhabit, sometimes for much longer even than the spans of genera—may be under preservation. And yet the experts of population genetics, although coming near this topic when they have treated gene preservation under the topic of random variation of environment, have paid selective maintenance by determinately dynamic processes little attention.[48]

He adds:

> Having myself played with the dynamics of not only HAMAX but other dynamical models—dynamics that, as seen from a distance, one might call states of stable preservation even if they always include a jittering of frequencies of some kind...—I have come to believe that theorists ought to pay this kind of preservation much more attention. Increasing numbers indeed begin to. But many still pursue only the older focus on full stabilities. And full stabilities have to be, of course, uncompromisingly hostile to PRQ sex.[49]

This introduces again a point that seems to underlie many of the referees' problems with the Parasite Red Queen paper. The hypothesis is based on an idea of non-equilibrium dynamics, while most of the other models at the time were traditional equilibrium models.

Although Hamilton tried hard to be impartial as to which one of the queens would 'win the Olympics' (his term), and did not see the hypotheses as mutually exclusive, it seems clear that he was emotionally attached to the Red Queen. At the end of the 1990s Bill's former student Laurence Hurst, editing a volume, invited him to write a comment to a review essay which suggested that the parasite Red Queen needed supplementing with the idea of mutational decay and advocated the development of mixed models. Hamilton declined. He was very busy preparing for a trip, he said, and anyway 'he didn't see the point of exploring mixed models until someone could show him that he was wrong'. (Kondrashov, who had also been invited to comment, originally took just the same kind of position, but later did contribute a critique.)[50]

An interesting factual question is of course: where are we now as a human population? Are we close to the Kondrashov threshold or not? Unfortunately, this seems difficult to determine. Hamilton in fact found one calculation showing that for Kondrashov's model to work, the amount of deleterious mutation would actually have to be more than two, rather than one, per person per generation.[51] In any case, although Hamilton in the second volume of *Narrow Roads* certainly worries about accumulating mutation load, he seems much more concerned about the degeneration of the human genome in the absence of selection ('the phase we are in now').[52] The timescale he operates on is large—he almost seems like a population geneticist watching us from Mars. And what he sees is an asteroid about to hit the Earth.[53]

19

In Tune with Nature

Bill Hamilton's enormous frustration with his Red Queen project and the uncomprehending referees did not mean that he put everything else on hold, waiting for the paper to be published. On the contrary, he was keen to start new projects. One of those projects, started in 1989, involved a study of a great scarab carrion beetle in an area close to Manaus, Brazil. Bill was fascinated by this beetle and the fact that a huge horn was present in both males and females. What was this horn being used for? His Finnish student Merja Otronen was studying the habits of this beetle and he went to Manaus to help her with early observations.[1]

Bill had a preliminary hunch about these beetles' general behaviour based on what he knew about their closest relative, a huge dung beetle in South Africa. Still, almost anything might be discovered:

> [A]lmost any social variation or oddity seems possible for this amazing group of animals—maybe they will really turn out to be the insects' first unisex provisioning parents, with all roles except for the laying of the eggs shared equally... What are our beetles going to show? At the moment we know little more than that they do bury carrion, and this was well known even by J H Fabre, from a hen's egg size ball of a smaller congeneric sent him from Argentina a hundred years ago. Then there is the claim in the literature that four of ours buried a 12 kg dog in one night; and then my own rather casual observations in S Brazil how they would come buzzing with sounds like small airborne motorbikes through the dusk of my little town to patches of fresh horse dung and even on one occasion dug holes under a decaying cabbage. All this goes on almost as if it were play... With ours digging tunnels seemed almost effortless. They

> plunge into the forest soil under a chicken almost as if swimming. Presently you see the whole chicken rocking about as they try its weight from below; and I suppose the principle of their burying is to keep finding the pressure points and keep shoveling the soil from under them. They are lovely metallic colour, mixtures of greens, yellows and purples.[2]

This is why it was so much more disappointing for Bill to see so little activity among the carrion beetles during the three weeks that he and his student spent watching them. The beetles seemed uninterested in mating or doing anything—was it the wrong season? He hoped that his student would have better luck for the rest of her four-month watch.[3]

Back in Oxford in 1990 Hamilton received an interesting proposition in regard to the Parasite Red Queen hypothesis. Dieter Ebert, who had become completely fascinated by it, wanted to do a post doc with him. As Ebert put it himself, the Red Queen hypothesis had attracted a lot of attention but very little empirical work had been done: 'Surprisingly, the widespread support that the hypothesis had earned by the early 1990s among many evolutionary ecologists was not based on hard data, but on its plausibility and the increasing recognition that parasites are indeed everywhere. What was actually known about the interactions among hosts and parasites boiled down to only a few studies.'[4]

Not many host-parasite systems, though, were suitable for experimental work. (One that was believed to be usable for testing the Red Queen hypothesis was a system with a eukaryote host and rapidly evolving parasites.) Ebert had been so taken by the whole idea of host-parasite interaction that after his PhD he abandoned his earlier study of the life-history of *Daphnia* (a water flea) for the exciting prospect of testing the Red Queen hypothesis experimentally. After some discussion about its viability, Bill and he decided to try to use *Daphnia* to develop an experimental host-parasite system. (This involved finding both *Daphnia* and suitable parasites, and cultivating them, which was quite a challenge.)[5]

Which predictions of the Red Queen hypothesis could be tested experimentally? Ebert started examining a literature which he thought would give suitable insights—there was a huge amount of information available about serial passage [transmission] experiments:

> Typically, in these experiments, a novel but related, host is artificially infected and the infection is then transmitted from one host individual to the next (e g by

> syringe transfer of blood)... [S]everal studies consistently reported that parasite virulence increased during serial passage experiments as a result of within-host competition, and that this increase in virulence depends on the host genotype. (Parasites passed though one host-type become 'attenuated', i.e. their pathogenic effects are reduced in hosts different from those in which they were passed). Attenuated parasites are useful vaccines...[6]

One of the questions that Ebert and Hamilton asked and answered based on their experiment was 'Why does virulence not increase under normal "non-passage" conditions?' The reason, according to them, was genetic diversity in natural populations. As Ebert explained:

> During serial passage experiments parasites are exposed to a narrow range of host genotypes. Infection of a novel host.... results in parasite attenuation, indicating that growth and virulence are adaptations to the host-genotype in which it evolved. Ebert and Hamilton [this refers to their 1996 paper] proposed that virulence does not usually escalate in natural populations because genetic diversity among hosts prevents the parasite from evolving host genotype-specific virulence.[7]

In other words, it seemed that the *Daphnia* model system experiment could provide important insights about the evolution of virulence.

But soon Bill was back to the Amazon again. In the early 1990s he spent a lot of time there together with Peter Henderson and various students, operating from the Mamirauá Ecological Reserve close to Tefé. The idea for that reserve had come from Marcio Ayres, a native Amazonian from Instituto de Pesquisas da Amazonia (INPA), the Brazilian research institute in Manaus. (Marcio had in fact been working with Warwick Kerr, its director.) Marcio was convinced that Lake Mamirauá, located at the floodplain confluence of the rivers Solimões and Japura, would be ideal for an ecological study of the Amazonian seasonal flooded forest. Both Hamilton and Henderson were captivated by the potential site as they surveyed it on a two-week trip on Marcio's boat. The result was the Mamirauá floating research station, which had its own special expedition boat. Later the 'Centro Itinerante de Educacão Ambiente e Cientifica Bill Hamilton', a school for ecological education, was added. Bill and Peter regarded the whole Mamirauá project as part of a great British tradition (after all, Henry Walter Bates had stayed in Tefé, then called Ega) and managed to obtain generous initial financial support from various British institutions and governmental agencies.[8]

On their exploratory expeditions in the area they divided the labour so that Henderson would concentrate on fish and ecology while Hamilton would be collecting and identifying plants. They also brought some students with them to the Amazon and were able to describe a great number of fish and plants in his way. For Hamilton, though, the main point of the trip was the bolder speculations regarding adaptation that the flooded forest invited. He believed that the plants and animals, being submerged in water at regular intervals, were in fact specially adapted for phenotypical flexibility and lifestyle adaptability. In their joint paper, Hamilton, Henderson, and the Oxford student Will Crampton speculated that the flooded forest might in fact be the cradle for macromutations of various kinds, such as the origin of land animals (via the lung fish) and trees from plants. This habitat does not encourage evolution through speciation, but through *plasticity*. 'Low speciation is not low macromutation; indeed, much remarkable floodplain adaptation is present. Genetically assimilable plasticity often precedes radical novelty.'[9]

And this applies both to the fish and plants:

> Many fish demonstrate striking plasticity in their ability to change their physiology, behaviour and body form in response to anoxia and factors of the floodcycle. In plants, as just one example, diverse rooted life-forms swiftly develop arenchyma and become floaters when they are flooded... Simplified, responsive organisms are a clay from which evolution may be moulding not so much abundance of species as novel forms.[10]

The fish examples are colourful:

> A lungfish that has to live as an adult in an expanse of drying mud as well as living at other times in a shallow lake may find itself, via genetic assimilation, within reach of macroevolutionary slopes that are unattainable to a pure shallow-lake specialist... Likewise, for the tambaqui, *Colossoma macropomum*, which varies from being a predator when young to a seed eater in sparsely populated flooded woodland as an adult. As an adult this fish displays a remarkable further plasticity in switching seasonally from feeding on seeds crushed by strong teeth at high water to planktivory using gill-rakers at low water. Such an example may help in part to explain a related fish group whose origin may indeed be in the floodplain. The biting piranhas (family *Serrasalmidae*) are found currently in many aquatic habitats of tropical South America...[11]

The paper goes on to describe how a particular group-living floodplain piranha species attack large prey in groups in overcrowded *varzea* (flooded forest) low-water lakes while it at high water scavenges flood-stricken victims in the forest, and then ventures the following conjecture: 'Plausibly ancestors of *Serrasalmidae* at some point specialized from the plastic repertoire of a tamaqui to become unique neotenous chunk-biting predators, thus initiating the radiation of form and behaviour that we see'.[12]

The exciting general principle, then, that Hamilton believed the unique flooded forest habitat could illustrate was: 'A habitat's demand for plasticity could well accelerate macroevolution via genetic assimilation within reach of macroevolutionary slopes (provided—the usual difficulty for this concept—plasticity can be sufficiently maintained as the process goes on)'.[13]

In regard to plants, Bill was especially enthralled by 'floating lawns', a collection of three dozen or so free-floating small species forming a plant community. The floating lawn was one of three main types of plants in the *varzea*. Some plants were rooted at the bottom but with floating parts that photosynthesized, other plants were completely under water but floating unrooted, while the floating lawns photosynthesized above the surface.[14]

This is how Peter Henderson summarized the overall ambition of their joint paper (a product of several years of research and discussion):

> In 'Evolution and Diversity in Amazonian Floodplain Communities', we wanted to consider large-scale major evolutionary innovations. We discussed events over millions of years and ranged over both the plant and animal kingdom on both land and in water. We wrote about some of the greatest events in biological evolution, the conquest of land by the vertebrates and the evolution of wood so that plants could become trees and forests could form.... I think our combination of ecological and evolutionary thought within a single paper is interesting and unusual and reflects our combination of complementary skills... If in this paper we rather overextended ourselves, I hope the reader will be generous and remember that it was written while the grandeur of the floodplain and the flash of sunsets over lake Tefé were still clear in our minds.[15]

In 1992–93 the Israeli evolutionist Eviatar ('Eibi') Nevo joined Hamilton for a project on plant genetics in the Amazon, with the idea of contrasting the white and black water of the Amazonian basin, the meeting point of the two rivers Solimões and Japura:

> With Bill, in 1992, we started a long-term population genetic program on the protein and DNA levels in the flooded rainforest of Amazonia, in the Lake Mamirauá Natural Preserve, near Tefé, Brazil.A major goal of this project was to compare and contrast the population-genetic structure, diversity and divergence of taxa, primarily of plants, at the three sharply differentiated Amazonian habitats of *terra firme, varzea,* and *igapo* [these words refer to, respectively, non-flooded habitat and seasonally flooded habitats inside and outside the outermost levees of the river]. The problems we aimed to resolve included the following: What are the modes of speciation in Amazonia? What is the rate and genetic differentiation during speciation in Amazonia? Is the 'adaptationist program' and natural selection tractable in Amazonia? Unfortunately, our dreams of studying all these intriguing problems did not come true for lack of resources.[16]

The plant project did not materialize, but the two colleagues had several discussions and projects going, and together made a remarkable discovery. They found what they called 'global adaptive genetic convergent patterns' between Amazonian and Israeli mole crickets.[17] This discovery might be described as having 'fallen in their laps', because the Amazonian mole crickets 'swarmed in the thousands during their nuptial flights' and fell onto the deck of the boat at night. They found two similar convergent phenomena: 1) high genetic polymorphism but very low heterozygosity, which indicated significant deviations from the Hardy-Weinberg law (which assumes random mating and probabilistic gene distribution); and 2) significant linkage disequilibria ('coupled' genes), at an unprecedented level for outbreeders, and also great intersite differences in Israel. To explain this unusual genetic pattern, found in both types of crickets, they developed a multifactor mathematical model combining three factors that they believed were responsible: niche viability selection, niche choice and assortative mating. 'Simulations based on this model showed that a combination of these three mechanisms may produce the observed distribution of alleles, via selection on a few loci, to affect the entire genome organization', Nevo reported, and added: 'So, quite remarkably and unexpectedly, shared underground lifestyles of gryllotalpids [mole crickets] in Israel and Amazonia resulted in a convergent genetic pattern.'[18] So here again, Bill was revisiting his 'underground theory' (explored in chapter 11), this time with the help of a population geneticist.

***

Bill Hamilton was more than a naturalist. He had an organic connection with the living world, which could sometimes take extreme forms. He wanted to understand how Nature worked, he wanted to become one with her. Nature was his source of inspiration and excitement, she was his true conversation partner, his confidant.

His perspective was a very long one, a true evolutionary scale, would-be geological, and this also gave him a fascination for topography. As mentioned earlier, Bill had learnt from an early age to 'read' chalk lands, such as the Downs surrounding his childhood home in Kent. He knew what to expect in terms of flora and fauna, and later on found quite similar nature in other parts of the world, such as Israel and Italy's Ligurian Alps. One of the great delights of flying for him was watching the changing landscape and being able to see how Nature's forces had been shaping the Earth.[19]

Bill had a truly unusual empathy with Nature. It was not just a strong delight in all living things, his extensive knowledge and interest in all kinds of organisms, or the awe he felt in the rainforest. No, he placed himself in the mind of Nature herself. He wanted to understand her logic, as it applied to the process of evolution, the fate of Earth and the fate of mankind. He believed that Nature had a message for us, if we only listened carefully. At a more concrete level, he empathized with plants, the various insects in his living quarters, wingless fig wasps, even an intestinal worm.[20]

Anthropomorphism was for Hamilton both a metaphysical commitment and a method of research. The reason he felt free to make comparisons, say, between birds and humans in regard to perception of beauty, was that he believed that we are really quite similar at a deep physiological level. This kind of reasoning helped him in his research, too. It was clearly crucial in his view of altruism. For readers whose training has made anthropomorphism taboo, reading statements such as 'the babassu palm embryo is happy to sacrifice itself' or 'why shouldn't plants be altruistic' may have sounded surprising, to say the least.[21]

The flip side of Bill's attraction to Nature was his relative lack of interest in people and his social awkwardness in new situations. But there was more to Bill's strange type of empathy. He wants us to know something more, and he has left us clues in his autobiographical introductions to his collected work in *Narrow Roads of Gene Land*. Not only does he tell us that he is not interested in chit chat with fellow Britons in bars—he prefers natives from the Amazon area—and that he prefers the Amazon to the luxury of Washingon hotels. That's fine, we could probably have guessed that by now. But when he describes how he was deeply moved by a brave yellow ragwort in the London asphalt, and immediately afterwards adds: 'A screaming child would not have touched my heart in the same way', this comes almost as a provocation. It is almost as if his original human capability for empathy had been rewired in order to apply to non-humans.[22]

This relative lack of empathy for humans appears to have fluctuated depending on the situation. For instance, he bought or made for his family and friends things that he thought they would enjoy, and he was often right. It is just that Bill often did not have a clue about how other people felt or interpreted situations, and this sometimes affected his close relationships. As soon as he could use himself as an example, there was no problem. So, for instance, he could understand students and their struggles very well, and helped many of them in significant ways, earning their eternal gratitude. Other times he used incorrect explanatory models and was surprised at the result. This was particularly the case in relation to women. He didn't have the habit of asking or discussing matters very much—instead, he used modelling. Bill's way of coping was modelling a woman on other women he knew, or sometimes using sociobiological models to explain their behaviour (particularly anger). Needless to say, they did not agree with such a far-fetched sounding explanation when they knew very well the immediate reason![23]

If one cannot read 'other minds' very well, another way of figuring out what is going on is close observation of others' behaviour. Bill was an excellent ethologist—see for instance his description of the luminaries among ethologists and anthropologists at the Washington, DC, conference in 1969. And of course Bill had a broad repertory of appropriate social behaviour, he was well brought up and could be very polite. It was probably from his father Archie that he got a particular model of jovial storytelling. Indeed, many know Bill as the entertaining host at dinner parties in his own home, or telling stories in the evenings with colleagues during an expedition. Others know Bill as a retiring figure, comparing him unfavourably, for instance, with someone like John Maynard Smith. But for Bill it very much depended on the company and the situation. In the Amazon and on trips, Bill could be Mr Congeniality. In everyday life when things were not going well Bill could sometimes retreat into morose silence, making his home atmosphere—and dinner table communication—quite awkward.[24]

Bill was aware of these features of his personality, but probably not of the effect it had on others. He talks about himself as a 'lonely bear, getting to become more human as better winds blew'. This ability for self-reflection, though, doesn't mean that he was able to snap out of his moods when they struck. But he knew intuitively what the right therapy was to break the downward spiral toward depression that easily followed when things did not go well and he was left to his own devices—some distraction, some new excitement, a trip. His family knew that he was working and left him alone, they did not know how to cheer him up, and the efforts of his Oxford colleagues and students were not sufficient to improve his mood. Bill knew that he had to regularly get away. In other words, he needed his trips to the Amazon and elsewhere as sheer therapy and to help with the renewal of his creative powers. There he became a new person, speaking a different language, meeting new challenges, and rediscovering other sides of his personality. It had to do with the exotic nature as much as with the expectations of the people he met. In Brazil he could take on a different personality which he had to suppress in the relatively stuffy atmosphere of Oxford—he could be himself, have fun, without being hampered by others' expectations of the famous Oxford scientist. Pictures from the Amazon almost invariably show him looking very happy. (Arguably, almost any type of physical exercise might have worked—when he was younger he decided once to work on a farm for this reason—but by now his requirements had become more complex.)[25]

Bill seems to have reflected on this quite deeply, because in a public lecture in 1993 he made some remarkable statements about himself. There are two types of people, he declared, 'thing' people and 'people' people. He himself belonged to the 'thing' people. And in fact, he pointed out, he would not have made the discoveries he had done if he had not been a 'thing' person. Societies need thing people, thing people are useful—they are often scientists.[26]

This is self-revelation of a quite specific kind. Is there something Bill wants to convey? It seems to me that he is trying to present aspects of himself that fit with (the rather impressionistically defined) Asperger's Syndrome, a mild form of autism. Autism, according to recent medicine, can be a matter of degree, and some forms are highly functional. This form is typical of scientists, and there is now a vogue of post mortem diagnosis of famous scientists with this disorder (Einstein and Newton are on the list). In other words, Hamilton is in good company.

In fact, here and there in *Narrow Roads of Gene Land* there are hints pointing in the direction of Asperger's Syndrome. We learn, significantly, that his own mother, a medical doctor, thought that he was autistic. We also read that his father, upon seeing Bill obsessively sitting for hours throwing a ball into a corner, told him 'Bill, are you batty or what?' Everything seems to fall into place: 'thing' orientation, social awkwardness and lack of social competence, strong interest in patterns, obsessive behaviour. Add to that other typical Asperger features that can be observed in Bill: his focus on detail and his occasionally minuscule handwriting (Newton's notes were also notoriously difficult to read and minuscule), not to mention his tendency to mumble.[27]

Children with this syndrome in a strong form are said not to have a 'theory of mind'—the ability to put oneself in the place of another and understand his/her intentions, thus being able to reason 'from the inside' (something that humans and primates are capable of). Some of the inability to develop proper social skills may have to do with inability to read non-verbal signals correctly. It has also been suggested that the brains of children with this particular disability have a particular feature stemming from early development: they are making too many connections. (This presumably leads to excessive focus on details.) The list of supposed Asperger's symptoms vary, but there are some recognized authorities in this area, such as Simon Baron-Cohen at Cambridge. Other psychologists find the diagnosis too vague and potentially more harmful than useful. The current enthusiasm to diagnose children—and famous people—with Asperger's may lessen over time.[28]

Two people who knew him well have suggested to me that Bill Hamilton had a touch of Asperger's: his student Bernie Crespi and the English sociologist Christopher Badcock. But others disagree—his sister Janet and his Finnish colleague, entomologist Rainer Rosengren.[29]

I think there are too many indications that Hamilton was not really a genuine Asperger's candidate, although he may have been fascinated with the idea, and it certainly correlated with some part of his personality. Let's take a closer look at different aspects of Hamilton's complex personality.

The standard description of Bill has him as mild-mannered and absent-minded, shy and self-effacing. The stories about Bill's lecturing style are legion. He is described as scribbling blackboards full of equations, talking towards the board in a voice bordering on the inaudible. Once he didn't have a pointer, so he used the microphone as a pointer instead—with predictable results. And more than once there was a near stampede to the door as soon as the lights went out so that he could show slides.[30]

What was going on here? I believe that although Bill *looked* as if he was lecturing, he was actually using the lecture as a vehicle for thought. Bill had the same attitude to his public lectures as he had to many other things—he was interested in new things, not a rehash of the old. This meant that in his lectures he was often operating at the horizon of his own knowledge. He was not on well-trodden ground, but rather on the edge of his current research—which meant that he could surprise himself if some unexpected thought emerged. That is why we have those reports of Bill stopping in the middle of his own presentation for up to two long minutes, trying to figure out a sudden puzzle.[31]

Here then we have something of a theoretical, often mumbling, oracle interpreting Nature's secrets for us in the strange language of population genetics. This is the man sitting among heaps of papers totally oblivious to the world, struggling to crack a particular puzzle. The other side of this shyness and reticence was suppressed feelings of anger and injury, sometimes resulting in surprising outbursts to trusted friends.[32]

While many saw Hamilton primarily as a mathematical modeller, Bill saw himself primarily as a naturalist. It was just that he wanted to express his conclusions about general principles in the most scientific language he knew—Neo-Darwinian population genetics. This nature-loving Bill is easily compatible with an emotional and sensitive Bill reciting poetry, citing Greek mythology, appreciating art, and writing essays and poems himself. The themes are often dark. As we saw, Bill's early poems and essays when he was a schoolboy seem to have been largely about such things as death and apocalypse, and insidious struggle in nature. This is also the Bill that later cries when he thinks about the fate of mankind, recognizing as his own duty the need to tell us all the truth.[33]

But there is another side to Bill's love for Nature: Bill the expedition man! This is a fun-loving and risk-taking fellow, a man climbing trees and jumping over fences, feeling and looking fit even in his 60s, loving to rough it in the wilderness and test his problem-solving skills on practical matters as they arise. This is a man who lets his hair grow, who likes to run around barefoot, who is looking for adventure, and if he doesn't have it naturally, creates it. Also, he is a first class raconteur—perhaps sitting with a glass of whiskey at the camp in the evening, telling almost unbelievable stories. This is the Bill known to those who followed him on expeditions.[34]

How, then, did he combine his various sides? I believe he actually kept cycling between two main modes throughout his life, the theoretical mode and the expedition one. The first one involves intense theoretical concentration—especially the mode that he was in during the time of his two main 'obsessions' (his own term), inclusive fitness and the parasite paradigm. I want to call these modes 'Kafka' and 'Bates'.

Kafka was the author that Bill read at Cambridge as an undergraduate as he struggled to understand Fisher's even more difficult *Genetical Theory of Natural Selection*. 'Kafka' will represent the frustration Bill often felt when trying to solve a problem, feeling trapped in wrong thinking and looking for a way out.[35]

In his collected works, Bill uses a Kafka quote that seems to have been helpful:

> It is not necessary that you leave the house. Remain at your table and listen. Do not even listen, only wait. Do not even wait, be wholly still and alone. The world will present itself to you for its unmasking, it can do no other, in ecstasy it will writhe at your feet.[36]

Bill interpreted this as encouragement to think more about what he already knew rather than gather new information.[37]

The second mode, 'Bates', stands for Henry Walter Bates, the early explorer of the Amazon and one of Bill's great heroes. This mode represents the pioneering spirit, daring and persistence in the face of adversity and especially perhaps the total freedom from convention that Bill associated with this man. As Bill wrote in a letter to a friend:

> I have always loved the book of Bates about the Amazon, and recently read something biographical that made me see him as even more of a fellow spirit. Someone wrote that part of what may have made him separate from Wallace was not any explicit quarrel, but rather that Bates' taste for 'going native' ran well ahead of Wallace's, and that even before they both left Belém, the respectable townspeople and expatriates began no longer raising their hats to Bates in the street because he looked so eccentric. Yet it is obvious that the people of the backwoods loved and accepted him and taught him correspondingly from their lore: being barefooted and in rags, as he says he often was at times, probably seemed nothing strange to those who were commonly the same way. Well if I could be as well liked and as tenacious in that little town (Ega as he called it, now Tefé) which he first put on the scientific map, I would be proud.[38]

How does cycling enter this story? Let's start with 'Kafka'. Bill was the type of person who tended to go deeper and deeper into a particular problem, totally absorbed, but it was exhausting and lonely, driving him into obsession and depression. This was the fullblown 'Kafka' mode. The best example is (what he himself called) his four-year 'absolute obsession' with inclusive fitness. But after a while he knew that he would have to snap out of this mode. He needed to get away. So here he entered into the 'Bates' mode. A new environment would be bound to stimulate him and give him new ideas. However, this meant the beginning of the next cycle: the insights gathered during his travel would necessitate a new phase of intense theoretical work, which in turn would exhaust him, which would again require distraction by travel, which...and so on. Throughout Hamilton's life we see examples of this alternation or cycling between the two modes.

We have an interesting testimony of the two Bills from Bill's last Brazilian student, who first accompanied him on expeditions in the Amazon and later saw him as a professor at Oxford. Servio Ribeiro joined Bill and his student Steve Harris on their expedition to Mamirauá in January 1991. One of Ribeiro's first descriptions of Bill turns into an instant Bill story. This is Bill in top form in 'Bates' mode:

> An impressive example of his energy was his performance when our boat almost sank in a large channel of the Japura river. We just had time to drive the boat onto the riverbank, where we spent 12 hours from night to day bailing water from the deck and trying to plug the leak. At a certain point, Bill jumped into the dirty water with a bed sheet in an attempt to block the hole in the bottom of the boat. While he was under the water another boat with an American scientist from the New York Botanical Gardens and a Brazilian scientist from Instituto de Pesquisas da Amazonia [INPA], approached us. At that time I did not speak English (Bill used to speak fluent Portuguese), and so I tried to explain the situation to what turned out to be a very selfish Brazilian person. As he didn't realize how important my boat mate was, he decided not to delay the trip of his foreign counterpart and left us behind, with no help. During this conversation Bill stayed under water much longer than expected. Later he learnt about the incident with the other boat and was so upset that he almost turned the story into a note to *Nature* about the lack of solidarity among scientists in the Amazon basin (the note was actually avoided in the last moment due to an apology from the American botanist, who did not speak Portuguese and so did not follow my conversation with his partner).[39]

Ribeiro describes how they then spent nine days at the Mamirauá floating station and went on a two-day trip to a complex set of channels ending up in the Amana Lake, where they stayed for six days, after which they returned to Mamirauá. Travelling with Bill meant being subjected to a lot of amazing experiences for a naturalist, because 'any form of life would be captured by his eyes, touch, or taste. In the position of a student I was invited to taste all kinds of fruits, but also strange things like the flowers of *Spilanthes acmella* (Heliantheae), which turned my tongue senseless due to its high concentration of alkaloids'.[40]

Hamilton was wonderful in his role as Servio's tutor: 'he tirelessly fed and stimulated my own attention toward nice isolated natural events'. Among these were a tropical salix species, one of the largest moths in the world, and *Azteca* nests and their structure. Bill also explained how seeing the rare black-faced uacari monkey in a malaria-free area supported his and Mario Ayres' hypothesis that the red-faced uacari had evolved to advertise resistance to malaria. During this trip Bill came up with a possible PhD project for Servio after noticing a particular fast-growing species of the *Tabebuia* tree that he had earlier seen in Bangalore, India. Maybe the high growth rate was connected to resistance to disease and herbivores? This hypothesis was worth testing. But a few years later, when Servio as a PhD student visited Bill in Oxford, he got a big surprise. 'I found myself in contact with a different Bill. This time speaking in English I was re-introduced to a spirited, though shy, man.' He also mentioned Bill's absorption in his own thoughts and the special warmth that Hamilton had for his Brazilian students.[41]

Here, then, the same person was able to experience both of Hamilton's styles—in this case in the order 'Bates' and then 'Kafka'—and how he and his Brazilian countrymen, through their mere presence, were able to change Hamilton's 'definition of the situation', transporting him once again to the Amazonian rainforest and to the happy version of himself that naturally emerged there.[42]

20

Truth at any Price

For Bill Hamilton, the early 1990s were full of surprises. The first was the news that his wife had managed to find a full-time job as a dentist. For the Hamilton family, always on a tight budget, that was great news. The problem was the job location: it was in the Orkney Islands!

Christine had not been able to work in the United States and had been looking forward to practising dentistry again, especially as the children grew older. But it had been difficult for her to find a permanent job anywhere near Oxford. The type of dentistry that she felt comfortable with was more community-orientated, not highly competitive, fast paced private practice. She had had some temporary positions, including one that involved treating drug addicts in a mobile dentistry clinic, but that clinic-bus had been raided and vandalized in 1991. In fact, the incident had made the local news.[1]

Christine really wanted to work. In the first place the family needed more money. (She knew, because she was handling the accounts.) With only Bill's salary they had not been able to buy a suitable house in expensive Oxford. Also, she was turning 50 and felt it was high time for her to start earning in order to be able to accumulate any pension at all. And finally, this was her profession and she wanted to practise it, especially after her older girls were away at school. As she wondered if any employer would consider her seriously for a job, a friend challenged her to go ahead and just apply for a job and see what happened. Almost as a joke Christine took the challenge and responded to a position as a dentist in the Orkney Islands.[2]

To her surprise she was invited up there for an interview. She did not take this very seriously, but still, she was curious so she went. However, when she saw the islands and the dramatically changing colours and moods of the water and sky she was overwhelmed—'it was so beautiful!'. The work as described seemed quite exciting too. She was to be a 'flying dentist', a role which involved darting from island to island by aeroplane, giving dental care to its less mobile inhabitants. And she would get a real salary! She really wanted to give the position a try, at least for a couple of years. Her youngest daughter Rowie could come with her, and she and Bill were quite used to being away from each other for months. So in May 1992 Bill drove her and Rowie up to the Orkneys, and helped her to settle in.[3]

This all sounded like a rational experiment, and intellectually Bill could see Christine's point. But he had not anticipated how he would feel being left home alone. (He was often surprised by his own emotional reactions.) According to people around him at the time, Bill responded badly. He was used to being taken care of in various ways, allowing him to concentrate on his research. He felt abandoned. This became apparent to everyone as he started letting his hair grow. And longer it grew, white and thick. At one point he was even sporting an Alice-in-Wonderland head band to keep it in place. A visiting post doc entering the departmental tea room at Oxford at this time wondered who on earth that funny-looking granny was before someone told him it was Bill Hamilton. There was a strange fascination with Bill's hair in his department, especially after his cryptic-sounding pronouncement 'I will cut my hair when Christine comes back'. Far from being an allusion to some fairy tale, this was the simple truth: it was Christine that used to cut his hair.[4]

Bill was looking miserable. His students and post docs tried to cheer him up, taking him out to dinner or going with him to the bookshops of Hay-on-Wye, a small town on the Welsh border where one could buy marvellous second-hand textbooks, one of Bill's great passions. Taking him to dinner also seemed a good idea after he was spotted attacking a big chunk of frozen meat with an axe in his kitchen. Indeed, there was some concern that he was too keen on the pheasants employed in Nigella Hillgarth's experiments (she was following up the Hamilton-Zuk hypothesis). But soon Bill was off again to the Amazon and the Mamirauá Reserve, and there nobody cared about his hair.[5]

An even bigger surprise was the number of major prizes that he received the following year. One person to whom Bill would write of his news was his close friend Yura Ulehla. (Yura belonged to those friends and colleagues of Hamilton's who were always positive and kept encouraging and assuring him of his abilities, whatever the situation. Hamilton desperately needed such people around him, or at least as correspondents.) This was how he broke the good news to his friend and his wife Blanca in his Christmas letter 1993:

> When I visited you a few years ago Yura said something to the effect that he would not be surprised for me to win something like a Nobel Prize in the not distant future. I was flattered by the idea of course but thought it pretty wild. But during the past year I have won what seems roughly the equivalent of two Nobel Prizes!!!!! The first was the Crafoord Prize of Sweden which I shared with Seymour Benzer, a Drosophila behaviour geneticist of Berkeley. This prize series was set up explicitly to fill in gaps in the subjects of the Nobel Prizes in science. The value is not quite as much but still very, very substantial. It is administered by the Swedish Academy of Sciences like the N Prizes. While I was still staggering from the surprise of hearing about that one last Spring I heard that I had also won the Kyoto Prize for Basic Sciences, which is part of a recently created series that I hadn't even heard of. And this time I was a sole winner in the field of basic sciences (with the topic this year roughly in my field—whole organism biology and genetics or roughly that) and the value was approximately 2.5 x as much as the other one!!! Having travelled to Sweden and Japan to collect them in late September and mid November respectively I am suddenly quite a rich man and wondering quite hard what to do with it all.
>
> Christine came with me to both prize givings and my sister Mary came to the second one in Japan as well. I tried hard to persuade my mother to one of them but she wouldn't... Both prize ceremonies were very grand and the Japanese one almost unimaginably so. I socially climbed between the continents from chatting to a mere King in Sweden to chatting to an Emperor in Japan. The latter, though, did not actually hand out the prize there, merely met us and talked for a half hour or so in the palace in Tokyo on the last day of our ten day visit. But on the way, at the prize ceremony, I met his cousin Prince Takamodo. These two were actually quite interested to hear about my work and I am on the point of fulfilling my promise to the Princess of sending them a set of my papers about which she especially wanted to know more.
>
> The Kyoto Prizes cover a pure science topic, an applied science, and the arts each year. This year the other winners were Jack Kilby who is the inventor of the printed circuit and former director of Texas Instruments, and Polish composer Withold Lutaslawski (or spelling to that effect)! I confess I hadn't heard of either of them before the prize announcement (equally I'm sure they hadn't heard of me). They were 70 and 80 respectively so I was the youngster of the batch.

> Lutaslawski is a modern composer who allows a lot of freedom to his instruments to play what they feel inspired [to] at least within a certain framework that he imposes. Not being musical I do not quite know how it works, but since coming back I have listened to one piece and it sounds not bad, and certainly not so wild as I had expected from its so called 'alatory' [sic] component. Anyway both were very nice people to be receiving prizes with, as was Seymour Benzer at the other party. Kilby had brought his sister and though she was quite a bit older she got along very well with my sister and Christine. He also had his two daughters and their whole families. No one apart from wives were being paid for by the Inamori Prize Foundation so what his party could be costing to keep in Kyoto's most expensive hotel I can hardly imagine. The suite allotted to Christine and me was costing about $750 per night (so that at the expense of looking a cheapskate more than a prizewinner I quickly moved Mary in with us at the cost of a few dollars for an extra bed in our living room). However the cost probably did not mean much to Kilby who is probably a millionaire many times over, since I think he makes a patent royalty on every circuit and computer chip ever printed—including after a legal struggle that seems to cause no rancour now on either side, most of those made in Japan.[6]

Hamilton goes on to describe the meticulous planning of the Japanese prize occasion and how they were taken care of by an extremely efficient female guide. Also, he is very impressed by the kindness of the Japanese people. He ends his letter thanking his friend 'for predicting and wishing and perhaps thereby causing my imminent unbelievable success'. And then in a handwritten PS he says he realizes that 'winning prizes like this probably includes quite a large element of getting to be popular among one's academic peers' and thanks Yura 'for helping to make me into a social being'. He also muses that all those seminars he has given over the years have borne fruit.[7]

The story about Hamilton getting the phone call about the Kyoto prize is a typical 'Bill tale'. Here is an account from someone who happened to witness it:

> That phone call on the Kyoto prize was funny because, despite the fact that he was being offered a very prestigious prize as an exceptional scientist, his reaction to the Japanese person on the phone was that of a little school boy, worriedly taking notes on the materials he was required to send (a photo of his head with certain dimensions, etc). And then immediately, within seconds of hanging up, with no gloating or celebrating, he was thinking out loud, calculating the number of pounds from the yen, and happily realizing that now he can buy that

house that Christine wants. Then back to worrying where to get the photo done, how to send it—as if he wouldn't get the prize after all if this wasn't done exactly and on time.[8]

This is clearly a very 'Billish' story. We learn here that there was enough prize money to buy a house and that Bill considered it at the time. Buying a house was also on Bill's mind in his earlier note to Staffan Ulfstrand, where he thanked him for supporting him for the Crafoord prize (Bill said he had just heard from someone that he had won)! '[M]aybe [I] can buy a house at last and lure my wife back from the Orkney Islands! Endless possibilities!', he told Ulfstrand in that letter.[9]

But by the end of 1993, Bill reconsidered the house buying idea—it did not seem practical. As he told Yura and his wife in that year's Christmas letter:

> Two years ago with Christine still here we would almost certainly have bought a likeable house in the Oxford area, a thing she was always yearning for (we have a very nice house that we rent but it seems that that is not at all the same as one we owned, and most women tell me this would be the same with them). But as you probably remember from my last Christmas letter Christine is now living in Orkney, and even before the news of the first prize had commenced buying, with my consent, a tiny farm with about 11 hectares on the main island. About a week ago she moved into it and although the dentist contract she is working on comes to an end next May I don't see much sign that she intends to return south when it does end, rather I have the impression that she wants to renew. So what with her there and two of our daughters flown, the prizes seem just a year late for any serious house buying and perhaps they are even too late to glue our centrifugal family back together. Don't think from this that we are at war in any way or even that we are necessarily permanently separated. E.g. I am flying up to Orkney for Christmas in about a week, and our girls 2 and 3 (Ruth and Rowena; Helen being in Brazil) will be converging there too. Nevertheless the family does seem to be slowly floating apart. I am not sure that I want to live in Orkney even in retirement. I am strongly urging Christine to become a wind farmer since wind is an abundant commodity up there.[10]

What was Christine's idea when she bought a house in Orkney? To her it made clear sense to buy instead of paying rent as long as she worked there. The idea of a large orchard was particularly appealing, because she was interested in conducting various horticultural experiments and there were government grants available for such endeavours. On their travels, she and Bill had been collecting seeds for trees, and had already been planting Chilean trees on the west coast of Ireland. Christine's father was an avid horticulturalist, and it was in Ireland that they had so far done their tree planting experiments. She also had thought that the Orkney house might become her and Bill's retirement home. And if he didn't want that, there was a Plan B: it could become a holiday home for the family. And it could always be rented out in the meantime. For Christine this whole thing made good economic sense. For her owning a house did not necessarily imply that she had decided to settle down there. For Bill, on the other hand, owning a house had a deep emotional meaning. It implied a serious commitment, it meant settling down. And that was partly why he had been ambivalent about buying a house in Oxford in the first place.[11]

Ironically, now that he finally could afford to buy a house for the family in Oxford it was making less and less sense. For some time already, he had not been very happy there. He still smarted from the lack of support or outright betrayal that he felt he had suffered in conjunction with his parasite theory of sex. He could not really see himself staying there after retirement. Why then buy a house there? He needed to find out first where he wanted to be. And that was a good question, though not imminent. (At this point Bill was 56 years old and the mandatory retirement age of 65 some nine years away.)

Finally, it may well be that the only place that Bill ever regarded as home was Oaklea. All other arrangements for him were secondary. An unwillingness to commit himself to any other home might explain why Bill from the very beginning was less eager than his wife to buy a house in Oxford.

***

Hamilton's reception of the Crafoord prize was given due attention in the Swedish press. Here was someone who was speaking up for altruism rather than selfishness, and who had scientifically shown its benefits. This, in a nutshell, is how the press conveyed Hamilton's contribution (various newspaper reports seemed to build much on the same sources).[12]

The emphasis on altruism fitted Swedish general opinion hand in glove. Sweden was a country where ideas about 'the selfish gene' had been received with particular hostility, as had Wilsonian sociobiology during the heat of the controversy. It seemed that in Sweden public opinion fell naturally on the side of the critics of sociobiology and just as in the United States, most people got to know about sociobiology through the polemical writings in the controversy, and many therefore 'read' *Sociobiology* and *The Selfish Gene* through the eyes of the critics. In a country where social democracy had reigned for so long (even if not always the politics of the party in power, it was the underpinning of the nation's value system), with its concern for social equality and social solidarity, and based on the idea of the public good any seeming justification of selfishness was anathema. The perceived nasty political connotations of sociobiology as painted by its critics held as much sway in Sweden as in the United States.[13]

For the socially oriented Swedish public, it was the idea of 'the selfish gene' that best captured the wrongheaded overall thinking of sociobiology. Such a title appeared to legitimize individual selfishness and unregulated capitalism in a country with a state religion of Lutheran Protestantism and an old state capitalist tradition. It essentially undermined everything that Swedishness was about. But here now came somebody with an alternative message—Bill Hamilton. With Hamilton, altruism was alive and well and the way to go scientifically too!

This kind of ideological intrapolation from perceived consequences back to the underlying science was probably what caused the Swedish press to present Hamilton's contribution as an *alternative* to the despised idea of the selfish gene. This was of course ironical, and that in a multiple sense. Scientifically Hamilton's idea of altruism and Dawkins' idea of the selfish gene were one and the same, and the person who originally came up with the gene's eye view (which Dawkins later used for his 'selfish gene' approach) was none other than Hamilton himself! Moreover, Hamilton had endorsed Dawkins' further elaboration of his idea and had strongly rebutted Lewontin's review of *The Selfish Gene*. So, if anything, Bill was a defender of 'selfish gene thinking' (as a heuristic tool in science, that is, not as social ideology). And that was also how other biologists typically interpreted the title of Dawkins' book.[14]

In addition, though, Hamilton had a conception of selfish *behaviour*, which he saw as an opposite to altruistic behaviour. This was quite a separate matter from the idea of selfish genes. As we have seen, in his original 1964 paper (and his dissertation) Hamilton considered four different types of logically possible social behaviour—selfishness, competition/cooperation, altruism, and 'stupidity' (later changed to 'spite'). (This is why his 1964 paper is called 'The Genetical Evolution of Social Behaviour', not 'The Genetical Evolution of Altruistic Behaviour'.)[15]

Therefore, when Bill was interviewed about his ideas by Swedish journalists he felt totally free to discuss selfish and altruistic behaviours—by which he was probably just reinforcing the popular misconception about selfish genes without knowing it. The press somehow ended up seeing altruism as a challenge to

Darwinian theory rather than as the intended solution to Darwin's big puzzle. As the journalists further pressed Bill about his own religious views, he explained that he was not religious but had a strong sense of morality.

Overall, Bill passed the Swedish trial with flying colours, and seems to have enjoyed his visit. There is a picture of the Swedish king presenting him with the Crafoord medal. He got to sit next to Queen Silvia at an official dinner and they seem to have hit it off extremely well. Of course, Silvia's mother was Brazilian, and one can only guess at the excitement of the Brazilian and Portuguese connection. (Afterwards, Bill was spotted buying various types of souvenirs with Queen Silvia's picture on it).[16]

Before the award ceremony Hamilton was giving a lecture of a more informal kind. He surprised his travel companion and fellow lecturer Laurence Hurst by talking about James Lovelock's Gaia hypothesis. This controversial idea was on his mind at the time and he was struggling with a way to reconcile it with evolutionary theory. Ever the rebel, he found it interesting, and he may particularly have enjoyed the thought that his Oxford colleagues did not find this theory a proper one to pursue. On the other hand, however much he would have liked to support the theory, Bill was unhappy with the fact that he could not make it into an evolutionary theory. In order to have an evolutionary theory, there was the necessity for some organism or entity that reproduced itself and in whose interest it would be to have a system like Gaia. He had not found that so far. Was it even possible? (He was, in fact, to hit on the solution a couple of years later.) In any case, this was one of Bill's better lectures, inspired and entertaining, although it represented the forefront of his current research.[17]

In conjunction with the Crafoord prize, an ecology tour in the beautiful Stockholm archipelago was arranged for Bill. He totally amazed the 'native' biologists with his extremely detailed knowledge of the Swedish flora and fauna—*their* supposed area of expertise.[18]

***

In the early 1990s Hamilton undertook a serious task. He had been approached by some individuals who wanted support from scientists in regard to the so-called polio vaccine (OPV) theory of the origin of AIDS. (That theory suggests that the origin of the AIDS epidemic was accidentally contaminated polio vaccine in Africa in the 1950s.) He took them seriously, and based on what he found, started to think that this theory merited more serious consideration and needed to be tested. However, the original article had appeared in the *Rolling*

Stone magazine, not a scientific journal, and the person in charge of the original vaccination campaign, still active, had threatened that magazine with a lawsuit. Surely that was not the proper response of a scientist! Hamilton decided to take it upon himself to examine the available information and write a critical analysis of the handling of the case. The original article had been dismissed, but Bill Hamilton could not be dismissed so easily! He sent in his letter 'AIDS theory vs. Lawsuit' to the Letter Editor of *Science*, Christine Gilbert, with a separate cover note on 17 January 1994.[19]

The following excerpt gives the gist of his cover note to the letter:

> Dear Madam,
> I am submitting the enclosed letter for publication as a letter in Science. It is rather long but I hope you will agree that its content, if sound, justifies this length. It concerns (a) a threat to the scientific approach to knowledge, and (b) a hypothesis of potentially enormous importance.
>
> ...
>
> I am aware that the very topic the letter treats has now quite a long history of rejection and even near ridicule in Science (as also in Nature), an attitude which has long seemed to me to be not at all justified by any evidence. Science's uncritical (and even unedited, as shown by the mess with references) publication of Koprowski's rebuttal, and some of your other pieces on the AIDS-polio issue with much the same flavour, are somehow characteristic of how your magazine that has risen to be, for doubtless excellent reasons during the rise, an establishment organ of Science.... Is this line however, justified by the general slant of the evidence in this case?
>
> ...
>
> I do not usually try to explain to editors why they should take any particular notice of what I submit but perhaps in this case, because Koprowski may be considered on a level with a Nobel prize winner (probably having had a near miss to join Salk and Sabin after his own magnificently successful polio vaccination campaigns), I should try. I will therefore mention that in 1992 and 1993 within twelve calendar months I won three large international prizes for my work in evolution theory. They were the Wander Prize of the University of Bern, the Crafoord Prize of the Swedish Academy of Sciences and the Kyoto Prize of the Inamori Foundation of Japan. The Crafoord Prizes are intended to fill subject gaps between the Nobel prizes and to be equivalent to them; the Kyoto Prize series has a similar aim. The total sum I received in the year was $385,000. I also mention being the winner of the Newcomb-Cleveland award from the AAAS

> for my paper in Science with R. Axelrod in 1982. My first Science article, which was on sex ratios in 1967, was prominent in my citation for the Kyoto prize, and I am still thankful to the journal for the publicity it gave that paper. If you look only at the papers I have published with you, on the whole you I think you will agree that I have a good record of being ahead of my time with evolutionary truths. I suspect my record will continue with an intuition I have about the evolution and species jump made by what is now HIV-1.
>
> In short I hope you will give the enclosed letter careful consideration.[20]

The response from *Science* was polite but negative. Hamilton persisted. He wrote a letter to the editor-in- chief, Daniel Koshland, asking him to reconsider:

> As you know the issue in the background here has very wide ramifications and implications for human safety as well as for the conduct of science in general. The contention is the possibility that the AIDS pandemic originated from contaminated vaccines in the early polio campaign in Africa in the late '50s. You may remember that in my submitting letter I gave two reasons why Science should very seriously consider publishing my text or something close to it. These were that I discussed '(a) threat to the scientific approach to knowledge (referring Koprowski's resort to a lawsuit aimed to intimidate authors and publishers away from any hypothesis that might reflect adversely on his work), and (b) a hypothesis of potentially enormous importance.' The second issue implies of course that if AIDS did indeed originate in the way suggested, a very thorough reconsideration of *all ways* in which medical procedures can conceivably facilitate zoonosic transfer of diseases in the future is essential. Bad as it is the AIDS epidemic could be just a dire warning. No issue currently facing humanity (other than possibly the reverse danger, overpopulation) can be more serious than the possibility of starting a disease as deadly as AIDS but with, say, the infectiousness of 'flu; and yet an event like this is just what we are inviting if procedures like whole organ transplants from animals into humans continue without assessment of the zoonosis risks involved.
>
> ...
>
> I mentioned my recent scientific honours in my first letter just to emphasise that I am not a crank since people who try to publish on this hypothesis are always being treated as if they were cranks (see Koprowski's letter plus my critique of it in my first letter; see also Science's various editorial comments). I am just a scientist with common sense plus what might be called old fashioned

standards plus a feeling that intelligent lay people should be encouraged to participate in scientific debates, not shut out. Among the standards I support is one that says that every idea has to be assessed on its rational merits and quite independently of vested interests, power structures, reputations and the like. There are innumerable medically oriented scientists who are far better qualified to comment than I am but, firstly, they are part of a clique that sees, consciously or unconsciously, a short term advantage in dismissing the hypothesis in question because it threatens the underlying prestige of their discipline, and secondly if they do speak out on the basis of common sense and what they find they are likely to be actually oppressed by their hierarchy. This of course has happened in the case of Dr. Eddy and the discovery of simian virus SV40 originating from the polio vaccinations, and I could cite others.

Here in my own department I am finding people far better qualified to investigate or to support than I am who say to me things like: 'Well, I can see the theory may have a case, but I'm afraid I can't touch any of that: our grant comes from the Medical Research Council...' or 'Labs that could test what you want in Britain are all in the same boat, they all get money from the MRC or drugs companies. I don't think you are going to find any of them wanting to be testing an old vaccine with a risk of turning up something. You just have to accept this is what the AIDS field is like...'

Surely you must realise that the development of this sort of situation in science is terrible—literally terrible for all mankind. Thinking only of the narrow escape in the SV40 case, leave alone of the possibly worse and determinedly underinvestigated case of AIDS, anyone should see that the situation ought to be terrifying us. Those scientists who are best placed to do so ought [to] help to combat it.

Yours sincerely

[signed]

Professor WD Hamilton.[21]

What Hamilton is invoking with this letter appears to be something like the traditional Mertonian norms for science, which emphasize the need for disinterested research, open-mindedness, and public knowledge, with an interesting extension of the norm of universalism—non-discrimination on the basis of nationality, gender, race, etc—to include people outside the scientific community. He is especially suggesting enforcing the norm of disinterestedness, which involves a consideration only for truth, not for other interests, such as economics, power, etc. (Most scientists, while not necessarily following these norms in practice, at least recognize them as important guidelines.) At the same time Hamilton gives us typical examples to the problems facing science as it is being increasingly connected to funding from outside sources.[22]

***

The general problem, then, was that certain things that would merit discussion in science could not be discussed. This resembled the problem of not discussing matters of great scientific interest simply because they were of a sensitive political nature. Hamilton had been thinking about these things a lot. As the first president of the Human Behaviour and Evolution Society, he had even had an opportunity to specify what he would regard as an ideal atmosphere for scientific discussion of difficult matters—in this case having to do with human nature. In his presidential address at the inaugural meeting of the society in Evanston 1989, with all the big names present—George Williams, EO Wilson, Richard Dawkins, Richard Alexander, and Irenäus Eibl-Eibesfeldt—he formulated this ideal atmosphere as follows:[23]

> My very strong hope is that the Society will continue long and always succeed in retaining the spirit in which it is born—a spirit of free interdisciplinary discussion of all the probabilities of human nature. I continue to suspect, however, that wishing long life for the Society and wishing it to have free, truthful discussion are not completely compatible. If asked whether I would prefer to have the Society based on some vague mish-mash of belief and ritual like Free Masonry, which would ensure its continuance for thousands of years, or to have it founded on acute, falsifiable doctrines in those of the late Shaker sect and hence prone to die out in a few years, I rather think I would prefer the one of being Shakers.
>
> ...
>
> Our view is undoubtedly giving us access to truths that can be useful to many professions—medicine, psychiatry, social planning, etc. However, there is justifiable concern in the world at large that the truths also teach cynicism and antisocial attitudes in places where myths have managed better. Hence, I still think that the clash of our truths with current myths and religions are inevitable for a very long time and perhaps that is even a Good Thing and to be endured in a stoic spirit rather than with resentment. Possibly clashes will indeed force us ultimately into paths of jargon and secrecy but I would like to see that resisted as long as possible.
>
> ...

> I hope that I am being unduly pessimistic about problems of our future, but I really think we face difficulties of a kind certain never to trouble a new Society discussing, say, the modern problems on non-linear dynamics. From my experiences as a graduate student when I was more keen and less cautious in applying my ideas of natural selection to humans, I feel that I foresaw correctly the controversy that would arise when the ideas about humans were brought to the surface, as happened in Wilson's 'Sociobiology'. For my own part I have generally kept clear of human statements and in fact of most of the matters which are now the focus of our Society. I kept clear mainly not out of lack of interest, but out of cowardice. However, as I said above, I think that such thoughts and doubts expressed above shouldn't inhibit discussion at our meetings; all people while they are with us should dare at least to try out ideas which might estrange colleagues and ruin careers elsewhere. Among us let all ideas be critically and courteously discussed; let us watch the poor stranger among these ideas who may be a king in disguise; let no idea be simply 'put down' because it seems odd or heretical. I think scientific societies are among the most brilliant of human achievements: I hope that ours, which is devoted so much to the study of conflict and unfairness in human life, will rise above 'all that out there' which we describe and will prove that it can be more peacable, fair and democratic than any organization has ever been before. Perhaps our Society alone is able to understand the forces ranged against it working both from inside and out—even if, as I suspect, it is other groups such as politicians, that may sense such forces better.[24]

We see yet again that Hamilton has the highest of hope for science as an arbiter of truth, and it seems, moral behaviour. But he also realizes that telling the truth is not always easy. Not only is it dangerous (something that he personally doesn't seem to worry about too much), but it is possible that myths and false beliefs might in fact work better to keep society together! Unfortunately, there are scientists too who take a popular view rather than facing the hard issues. Hamilton labels these demagogues, and even half-seriously suggests a scale for such scientific demagoguery based on 'demog' as unit. He feels that his own serious messages and warnings are not pondered as gravely as they ought to be, while scientists with optimistic-sounding messages are listened to by the masses.[25]

The big contradiction experienced by Bill Hamilton throughout his life was the contrast between what his evolutionary reasoning told him was right and what people seemed to want. People didn't seem to realize or care what was in mankind's best self-interest from a long term evolutionary point of view! While various 'demagogues' promised easy sinecures to the masses, serious scientists who came with warnings were not listened to (well, he had to admit that being a Cassandra could sometimes make money, too). Even science itself, not to mention medicine, seemed to be going for short-term economic interests. Meanwhile, it was important not to suppress the Truth.[26]

Hamilton, in his self-chosen role as an evolutionary rationalist, took upon himself the role of guardian of mankind, a function akin to that of the State looking out for the long-term national interest far beyond the present generation. At the same time, though (still playing the State), he realized that it might not be socially and politically expedient to speak the (evolutionary) truth to everybody all the time. Inconveniently for his own convictions, various types of myths in fact seemed to work as 'social glue!' Was untruth then the price of social peace?[27]

Hamilton's whole life seemed geared toward resolving this conflict. And for him, the problem had to be resolved with the help of science. If it was the case that people reacted against evolutionary theory because of the perceived negative social implications of what it was saying, well, then, the point would be to demonstrate that evolutionary theory can do good things, too!

Let us give the last word to Hamilton himself. This was what he said in 1996:

> [T]he implications of a fully rational and evolutionary theory of behaviour, and that includes human behaviour too, is not such a nasty thing as it may seem at first. If you believe that we evolved out of animals—are animals—and have the same kinds of drives, it doesn't mean that we have to be selfish and inhumane. When you fully work out the consequences of the rules of kinship and of reciprocation, and ensure maintenance of the standards implied, you will see that the outcome is in fact quite a moderate kind of behaviour, avoiding evil and as good in holding the society together as are the religious myths. Indeed, under a rational theory we should be able to do better for human happiness by avoiding various naive errors.[28]

21

Creative Strategies

After sending off his letter to *Science* Hamilton was still upset. The extreme contrast between what he believed science stood for—the pursuit of truth, no matter what—and the apparent attitude of *Science*, a leading journal, dismayed him. Now he, an international prize-winner comparable to a Nobel laureate, had asked *Science* to reconsider its policy. But would he be able to get the journal to open up the discussion of the OPV hypothesis and the bigger problem of virus transmission from animals to humans? Would *Science*'s Letter Editor and Editor-in-Chief recognize their grave responsibility to science and humankind? With this move, he had put his own scientific standing on the line, but he felt he had no choice. Where was today's science when it came to truth and the benefit of mankind? Where were things going?

At a personal level, it was getting gloomier and gloomier for Bill to return from his long expeditions to his house in Wytham, now looking more like a bachelor pad. The contrast between the warm camaraderie in the Amazon and his lonely existence at home started affecting him badly. At home he had usually been absorbed in his theoretical pursuits, but there had always been a background of domesticity. Now he felt this taken-for-granted dimension acutely missing. He had also started wondering if Christine was really coming back after the expiration of her two-year contract, or if she would want to renew.

What happened next was again like his cherished cartoon from the *American Scientist* where the math professor scribbling a formula at one point states: 'Here a miracle occurs'. Bill himself had applied this parallel to the help given to him by Bob Axelrod's (see chapter 16). In this case the miracle was the unexpected letter that Bill found in his mail at the end of February 1994 on his return from a two-week trip to the United States.

Hamilton was a scientific inductivist, for whom any unexpected sight in nature could stimulate an instant flow of creative insights and comparisons. Just as at the University of Michigan he had associated a red cardinal singing in the winter snow with the thrush singing for Thomas Hardy, which in turn helped to inspire his parasite paradigm, so this letter now fuelled his creative imagination. (He may even have seen it as a 'sign' of sorts; there was an occasional mystical side to Bill.)

The letter—actually just a polite, handwritten note—was from the person who had interviewed him for *La Stampa* in 1988 about his parasite avoidance theory of sex. Although she worked as a freelance journalist, she held a doctorate in biochemistry and had later, while working as a school teacher, become interested in animal behaviour and written some popular books. The note accompanied a photo of Bill from the recent Castiglioncello conference in Italy that one of the organizers, a good friend of hers, had suggested she send on to Bill. Bill remembered the letter writer well—Maria Luisa Bozzi, someone with great interest in evolutionary biology and an excellent listener. He had thoroughly enjoyed being interviewed by her some six years earlier.[1]

Here is how Bill described his situation in that year's Christmas letter to friends:

> It was the prospect of the < 5% situation [his estimate of how much he had been seeing of Christine since she took the job in the Orkneys] continuing indefinitely that, early last year, made me break out. At the end of the 2 years that she originally specified for her adventure in independent living there was still no sign, to me anyway, that Christine was even thinking about returning.[2]
>
> When I asked her about her return in Autumn last year when we were on the trip to Japan she replied that I must realize that jobs were not at all easy to obtain in Oxford. Yet of course I have always emphasized that there is no real necessity for her to have a job at all; that is just a matter of her preference. All the two years it seemed to me that she was actually happier in Orkney than she had been in 'stuffy Oxford' as she had felt it to be (also perhaps living along with her 'stuffy old professor' as I rather felt myself to be). While she was there we had between us bought for her a house and a small-holding with farm buildings and she had begun to plant trees, concentrating on the species she and I had brought seeds of back from Chile in 1975 for trial in the west of Ireland. I followed this endeavour with enthusiasm from one side of my nature—I would really like to see those trees do well and knew that her father was always interested in tree

planting—and dismay on another, seeing the trees and house as further proof she didn't care about living with me. After I told her that I had a new companion she said her preparations for return had actually been imminent, she had been about to begin discussing this with me, the trees and farm were just a project and a holiday house, and finally that it was treacherous to act as I did without at least warning her. I see some justice in the accusation. I had pondered what to do, whether to warn her how she was slipping out of my life (as I, seemingly from hers—for weeks or months neither of us would even telephone), but eventually decided that . . . she was happier at a distance and living in a culture that seemed to suit her better than living attached to me, to our rented house, neglected by my snobbish colleagues, etc.—all the things that had always so riled her. Anyway, I did act, finding a few unexpected free days at Easter due to a conference proving shorter than it was billed to be, I flew to Turin instead of to the W of Ireland.[3]

Luisa had interviewed me 6 years ago for La Stampa, the daily she writes for, following our first encounter at a conference in Florence. Last year there was a wasp conference at Castiglioncello . . . There I met her again and we chatted briefly. More fateful than this was a close friend of hers taking a photo that happened to include me seated in the lecture theatre. For some reason the friend didn't send the photo to me herself but gave it to Luisa saying that she knew me better: if she thought Hamilton would like it, why not send it?[4]

Luisa did so, with a note with best wishes and a remark that she hoped we might meet again soon. Because the letter arrived close to St Valentine's Day, I read into this (or maybe my loneliness read into it) more than just a professional, friendly regard and I replied with a long letter—a kind you know of, this for example—that wasn't at all professional, a philosophical monologue of my recent life and travels and doings . . . This gave her a great surprise but astonishingly it also struck a chord in a 12-years lonely woman.[5]

Actually, the really surprising thing was that in the letter, Bill asked Luisa if she would like to come with him on an expedition to the Amazon in July that year—either to gather material for an article or just for a holiday. She agreed but suggested that they meet before that (which was why he went to Italy for Easter).[6]

In his letter to his friends, Bill went on to describe how on a later trip to Italy he was able to observe and photograph the parasitical *Polistes* that he had long wanted to see (the only social parasite in this genus) and how another friend of

Luisa's who studied wasps at the University of Florence facilitated this discovery by kindly lending them her mountain house in the alps at the Cesana pass. Meanwhile the first friend lent them her apartment in Florence. This was to be the beginning of a complicated commuting life between England and Italy. As Hamilton put it, Luisa's frequent commitments in Italy (she had a son, who was just starting university, and an elderly mother) made him suspect from the beginning that he would have to put up with her spending no more than 50 per cent of her time with him. 'This is less that I hoped for but still much better than the < 5% of the previous two years', Bill wrote.[7]

The Amazon trip that year went up the Japura, one of the large tributaries of the Amazon. Bill did one trip with Luisa in July and one by himself in October.

> The two trips were up the same tributary and have ended a simmering controversy amongst our Project personnel as to whether the Japura was a 'white-water' river or a 'black'—and they settled it, rather drably, in favour of it being a bit of both: it is 'white' in sediment load but 'black' in terms of conductivity and pH. Though disappointing in terms of deciding definitely which of us, Peter Henderson or myself, owes the other a bottle of champagne, this situation is very odd and interesting in the sense that it means the silt is present but anything in it or accompanying it that can be of much use to plants is lacking... Seemingly it is all completely extracted in the course of the water's passage from the Andes. It is almost equally odd that we should be the first to discover all this in Amazonia's fourth biggest tributary.[8]

Since the trip with Luisa was in July, it was not the flood season, but that year the 'low' was not very noticeable. There was even some flooding, and they spotted a lot of people packed into a house on the Japura riverbank, sitting on window sills with their feet in water. What were they all doing? It turned out they were watching the final moments of the Brazil-Sweden World Cup match, to which the TV owner had invited everybody in the village (the TV itself was carefully kept out of the water). Bill and Luisa climbed through the window from their canoe and joined the crowd.[9]

The following year Hamilton was invited to New Zealand as a Rutherford lecturer. This meant he could finally visit his many relatives there (he had first cousins on Bettina's and Archie's side). In New Zealand Bill reportedly gave a very inspired lecture on the Amazon (which goes to show that he was a good lecturer when he was clear about his subject matter). He was also able to visit the southernmost island, a nature reserve and parrot sanctuary. And he is said to have tried to give the government advice on planting trees, telling them they should avoid monocultures. New Zealand is free of many tree parasites, but he was worried what would happen if any pest arrived.[10]

Among longer trips in the mid-1990s was one to the Valdez peninsula in Argentina and to Patagonia. Bill had always wanted to see the interesting flora and fauna in these places and the trip had stimulated his theoretical reasoning too. He reports having seen marine mammals: Southern Right Whales, elephant seals, guanacos, and maras.[11]

In 1997 he participated in the trip that marked the thirtieth anniversary of the Mato Grosso expedition, which now included a visit to the new university of Nova Xanatinga. In Nova Xanatinga, on this occasion, Hamilton was made a freeman.[12]

The mid-1990s were also a time for honours and prizes. After receiving an honorary doctorate from the University of Guelph in Canada, Hamilton received the French Fyssen Prize in 1995. His prize lecture brought in Lamarck by invoking both his favourite JH Fabre and the 'semi-Lamarckian' Baldwin effect. Baldwin was on his mind these days, as he was theorizing about the adaptive flexibility of the organisms in the Amazonian flooded forest.[13]

Another trip was to Finland, where in 1997, Bill was elected as an external member to the Academy of Finland. He considered this honor to be very grand, in fact more so than a Nobel Prize. 'Since it [the Academy] has statutorily only 11 Finnish members and at present only 14 foreign members, I count them to have given me a much greater honor that the Nobel prize!'[14]

He was very pleased to chat with the Finnish president Matti Ahtisaari at the prize dinner and photos from that occasion show him looking happy. On this occasion, at which guests were invited to display honours and medals, Hamilton sported his Kyoto Prize, an impressive huge golden medal hanging on a ribbon around his neck, attracting much commentary. But an observation of one of his friends in regard to Bill's attitude to prizes seemed to be true: while he was very pleased to be recognized, he did not dwell on the honour, but typically took off for a field trip soon after.[15]

In the Finnish case, Bill went on an exciting trip to the countryside to watch a dawn spectacle of black grouse displayed on a bog, arranged by Jyväskylä University biologist Rauno Alatalo. He had also hoped to see a black woodpecker, but the tape-recorded shrieks and pecking sounds emitted from the loudspeakers connected to Alatalo's car did not convince the woodpeckers. He was deeply disappointed when the weather did not allow him to skate on the Finnish lakes, something he had hoped for. Skating was a near obsession for Hamilton, and in 1992 he had lugged his huge skates to a February conference at Tvärminne research station on the Finnish Bay, where the organizer Pekka Pamilo and the other conference members held their breath as they watched him sea-skate and approach the weaker ice caused by currents in the archipelago. (The organizer was imagined to have made quick calculations about kin relationships and a rescuer's chances of drowning.)[16]

As for research during this time, the Amazon project with Henderson was clearly the biggest. During the late 1990s Hamilton also started a number of new projects alone or with others, including his Master's student, Francisco Ubeda. Most famous of these are his closer involvement with the Gaia hypothesis which turned into a study of clouds as he started his research with Tim Lenton, and his and Tim Brown's theory of colour change in autumn leaves. Both cases involved evolutionary theorizing and collection of suitable empirical evidence, and in both cases Bill used the well-tested strategies that he had earlier employed in the cases of inclusive fitness and the parasite avoidance theory of sex, which once more drove him to libraries to find relevant literature.[17]

In both cases Hamilton was successful—with a little help from his friends. In the Gaia case, the problem was that although Bill would have liked to think in Gaia terms and thought the theory exciting at an abstract level, he could not figure out how it could involve the evolutionary benefit for an individual organism. That was the necessary criterion for Gaia to be connected to evolutionary theory, and the main the reason why his Oxford colleagues did not consider it seriously. Hamilton had corresponded briefly with James Lovelock and Lovelock had come to see him in Oxford in 1997. No resolution had been found at that occasion.[18]

Things were looking up as Hamilton found that Tim Lenton at the University of East Anglia had been studying micro-organisms that were able to generate turbulence at the top of the ocean water and propel themselves up in the air to be picked up by the winds. From Hamilton's viewpoint here, finally, was an organism that entered Gaia's self-regulating system for the reason of promoting its own reproductive self-interest. Evolution was, after all, able to be combined with Gaia thanks to Lenton and the sulphuric gas (DMS, dimethyl sulfide) employed by the bacteria for their 'jet' manoeuvres. The result was their joint paper 'Spora and Gaia: How Microbes Fly with Their Clouds', which again attracted great attention (and maybe some disbelief). But Lovelock was pleased. He and Hamilton shared the common characteristic of being very serious scientists but each with a playful imagination and unconventional attitudes.[19]

In fact, it was Jim Lovelock who had put Tim Lenton in contact with Bill, telling him to send Bill a copy of a review article on Gaia and natural selection that he had submitted to *Nature* but which had not yet been accepted. Bill read that article and became fascinated with the DMS (dimethyl sulfide) gas produced by phytoplankton. Bill saw this as a ray of hope for connecting Gaia with evolutionary biology, thinking of it as 'adaptive DMS'. He engaged in email correspondence with Lenton, read more of the literature and got excited. 'Bill had a wonderful capacity for finding needles of interest in haystacks of otherwise unpromising papers', Lenton comments. Bill then surprised him completely by sending him the first draft of the paper 'Spora and Gaia: Do Microbes fly with their Clouds?' The question mark at the end was later eliminated as Bill and Tim gained more confidence. The article was published in *Ethology, Ecology and Evolution* and generated great interest, indeed, the story made the front page of the *New Scientist*. This in turn led to an interesting chain of events, involving the independent film maker Anna Maria Talavera, who persuaded William ('Chad') Marshall to test one of the claims in the paper by collecting airborne samples of the ocean spray and analysing them. The test showed how easily plankton actually end up in the atmosphere. Some cloud samples were dramatically taken from an aeroplane and captured on film.[20]

The explanation of why autumn leaves turn red and yellow was one of those ideas that was routinely labeled 'wacky' or 'mad' by journal referees and readers of journals alike. But Hamilton was very serious, and so was young Tim Brown, whose PhD thesis this came to be. Brown, as he worked with Hamilton, just couldn't believe the sheer cheek of the whole idea! The basic point was to see the colour change as a signal of a plant to insects, telling them to stay away (or, as a clever heading in the *New Scientist* had it: 'Leaf me alone!'). Hamilton and Brown's explanation, in other words, actively avoided chemical considerations and focused on the ultimate question of why this happened at all. The research project was simple in principle: the point would be for Brown to rate the showiness of various trees (from garden catalogues) and correlate this with the trees' proneness to parasites (in this case aphids), obtainable from other catalogues. As it turned out, the hypothesis was supported, and created a stir. But the research was interrupted many times because of interference from other projects for both authors, and finally published in 2000.[21]

Finally, around this time, too, Bill Hamilton's new popular reputation may have triggered a Cambridge professor's curiosity. John McCullough wanted to see how the substantial killings of relatives in English history in relation to the fight for succession to the throne would hold up to Hamilton's Rule. The conclusion was that Hamilton's Rule did hold up even in this case. When one took into account the genetic relationship of the pretender to the throne to each one of his murdered competitors, the calculation showed that the resulting number always was smaller than 1, that is, the total number of killed relatives discounted by their various degrees of relatedness was never equal to him killing himself.[22]

***

Bill had an uncanny ability to identify important research problems and come up with plausible explanations. As he describes it, the idea may originally have come from some observation he made in nature, or some curiosity that someone told him, or he may have found it himself in some old textbook. The plants and animals of the Amazon were his stable source of stimulating new ideas and the information that he serendipitously found in libraries, while searching for material to substantiate some other idea played their part. He skimmed *Nature, Science* and the *New Scientist* to catch up on new ideas and to stimulate his own idea flow.[23]

Bill was usually on the lookout for unusual behaviours, evolutionary paradoxes, seeming exceptions to the rule. It was especially the seeming exceptions that thrilled him, and he went out of his way to learn more about the particular phenomenon—be it from a colleague, a student, or an Amazonian native who had found something interesting. (One exception that excited him but that still eluded solution was the puzzle of a species of parrots where both the male and the female are brightly coloured).

Sometimes the problem to be solved was sitting right under his nose, as he was studying the writings of Fisher, Haldane, or some other famous author. The following observation comes from Richard Alexander:

> Bill's originality of mind often turned the barely articulated ideas of distinguished predecessors—ideas overlooked or neglected by all the rest of us—into magnificent theoretical edifices affecting our view of all life.[24]

Students and colleagues accompanying Hamilton on nature walks, be these in Ann Arbor, Wytham Woods, or some place far away like Taiwan, saw his mind in action. On these occasions, however, he didn't start with the individual organisms themselves, but presented a general theoretical preamble for pedagogical purposes.

> On such outings, the talk would move in sudden but oddly seamless leaps, first to the contemplation of general theories, then to a recitation of the particular names and habits of startled arthropods that suddenly found themselves brightly lit and closely inspected by admiring humans; and then, once the stone or log was returned to its former position, back to theory.[25]

We see how Bill's thinking is quickly moving between general theory and particular, careful observation. It is always a question of fit, and if an organism's behaviour doesn't seem to be explainable by existing theory, well, the theory will simply need to be changed! For Hamilton, every new encounter with nature was an adventure and a potential opportunity to challenge prevailing orthodoxy.

Of course, while giving insights into Bill's reasoning style, those nature lectures were intended to educate and entertain his audience. Perhaps there was something of a showman—or possibly a preacher—in Bill during these performances, he so much enjoyed sharing his information and showing off his knowledge and reasoning style in a friendly way. And there was certainly the added thrill of a live performance, too, because who knew what exactly would crawl out from a rotten tree, or whether he could actually guess, at least roughly, the family of a totally unknown tropical plant!

Because of the way Bill initially handled information, and all the associations and plausible or even far-fetched explanations that any new stimulus spontaneously generated in his mind, he ended up with a stunning number of new ideas and would-be nuclei of new theories. It often seemed to people that he had theories for everything.

Bob Trivers remembered once being startled when he heard a group of graduate students at the University of Michigan discussing Bill:

> One fellow said: 'The thing with Hamilton is that 20% of what he says is brilliant and 80% off the wall, but you don't know which is which so you have to pay attention to all of it.' I thought [wrote Trivers], 'More like 99% and 1% and be real careful about the 1%!'[26]

So what did Bill himself have to say about these estimates of the soundness of his ideas? (Trivers was curious enough to ask him.) 'Bill, on the other hand, agreed with the student and pointed out that the remarks were not based on his published papers but on his everyday comments.'[27]

His doctoral student and friend Nancy Moran explained the seeming oddity of Bill Hamilton's ideas in the following way:

> He believed that if you have a weird idea and it seems wrong at first, you should give it a chance before you throw it out.[28]

Is this related to a sense of economy—don't throw things away—together with a trust in the subconscious ability of an intuitive scientist to find something true, even if he doesn't recognize this at first?

But let's consider Marlene Zuk's calmly numerical approach to the ratio of brilliant versus off-the-wall ideas. Bill Hamilton was successful simply because he had more *good* ideas than anyone else!

> If we assume that only 10% of new ideas are actually good ones, and if the regular fellow has only some 4 or 5 ideas during his or her academic career, chances are that these are not first-rate. However, if one like Bill has 100 new ideas, by this rule ten of these are good, and some of these brilliant![29]

But idea generation is one thing, selection of those ideas that are worthwhile to work on is quite another. How did Bill discard ideas, how did he know what to keep, and what to leave by the wayside? Bill was good at 'cross-tabulating' various pieces of information in his head for a preliminary check of the plausibility of a particular new idea he might have (here he was comparing and contrasting all kinds of different details he knew about animals, plants, bacteria, fungi, etc). Then he would go to the library in order to mobilize as much positive evidence as possible for his case. Finally, he would mathematize his conclusions and build as general a model as possible, after which he would test it by simulation.[30]

Bill wanted his models to be as general but also as realistic as possible. This meant that he wanted to operate with assumptions that led at least to an approximately right result. This (as we have seen) sometimes got him into trouble with more exacting mathematically trained colleagues, such as population geneticists.[31]

Bill used mathematics for developing his models, but there was an additional point to this. He explained this as follows:

> Often I use mathematics because I need to straighten out my own ideas. I have a somewhat illogical brain, and unless I put it through the mill of mathematics, I can continue to believe in the impossible for a long time.[32]

This of course was not only a help for Bill himself, but made it possible for others to follow his reasoning.

But Bill was happiest when he was finally able to test his theories and interpret the results visually. He was an extremely visual person, good at detecting patterns, and would surely have enjoyed some of the software that is currently available. He believed strongly in the supportive power of simulation, but this tool was not typically used among mathematically trained biologists.[33] This was probably a central factor in the difficulty that Hamilton and his collaborators had in getting their parasite avoidance theory of sex accepted. Indeed, as he himself later commented: simulation is not the same as proof.[34]

Hamilton's collaborator Akira Sasaki described his fascination with visual displays as follows:

> Bill was attracted by the very rich behaviours of the simulations played on the computer screen. They seemed to inspire his thought. I was happy to think that the way he was attracted to the patterns my program generated might look like the way he was charmed by a bunch of aphids in different colours on a leaf when he found them in Nagano. I remember many scenes in the study of Bill's house, with Bill observing every detail of what was going on in the display, and whispering about their implications.[35]

Hamilton had very high standards for papers that he wanted to publish. For him it was most important to satisfy himself first that he was right. And it was exactly because he had spent so much effort and was so convinced that there was no major error that he got so upset with referees who didn't seem to understand what he was trying to convey. In this respect Bill was different from some scientists, who may have tended to see science as more of a 'conversation'—that is, an enterprise where you publish something as soon as you think you have something to say and you then let others find the errors (this was, for example, the position of Nobel laureate David Baltimore in 1988). Bill said he was 'more "dinosaur" or "bird" against "reptile",' nurturing hypotheses to independence instead of 'leaving them to fend for themselves'. (This was probably also why he often felt uneasy about having to have a paper ready for conferences.)[36]

He was indeed very careful in his approach, probably more so than many current scientists are under the increasing pressure to publish or perish. (This could have been one of the reasons for the fact that he did not produce more papers than he did.) Even his 1964 inclusive fitness paper includes a discussion of possible counterarguments and problems with the theory. In other words, Bill himself had already thought of and dealt with many of those things that later commentators might criticize. (Incidentally, and typical of Bill, when he did become aware of errors, he corrected these himself (such as an error about the sex ratio, corrected in the reproduction of his 1964 paper in Williams' 1971 book).[37]

This extreme care—and wish to avoid criticism—was probably partly a feature of Hamilton's personality. In his childhood home, Bill was reportedly keen on doing things 'properly', while his younger brother Robert was the quick but somewhat sloppy fixer of anything that needed repair. Care and error-checking also seem like natural values of the boys' engineer father Archie.

But there is more to Hamilton's handling of criticism. His colleague and friend Peter Grant referred to Bill 'dismissing with a shrug contrary views'[38] This is surprising, since we know how much he disliked being criticized and how upset he sometimes was at the comments by referees. On what occasions then might he have displayed this kind of nonchalant attitude? This probably happened when he had finally made up his mind, tested his theories to his own satisfaction, become absolutely sure that he was right, and simply could not be moved in his belief. There is no doubt that Bill was extremely stubborn.

His friend Bob Trivers had the following vivid vision of Bill's stubbornness—and powers of persuasion:

> I remember thinking at some point in my relationship with Bill that if the argument ever became physical, the contest I would least like to be engaged in against Hamilton was a shoving contest. I felt that he would dig in his heels, that you would be unable to move him, and that he would lean forward and shove you slowly and stubbornly to wherever he wanted to get you.[39]

This is similar to Marlene Zuk's description of Bill's behaviour during an excursion in northern Italy. They were walking along a river, but at one point Bill absolutely insisted that they cross the river and walk back on the other side. This seemed a rather wild idea, since there was no bridge in sight. No matter, Bill would build a bridge for them (a party of four) to cross to the other side. He happily set about bringing together various branches and some debris that he found, and despite loud protests that it was totally unnecessary to go to the other side, set out to build a workable bridge. When it was ready, he took the members of the party over, one at a time. Marlene Zuk saw this whole episode as a parable for Bill as a scientist: he had a wild idea that others were highly sceptical about, but he persisted, and slowly brought them over to 'his side'.[40]

Hamilton even provided some advice to his readers about how to sustain belief in your own ideas. He said he chose to follow the humorist James Thurber, who 'found it best not to listen to his opponent at all because this enabled him to keep his mind calm and keenly focused on what he was going to say when his opponent stopped talking'.[41]

Is taking Thurber's advice really acceptable for a scientist? According to Bill, it is. In fact, he recommended that scientists become more of advocates of their own ideas.

> This may sound very un-Popperian and even a dishonest approach to science, and in a way it's true; but... one of the wonderful properties of science is its self-correction: at the same time there is a lot to be said for everyone doing their own advocacy. Very possibly the truth comes faster that way—the courts, at least, seem to think so.[42]

As a warning counterexample, Hamilton held up Gregor Mendel who failed to follow through after his discovery because he listened too much to others. 'Had he not listened to the Swiss botanist Karl Nägeli, for example, who had recommended Hieracium—hawkweed daisies—for his further studies, he himself would probably not have started on that cryptically sexless group of plants and would not have become confused.[43]

On the other hand, while ignoring criticism, one should not ignore evidence, Hamilton warned:

> A wise scientist never ignores evidence, of course, his theories soon forsake him if he does; but in science, as in law, advocacy is not a matter of lying. You put the best case that you can for the theory you believe a likely winner; but likewise you must do nothing to impede others who work similarly for the theories they favour. (In case of interest, both my grandfathers were lawyers).[44]

Aha! This can perhaps explain some of Bill's general tendency to rationalize and his skill in finding plausible arguments that could in principle connect his data with theoretical frameworks. (These arguments are then typically translated into mathematical form, and a model is built and tested by simulation.)

While Hamilton in his autobiographical notes often probed into his own feelings and motives, sometimes trying to catch even those that were hidden from himself, not much space was explicitly devoted to creativity. But at one point, Bill discussed such matters in principle and described how he typically went about things. It is clear that Bill was conscious of the importance of creativity for a scientist, and deliberately sought ways to reflect on his own behaviour:

> Seeing deep correspondences in seemingly unrelated things is the essence of science and is vital in mathematics and philosophy as well. Newton saw that the fall of an apple is the motion of a planet... When two phenomena give me even a hint of similarity I try as a matter of course swapping modes of thought applied to them, forcing myself to contemplate each one in the light of the other. This playful effort continues until distinctions and special pleadings stick so much clay to my feet that the back and forth with ideas is no longer fun. Or else if new and unsuspected correspondences accumulate while distinctions prove rare or trivial (a situation that I am undoubtedly helped to notice through word usage bringing various other meanings and situations swiftly and repeatedly to mind) then I grow serious and skip even faster, sensing a find that may be of beauty and use, a new tool with which to understand the world.[45]

This gives an interesting insight into the working of Hamilton's mind (and probably that of other scientists too). His specific point in this passage is that everyday word usage helps this quick comparison in science, and is therefore an important tool for scientific creativity. Certain commonly understood words should not be censored away in scientific discourse just because they are seen as politically incorrect.

Bill's whole life can be described as a search for the right conditions for creativity. He sought out people, environments, and various props that would serve this purpose. In fact, he was used to designing his own life and environment already from early childhood, choosing his hobbies and friends, and avoiding things that he didn't like to do. (Perhaps his mother was something of an accomplice in this respect, because young Bill could avoid chores by walking off on butterfly expeditions.)

Bill seems to have had a partly conscious, partly unconscious approach to his creativity. He cast a relatively broad net, because he could not know where the next creative instance would come from. But when he was in a situation or a conversation, he realized relatively soon its creative potential. According to one of his sisters, he used to give a new acquaintance some 10 minutes of his time to check out the future chances of something coming out of it. Casting a broad net also meant that he did not unduly censor the type of person that he talked to—for instance, at conferences, he often preferred to talk to students rather than learned colleagues (and many students remember him fondly for these conversations).

He also had a predilection for eccentric friends—think for instance of George Price. He may have had a theory that eccentricity goes together with creativity but it is also true that he had a genuine appreciation for more unusual characters, and a special spot in his heart for underdogs.[46]

But Bill was also sensitive to inspiring environments. Although he could work in whatever surroundings when concentrating on theoretical models—for instance, sitting in the middle of his living room surrounded by a sea of papers—when it came to composition, the environment took on stronger meaning. This is why he much appreciated the stays in various country houses and mountain lodges of friends that Luisa Bozzi was able to arrange, perhaps especially the cosy converted farmhouse of biologists Laura Beani and Francisco Dessi in Impruneta near Florence. Some chapters of *Narrow Roads of Gene Land* were written in that house's old olive oil pressing room, in front of a particular window looking out over the rolling Tuscan landscape. 'This is where I want to sit', Bill had told his hostess. That view, framed by the door-like window, appeared just like a painting. And, except for the occasional cypress, the view may also have transported him back home to Oaklea...[47]

He also actively sought out stimulating environments. Here certainly his expeditions provided the needed novelty for keeping his creative juices going. But libraries, too, were some of his favourite hangouts. As he describes it, sometimes it seems he was as much stimulated by the librarians as he was by the books themselves. And he did appreciate their presence. Some of these librarians were beautiful—that was enough. Others were helpful for his research, and a third category presented a challenge by their mere presence. They provided a stimulus for his ingenuity—recall his penchant for 'passing' as a regular customer with library rights when not a card holder and posing as a plumber in order to gain access to the reported unlikely habitat of an interesting beetle species: a hospital kitchen.)[48]

So, Bill enjoyed pushing himself into difficult situations, to test his own creativity and problem-solving capability. We know that already from his travels with brother Robert. Being thrown in prison in Greece for overturning a car was an accident, but Bill's student Bernie Crespi was almost sure that Bill and Robert deliberately got lost in the jungle on an expedition in Borneo. This allowed them to sleep in a cave with huge spiders, which they seemed to enjoy. Poor Bernie didn't get a wink of sleep, while Robert's wife Tiang, a member of the party, had wisely found more 'bourgeois' accomodation in a nearby village.[49]

Hamilton was also actively looking for materials that would stimulate his thinking—be these old textbooks, maps, paintings. Old textbooks were an incredible source of information for him and he savoured the quaint language. As he noted, there might be throwaway phrases, or interesting observations of animal behaviour worthy of being followed up. It was his belief that issues of scientific interest are often abandoned long before they are actually exhausted.[50]

More generally, a love of reading was something that his mother Bettina had installed in Bill and his older sister, and she had also exposed them to poetry and the arts. This kind of general liberal education continued at Bill's elementary school and during his early years at Tonbridge. Bill later educated himself in the great art museums of Europe and sustained lively discussions especially with his doctor-artist sister Mary on art and his Uncle Charles on modern poetry (Bill did not much care for modern poetry, but was polite about it, considering that Charles was the editor of the literary journal *Landfall*).[51]

I believe that all this exposure provided Hamilton's mind with good potential for making interesting associations. Indeed, his ideas often came from unexpected sources. For instance, we learn that he got the idea for his model of the 'seething cycling' between hosts and parasites from Tolstoy's novel characters.[52] Ideas can come from anywhere—what counts is what you do with them and if you can get other scientists to agree with you. Karl Popper would have agreed with Bill's varied sources of ideas. Popper pointed to the important distinction between where ideas come from and how they are scientifically supported (the difference between the context of discovery and the context of justification). Indeed, Bill really was something of a Popperian, believing in conjectures and refutations of scientific ideas within a quasi-Darwinian process.[53]

22

Through a Glass Darkly

Bill Hamilton's collected works, the two volumes of *Narrow Roads of Gene Land*, have been mentioned several times already. In fact, Hamilton spent a good deal of the 1990s on the organization and publication of these volumes. This chapter will take closer look at the background of this project and the nature of the resulting books, especially the second one.

The initiative to publish Bill Hamilton's collected works came from Michael Rodgers at Oxford University Press, a long-time admirer of Hamilton. The original idea was to produce a set of collected works with brief introductions by the author. But after some time, the introductions were turned into essays instead. The result was a collection of autobiographical commentary for each of Hamilton's papers, each describing the context of that particular paper: what he was thinking at the time, what else he was working on, the reception of the paper, and finally, his current reflections on that older contribution.[1]

We can in fact go one more step backwards, to Sarah Hrdy, the editor of a series in Evolution and Human Behaviour for Aldine, who had discussed this kind of book with Hamilton some 15 years earlier. It was this book project that was later taken over by Rodgers and turned into *Narrow Roads of Gene Land*.[2] The result was two volumes, the first focusing on Hamilton's papers on the evolution of social behaviour, and the second devoted to his later theme on sex and parasites, published in 1996 and 2001, respectively. Bill himself had planned a third volume, which he said would be devoted to what he called 'moral genetics', and to which he made a few cryptic allusions here and there in volume 2. That volume was not to materialize but there does exist a volume 3, edited by evolutionary biologist Mark Ridley, consisting of Hamilton's most recent co-authored papers, with introductions by his collaborators and students.[3]

It took a lot of effort for Bill to write these prefaces. Indeed, a request to account for the context of writing a specific paper, the various trips, and speaking invitations that may have been involved, not to mention what one was thinking at the time, is a daunting task for anyone (think about writing a report on your own activities only for the last academic year!). He took this task very seriously, and we do get a detailed recollection, sometimes so vivid that it is possible to conjure up an image of Bill at the time. We learn not only what he thought, but also what he felt, especially about reviewers and editors not understanding his contributions. Many of the recollections are indeed written from an emotional point of view: for instance, we see Bill sitting in Waterloo Station scribbling miserably, trying to solve the problem of altruism. But there is a quality to these prefaces which makes them read more like personal letters. For some of his colleagues reading these commentaries may even have felt like Hamilton was talking to them personally.[4]

What is striking about Bill's letters in general is their seriousness, be it about his scientific expeditions or insights, or places visited on his travels, and this is almost regardless of whether he is writing to his family or to a scientific colleague. We find the same style in these autobiographical notes. This does not mean that there are not digressions of various types and humorous asides, but even the digressions are serious, in the sense that they typically contain observations and reflections that might be on their way to theories. And looking up the footnotes to these various chapter introductions, we find more serious detail—as well as more digression.[5]

This is very much Bill's style: he wants to share with the reader as much as he can and in as much detail as he is capable of remembering. Bill is something of a boy scout in his quest for comprehensiveness—and in his surprising self-disclosure. Sometimes one sees a struggle between the subjective emotion that he relives as he writes (for example, with regard to how lonely he felt as a graduate student), and his rational mind telling him that there may have been objective reasons for a particular state of affairs (say, the absence of a formal seminar structure, his own lack of interest in mingling with fellow students, or whatever).

In volume 2, the emotional recall becomes more prominent, which gives that volume a rather different flavour. The innocent reader, expecting to learn about the scientific development of Bill Hamilton, is often dragged into the discussion, which is sometimes quite uncomfortable. The topics emerging in many of these prefaces in volume 2 are difficult moral and ethical ones. The reader may not feel at all prepared to suddenly argue with the author about issues of life and death. Other prefaces in volume 2 leave the reader feeling as though he or she has taken a long car trip with the author as he recollects the various instances in which he was—or felt—misunderstood or mistreated in academia. There is also digression and footnoting in these cases, except that this time the subject matter is often controversial.

But Bill needs us, his readers. He desperately wants our sympathy for those situations where he has spoken up about important matters and been rebutted by the Establishment—be this that of science, medicine, or the Church. After all, he was only telling the truth! He wants our understanding and validation. In moments like this, what he does not want to hear is that truth can be told in many ways, and that he misjudged the way the other party would react—perhaps because that other party 'should' have reacted differently?

Of course, these introductory essays were written in parallel with papers and conferences, Hamilton's usual expeditions to the Amazon, and other travels. In fact, it was because he took the writing of autobiography as seriously as any other task, that these recollections and commentaries came to compete for his time with emerging, new and exciting ideas. That accounts for the relative sluggishness of production.[6]

Why *Narrow Roads of Gene Land*? Bill chose this title with care and deliberation. As he explained, he regarded the work as an exploratory journey similar to the one taken by foot by the Japanese poet Matsuo Basho in the 17th century. Matsuo's diary of that journey, *Narrow Roads of Oku*, is a mixture of prose and poetry, as he describes the landscape and the people he encountered and reflects on life. Bill loved Basho's brief haiku poems, which according to him included 'some of the deepest and most poignant thoughts ever recorded'. For Bill, *Narrow Roads of Gene Land* was his answer to Basho: the chapters would represent Basho's poems, and the introductions correspond to Basho's lively narrative, with anecdotes and reflections interspersed with more serious observations and explanations. Bill saw a natural connection between scientists and artists.[7]

It seems that Bill wanted these volumes to represent him as fully as possible, thus he cast himself in a set of multiple roles—theorist, naturalist, explorer, poet, artist, teacher, buddy, raconteur, rebel, martyr, prophet. He designed the covers of the two volumes himself and carefully picked out the frontispieces and poetic introductions to each chapter. The first volume features a wasp apparently happily walking a narrow country road with blue sky and just a few clouds in the background. The second one sports brightly coloured birds and some of Hamilton's famous three-dimensional 'spaghetti' curves of parasite-host co-evolution, encased in perspex cubes.[8]

It was an imaginary journey both in science and in time, and a journey of self-discovery and self-revelation as well. This may have been both a good and a bad thing. Good, because it forced Bill to get in touch with his feelings as well as to reflect on situations as it were 'from the outside'—after all, here he had to present himself to an audience—bad, because these recollections brought back to Bill the memories of earlier slights and difficulties, opening up old wounds. There is a lot of repressed anger, sometimes thinly camouflaged, surfacing in these memoirs, as well as occasional suspicions. Bill shares it all with us.

For Bill, it seems only too easy to fall back on his deeply engraved early model of misunderstood, ignored, and slighted graduate student, a schema that he later uses for interpreting events also in his mature scientific life. Also, we get a sense of Bill's doggedness on finding that he devotes two appendices, in addition to two regular chapters, to further convince the reader that he is right about the Parasite Red Queen theory of sex as well as the Hamilton-Zuk 'sosigonic' criterion for mate choice.[9]

The forced self-analysis of autobiography at a time when he had achieved considerable scientific fame may also have encouraged Hamilton to tell it as he saw it, which is why volume 2 does not mince words. It seems he intended volume 2 to act not only as an overview of his own work, but also as needed shock therapy for a complacent world, assuming for himself the mantle of an unwanted prophet, or a rebel against the System, roles that he could identify with both intellectually and emotionally.

***

Bill undoubtedly had a rather dark view of human nature—or shall we say, he felt the need to emphasize the nasty as well as the nice features of human behaviour against what he saw as attempts to cover up the truth. The place where he goes into most detail in this respect is his contribution to Robin Fox's anthropology conference in 1973. Reading that paper, we learn that humans are xenophobic, racist, aggressive, and cunning, and that there is good reason to believe that there is a deep evolutionary basis for such features. He draws parallels with chimps and other animals, but also observations from anthropology, and even what he calls 'using my own intuition'. According to Hamilton, individuals are not blank slates onto which human cultures write. The genetic system has 'prepared' each individual, providing tentative outlines for the development of certain features. The evolutionary rationale for such features is the human need to know who is a friend or foe, and with whom one can reliably develop coalitions.[10]

One message that Bill especially wanted to deliver was that warfare was not a human aberration, something that could just be dismissed. On the contrary, warfare for him was a driving force of human evolution. It organized society, brought new resources and new genes to the population, and constituted an important test of foresight and intelligence. In early times, inter-group aggression and genocide was probably the typical model for interaction between groups of early hominids. Although Hamilton agreed that agriculture brought about more peaceful societies, and marvelled at the Neanderthal burial in Shandikhar, seen to represent early cultural practices, cultural explanations did not seem to enter his general reasoning in any significant way.[11]

Hamilton's basic position in regard to the relationship between nature and culture was (in Edward O Wilson's later famous formulation) that Nature 'keeps culture on a leash'. To those who believed that human nature was endlessly flexible, he noted that not all kinship systems that could be logically thought of are actually found in the anthropological record. In general, his position was that 'culture is a braggart'. The genes actually ran the show, but culture was taking credit.[12] For him the main function of culture could well be to reformulate in a palatable way certain genetic imperatives. For instance, in conjunction with his parasite paradigm he actually gave human culture the function of 'keeping the memory of beauty', beauty being a way of assessing an individual's health.[13]

Perhaps surprisingly, in volume 2 Hamilton gives great play to the idea of memes, those units of culture that according to Dawkins spread and reproduce just like selfish genes. Reflecting on current societies, he suggests that 'new memes that we are creating and are themselves vigorously "Darwinizing" and thus self-improving may acquire strength to take us over almost completely, and much in the way a "brain worm" takes over behaviour in an ant'.[14] There is a danger with such intellectual ideas coming to increasing prominence in our lives, he continues. For instance, a meme to restrict reproduction may well take better hold among more rather than less intelligent people, and genes for lower intelligence may thus come to dominate.[15] In other words, culture, in the form of memes counts after all! There are also some formulations where Bill admits that some behaviour or taste can be of genetic or cultural origin, or both.[16]

Paradoxically, perhaps, Bill saw a particular tension between nature and one aspect of culture—the challenge for mankind presented by the fast developing science and technology. His concerns were largely similar to those of his fellow biologists in the 1960s and '70s who were asking the question, Can Mankind Cope?—including fellow ethologists Konrad Lorenz, Niko Tinbergen, and Edward O Wilson. This general problematic spread well into the final decades of the 20th century. 'Can mankind cope?' had to do with such things as the danger of nuclear armament, overpopulation, and environmental deterioration. In addition, Hamilton has some unusual-sounding concerns of his own, to which we shall return later.[17]

For Hamilton, as for many other biologists, one response to the threat of runaway technology and its potential abuse, or to the escalation of social violence, would be better knowledge of human nature. Knowledge of who we really are, warts and all, and stripping away convenient myths of different kinds, would lead to more rational social decision making and policy measures. Hamilton was part of the neo-Darwinian revolution and the gradual extension of the Modern Synthesis to the human species. He left it to others, though, to directly communicate with the public in the form of popular books such as *The Selfish Gene, Sociobiology* and *On Human Nature*. (The trend of popular books describing the biological basis of human nature had been started by ethologists and ethologically oriented anthropologists in the 1960s.)[18]

But Hamilton also kept thinking about possible undesirable consequences of evolutionary developments and policy measures that could counteract these. Overpopulation had been a theme for him ever since he started his demographic studies, if not before. He was particularly concerned about the phenomenon of uneven growth in population of two neighbouring countries—according to him a potentially explosive situation, since an innate territorial instinct would want to assert itself among the more populous and lead to war. When as a student he suggested to population specialists at a mid-1960s meeting of the Royal Society that they pay attention to such situations, the experts appeared shocked. Later (in a letter to the *New Scientist*) he came up with the idea of 'reproduction permits' to keep the world population in check—everybody would get two permits and these would be freely transferable.[19]

At the end of the 20th century, though, it was the growth of medical technology in particular, the promise of various genetic therapies, and the advent of such things as cloning and xenotransplantation that came to preoccupy the policy-oriented aspect of Hamilton's scientific persona. These concerns were intimately linked to his scientific interests at the time: host-parasite co-evolution, sexual signalling, and the origin of sexual reproduction. For all of these a common component was health and the maintenance of health. So, looking at the proliferation of new hospitals and medical procedures, what could he as a biologist say about the general trend in medicine?[20]

Hamilton continued taking upon himself the role of the negative prophet, conveying to mankind uncomfortable truths. But as a biologist, and a population biologist at that, he felt he had to speak up. Everybody seemed so pleased with the new promises of medicine, with the possibility to cure illnesses, with gene therapies—nobody seemed to ask what this actually meant for us humans, not in the short term but in the long run. And here Hamilton's 'geological', long-term perspective on evolution produced quite a different picture than some of his colleagues' sanquine-sounding public proclamations. Were serious scientists like John Maynard Smith and Steve Jones perhaps victims of wishful thinking when they too bought into the talk about a rosy future of various gene therapies? Didn't they see where all this was leading? If so, why didn't they say anything? For Hamilton the prognosis was crystal clear. Humankind was becoming increasingly dependent on medical technologies, and these were progressing rapidly. What would happen if all these technologies for some reason broke down? In Bill's catastrophe scenario—a comet hitting the Earth, a sudden epidemic, a nuclear holocaust—the medical support system would not be able to sustain itself—or us.[21]

Meanwhile, an alarming long-term process was taking place: an ongoing decrease in the quality of the human genome as a whole. According to Bill's grim calculus, in every one or two generations there is at least one bad mutation in a genome, and these mutations accumulate over time. This was not Bill's own data, but data he had from luminaries such as Fisher, Haldane, Muller, and later Kondrashov. Even with the most cautious calculation there was no doubt that the human genome was slowly deteriorating.[22]

This deterioration was the real thing to fix, Hamilton argued. He analysed the various solutions suggested. For gene therapy to have real lasting effect for mankind it would have to be germ-line therapy, not somatic therapy. The point would be to excise faulty genes from the human genome. But what were the real chances of doing that? Replacing genes is a methodologically difficult procedure, and each success would only come after lots of failed experiments. And even if we succeeded in one generation, this would not be the end of it. There are always new mutations—you would have to go on constantly chasing and fixing mutations, just like some type of genetic Maxwell's Demon. Not possible, was Hamilton's verdict.[23]

Meanwhile our reliance on increased medicalization would only speed up the deterioration of our genome. More individuals with defective genomes would survive and be able to procreate, thus perpetuating existing genetic defects. This route also meant that we would be even more dependent on medical interference in our lives—Bill visualized the end result as a big Planetary Hospital taking care of humankind. The problem with this was that, in the case of a catastrophe wiping out a great deal of the population, the survivors would not be healthy enough to sustain a viable population. The only hope would be to have people in remote places who somehow had evaded the global medicalization programme.[24]

At the same time, Hamilton feared, the route toward medicalization implied authoritarianism and totalitarianism and an end to the freedom of the individual. We would have let technology take over our free will. On the horizon he saw nothing less than a *superorganism* dominating scattered individuals and running their lives. He even envisioned this superorganism as connected up with artificial intelligence, which meant that in the end, decisions would not be made by humans any longer, but by machines. Time is running out, Bill warned, people have to do something now, as long as we are still in possession of our individual intelligences and can think![25]

No doubt we are here getting into a science fiction scenario, but there is a clear sense of urgency. Hamilton's old spectre and childhood fear—totalitarianism—reappears, here in the guise of an authoritarian superorganism. The medical establishment represents the controlling System that has to be fought by brave individuals. For Bill, the model for fighting the System are individuals such as Bruno and Galileo. They broke the rules, they suffered for telling the truth in a world dominated by myths perpetuated by the power holders.[26]

But why should we believe in the coming of the superorganism, the Planetary Hospital? What is the rationale for this? Hamilton's dark scenario is directly derived from the history of evolution: the takeover of bacteria by eukaryotes, which became the mitochondria in eukaryotic cells. Just as freely living bacteria were trapped in eukaryotic cells, individuals will be lured to give up their freedom by subjugating themselves to the will of the Hospital. What will be lost if that happens is human variety, he warns, something that is crucially important biologically, morally and socially! This loss of human individuality and variety will come about because of the standardization needs of the Hospital. There will be a standardized model human genome against which 'deviant' genomes will be calibrated and assessments made as to the possible acceptability of the deviance. More likely than not, corrections will be made.[27]

What is the alternative to this futuristic scenario? Hamilton has a simple answer: natural selection. Just let natural selection take its course, don't interfere! This is a rather surprising-sounding statement, considering the undoubted medical progress that has been made and seems possible. But Hamilton's moral calculus is not concerned with increasing the health and happiness of currently living individuals—he is looking at the big picture. His calculus takes into account mankind as a whole, for any forseeable future, including the rights of so far unborn members of human society. According to him, it would be immoral to increase the health of our current generation at the cost of future generations. Is this truly his position? He asks this very question of himself and puts himself to a test. He imagines a hypothetical crisis situation where he would be faced with the choice to either sacrifice himself and his immediate family and thus save mankind, or, alternatively, save himself and his kin while letting mankind perish. He chooses the first option.[28]

Natural selection operates well enough by itself, Hamilton notes, often spontaneous abortions and miscarriages are of genetically defective individuals (those who carry a double dose of a recessive gene or many such genes). In historical tradition, malformed or weak-looking infants were 'exposed' or killed. Accordingly, in the case of 'bad genes' Bill approves of abortion and infanticide as 'natural'. Bill is truly obsessed with the long-term fitness of mankind. He has given up on genetic engineering of various kinds and believes that only 'natural' methods are to be recommended.[29]

It gets confusing, though, when Bill realizes that many highly gifted individuals have been clearly 'unfit'. Newton was born prematurely, Toulouse-Lautrec appears to have been born with genetic defects, Stephen Hawking is a genius, and so on. At this thought, his view seems to change towards a scenario where these kinds of individuals could despite all be accommodated. Bill's answer is that following the Parasite Red Queen theory, these individuals may actually be carriers of important genotypes, now latent, which in the next cycle will show their mettle. In other words, perhaps we had better retain all kinds of genotypes? Bill suggests that it might be possible to assess whether we are dealing with expressions of useful polymorphisms or merely isolated defective genes; frequency counts may give us guidelines.[30]

Bill also ran into trouble with women particularly when advocating the 'unnaturalness' and therefore biological undesirablity of Caesarean sections. It didn't go down well for instance when he presented this to a Harvard seminar organized by his good friend Naomi Pierce who had recently given birth to twins by caesarean. According to Bill, if all births take place by this method, circumventing natural selection, the resulting slim-hipped women will no longer be able to give birth naturally—a clear handicap for a population in the case of a major disaster. However, this is not the only possible way to look at the situation. The model of parent-offspring conflict would seem to be more immediately applicable.[31]

But what about the inherent possibilities in new genetic manipulations? Would it not be possible with the help of new genetic technology to create better humans, some kind of superhumans? Bill also briefly considers this, but dismisses the idea. If superhumans could be created, they would be unhealthy, he declares. He admits an earlier interest in creating better humans, but now realizes that the wrong features would be selected for. It is not intelligence but *health* that matters in the end! 'It was through my own research on the PRQ that I changed my mind' he declares.[32]

Indeed, Bill was interested in positive eugenics in his youth. So was 'everybody' I was told by John Maynard Smith, pointing to 1930s' volumes by Haldane. But with the Nazis and the events of the Second World War, the idea got corrupted and eugenics as an ambition went underground. As mentioned earlier (chapter 5), reading those 1930s' eugenics-inspired writings while growing up, Bill was clearly taken with some more fantastic ideas about procreation and society (this cost him his first girlfriend, as we saw in chapter 7). He also briefly joined Mensa (the organization for persons with a high IQ), a potential forum for assortative mating, but declared that experience very boring, people just talking about themselves.[33]

Even while advocating natural selection, though, Bill cannot quite give up his wish to improve the stock, and his fancy takes flight. His advice to marriage candidates is to find partners with good genes—that is, genes which have shown resistance to various illnesses. His hypothetical International Marriage Bureau would particularly recommend as partners individuals from populations that were resistant to such things as Tay-Sachs disease and HIV. (Indeed, he had the idea for an International Marriage Bureau in the early 1960s, but matching in that bureau was based on assortative mating, not on health considerations.)[34]

This is Bill in a partly prophetic and serious, partly playful mood. One might argue that his conclusions are only as good as the assumptions that go into his various scenarios. For instance, and especially: is it true that the genome is deteriorating? And if so, is this an immediate cause for alarm? Here is Hamilton's prediction: '[I]n two generations the damage being done to the human genome by the ante- and postnatal life-saving efforts of modern medicine will be obvious to all and be a big talking point of science and politics.[35] What are the predictions of other scientists? This is an issue that is not much discussed, I found. A few scientists gave me the old Muller calculation, but most said they did not know. David Haig thought that Hamilton presented too rapid a timetable for the coming damage to the genome—unless he was referring to the rise of a political debate about who will get the benefits of modern medicine.[36] It seems Hamilton was referring to the latter, because elsewhere he warned that it would be naive to believe that everybody would be able to be given treatment. There would have to be criteria and standards—but who was to set the standards? Better to start the discussion now as long as we are still in possession of our intelligence, he suggested.

Another puzzle: why does Bill in his texts prefer to use the word 'infanticide' instead of terms such as genetic counselling or abortion? There is something seemingly obsessive about this repetitive use—infanticide does not appear to be the way to go in modern societies, and the term sounds offensive to normal ears. Is it just that he is just getting obsessive about this particular term? Or is it an expression of Bill the contrarian, deliberately shocking people—as when once at the Oaklea dinner table he brought forth a big insect, quietly placed it next to his plate as a decoration, and after a while, to everybody's surprise, proceeded to eat it.[37]

Whatever the reason, it was this 'just let natural selection take its course' argument that he brought to the Pontifical Academy at the Vatican in October 1998. This scientific academy regularly organizes conferences around big themes, bringing in famous academics from different fields. This time they had been brought in to reflect on the meaning of 'Nature' in the 21st century. The meeting was to last three days, with a day each devoted to physics, biology and culture. Among the others who had been invited to the Pontifical Academy for the same meeting were Luigi Cavalli-Sforza, Steven Rose, and Stephen J Gould, the last two having become famous as part of the sociobiology controversy. Bill clearly felt challenged and wished to use the occasion to defend sociobiology as well.[38]

Bill's report about what happened at the Vatican constitutes the Introduction to volume 2 of *Narrow Roads*. He describes how the Pontifical Academy for some reason had originally wanted him to talk on the topic of developmental genetics. Since this was not his area of expertise, he asked to speak about a topic of his choice, and it was agreed that his topic would deal with the future evolution of humans.

The draft of Bill's talk was circulated before the meeting. It was a long paper, clearly unsuitable for presentation. It is unclear how Bill cut it down (he tells us that he was interrupted anyway). It began as the sincere attempt of a biologist to assert the truth and value of Darwinism, and an outline of the problems facing mankind, but soon took on the themes of the present chapter. In other words, the readers got a dose of counter-advocacy against genetic engineering, and advocacy for letting natural selection have its way, while defending abortion and infanticide. In his talk Hamilton had prepared three moral stories, comparing the deaths of a faulty neonate (his own brother Jimmy, born with a congenital defect, allowing him to live only a couple of weeks), his 18-year-old brother Alexander, and his puppy dog Freya. Bill's provocative point here was to present the dog as having more value than the neonate. The paper also included a point about Bruno and Galileo as defenders of the truth, while earlier and future superorganisms were in the business of sustaining myths (here Bill seemed to delight in using a semi-transparent Aesop language). Finally, for his defence of sociobiology against authors of books such as *Not In Our Genes*, Bill had prepared a sheet with references to authors pointing out errors in the critics' arguments.[39]

He did not get very far into his paper because he was interrupted, he complained. (An observer told me that this was not the case.) In any case, time was limited. Also, at this point the presentation time for Bill's paper had already been postponed a couple of times. He had started to feel he was being quite badly treated, while he saw Steven Rose getting very chummy with the organizers. Here is how he describes the reception of his paper:

> [J]ust before I took the floor the chairman warned me that I had only 20 minutes for my talk and that this would have to include the questions (the original allocation had been 45 minutes): we were desperately tight for time...It hardly needs saying that under continuous contractions of my time relative to what I had prepared for, my presentation was hurried and truncated. There was time for just one question and two comments; both the latter were hostile and dismissive of my

> points and there was virtually no time for me to respond to them. My appointed commentator then took the floor and told us very amiably how he so disagreed with my paper he could not try to address it; instead he would talk about historical issues and art that had interested him; these had been raised in another talk.[40]

Bill informs us that he was not only given much less time than he had expected, but also that there just before his paper had been a 'free' time slot (since Gould had not come), which had been filled up with small speeches.[41] In fact (as we shall see), there were at least two objective reasons for the moving around of Hamilton's presentation. Cavalli-Sforza could only attend the first day and needed to speak early on, and later adjustments could be attributed to reorganization because the conference was running late. What was really surprising, though, was that Cavalli-Sforza in his paper addressed biological points quite similar to Bill's in a brief, carefully worded way. Bill's longer paper may not have compared well with the well-measured one of Cavalli-Sforza. Despite his own paper's provocative content, however, Bill had not been emotionally prepared for the dismissal he felt that he got at the Vatican. Afterwards he found some consolation by going to Campo dei Fiori to pay homage to the statue of Giordano Bruno.[42]

How did others experience the conference? Here is an excerpt of a breezy report of the meeting by another invited participant, linguistic anthropologist Stephen Levinson:

> The Academy, originally founded in 1603 as the first of the Scientific Academies, is one of the most exclusive academic clubs in the world, with only 80 members, many of them with Nobel prizes.
>
> ...
>
> In this extraordinary setting we then got down to an extraordinary meeting. Our duty was to contemplate changing views of nature at the turn of the millenium. Day 1 was meant to be physics, day 2 biology and day 3 culture. But Cavalli-Sforza, a Pontifical Academician, scheduled for day 3, announced it was now or never for him. He then delivered a paper arguing for population control and specifically for the abortion of genetically handicapped individuals, letting cultural selection do painlessly what natural selection would have done with anguish. This caused some furore. Carlo Rubbia (Nobel laureate, head of CERN) said that if he was the foetus he would object, and Stephen [sic] Rose (British neuroscientist) pointed out that most of those present, since they wore glasses, would have been selected out by the same criteria. We escaped into physics...

> The next day [the third day], we had to zoom through the rest of biology and on to culture. [They were still dealing with biology, since the conference was running late.] Now for our second bombshell. William Hamilton, discoverer of kin-selection (which at last explained how selfless worker bees and selfish genes can cooccur), got up to give a shocker. He showed slide after slide of medical schools towering over campuses, claiming that they would soon take over the world. All because modern medicine has stopped natural selection in its tracks, so that deleterious genes are propagating. He advocated not only abortion, but mercy killing of infants. Most gallantly, the President of the Academy welcomed his opinions, and sought earnest discussion.
>
> ....
>
> Hubert Markl, president of the Max Planck Society, had the final talk, on the 'Future of Nature'. Pretty grim, in short. And population certainly had something to do with it. Science had created the problem, and it had better get us out of it.[43]

The reason they had to 'zoom through' biology was the very long wait to see the Pope the day before, which by day three had been mainly responsible for the accumulated half-day delay in the overall conference schedule. My informant also reports that Hamilton attacked the critics of sociobiology, Steven Rose and Stephen J Gould. (Presumably, though, that attack had to be very fast, since he did not have time to distribute the counter-critical reading list he had planned.)

Yet another participant, an ethicist, was disturbed by Hamilton's presentation, especially some of his formulations. He found it odd to hear Hamilton, after all an Oxford professor, voice views of this kind.[44]

The irony in the Vatican episode is remarkable. Here Hamilton, seeing himself as a rebel, goes to the Vatican to tell the truth, while his audience may potentially have seen him as speaking for the academic Establishment!

At the same time what he said was not so far from that of an authority of the Pontifical Academy itself—Cavalli-Sforza—who was also intent on telling uncomfortable truths, although in more pedagogical language and enjoying obvious insider status. As a speaker on the last day, Hamilton may in fact have had to experience the audience's combined reaction to his own and Cavalli-Sforza's talks.

***

Hamilton was used to seeing himself as someone whose duty it was to tell the world unwanted truths. Here was now an urgent and deeply disturbing one: we are in serious danger if we believe we can take care of the human future with the help of biomedicine but forget about natural selection. The truth that he discovered as a graduate student—that altruism was possible—was also rife with moral messages, but on the face of it, seemed a positive contribution (it was only later that some wondered if this implied a gradient from cooperation to hostility across human groups, depending on how much of their genes they shared).

It is clear that there was an important moral dimension in Bill's whole being, one that did not always make itself known in Bill's more technical writing. His idea of altruism was not just an answer to a scientific puzzle, it also satisfied a deep need to know that unselfishness was possible in nature. For Bill, the scientist, scientific proof was needed that there would be a point to self-sacrifice. Self-sacrifice was a thought central to him. It came up again and again throughout his life. An important early inspiration for the theory of altruism came from the honey bee sacrificing her life by stinging the intruder of a hive.

Indeed, here and there in Bill's writings the theme of self-sacrifice comes up, always with a positive connotation. In his discussions of the babassu nut and its six embryos of which only one can become a full-grown individual, or the imagined point of view of an embryo, or even his own brother Jimmy, it is remarkable how the theme of self-sacrifice surfaces. In regard to conscious human self-sacrifice, Hamilton's leading parable was the story about the Burghers of Calais, a group of citizens volunteering to give themselves to the enemy in order to save their town. And when it came to science, as we have seen, his heroes were Bruno and Galileo.[45]

Where might this kind of talk of sacrifice and suicide come from, also connected to resistance to authority, and to telling the uncomfortable truth? I attribute this to Bill's romantic vision and heroic imagination, but also to those early moral teachings.[46]

I therefore want to add one more twist to the meaning of *Narrow Roads of Gene Land*. Yes, Bill says it is modelled on the poet Matsuo Basho's diary. In Bill's case, of course, the landscape is an imagined scientific one, seen through genetic spectacles. The vision therefore is necessarily narrow and focused. And the genetic vision is what Bill wanted to pursue, leaving such considerations as culture to the side. But 'narrow' also has another connotation: it is the road to truth. It is important to stick to the narrow (also potentially lonely and difficult) road, not to be tempted onto the convenient broad one: 'Enter ye in at the strait gate: for wide is the gate and broad is the way that leadeth to destruction, and many there be which go in thereat: Because strait is the gate, and narrow is the way, which leadeth unto life, and few there be that find it' (Matt. 7:13,14). Hamilton is prepared for a moral journey, encountering and overcoming difficulties on the way, and even the possibility of dying for his belief.

Bob Trivers' wife Lorna was right when she observed that much of what Hamilton said had double and even triple meanings. Indeed, as Trivers noted 'Bill thought in chords'. *Narrow Roads of Gene Land* is one of those complex chords.[47]

Volume 2 was published posthumously. Bill had not yet had time to make his final changes and corrections. An obvious question is: how much would the book have changed had Bill had a chance to do so? This is the question I posed both to the editor (and Bill's Oxford colleague), Mark Ridley, and to the copy editor, Sarah Bunney. The answer was: not much! Ridley's impression from one chapter that he attempted to change at an early point was that Bill was not receptive to suggestions. The same was true of Bunney. She said she generally managed to make some small stylistic changes, but Bill had a habit of changing those back.[48]

Did Bill listen to anybody at all? His good friend Nancy Moran spent considerable time with the manuscript, trying hard to suggest reformulations and word substitutions. 'You can't say that', she would tell him, 'if you say that, people will think that you believe such and such...' but Bill paid no attention. He wanted to say what he wanted to say, and in the exact form he wanted to say it. He was probably preparing himself for battle and controversy. Perhaps he felt that the time had come for him to engage in serious debate. In the Unites States he had largely missed being a visible debater in the sociobiology controversy, defending a cause; others had been fighting it out. But now he finally felt ready to do something important for mankind. He had an urgent warning and needed to be heard. He wanted to open up discussion about matters not usually discussed. Little did he know that he would not be participating himself.

23

The Final Defiance

In the 1990s Bill Hamilton came increasingly to focus on the fate of Man, worrying about the preparedness of humankind for what he termed an asteroid impact. But he was himself perhaps less prepared for another type of impact and its consequences: the death of his mother.

In his Christmas letter to Yura in 1996 Hamilton describes his mother's death as 'the greatest event of the last year'. To be close to their mother in her final days, the family, including Bill's sister Margaret and brother Robert and their families from New Zealand had gathered at Oaklea. Hamilton had the idea of celebrating his mother by building a chair rigged with handles like a stretcher to carry her around the garden that she loved. He described this tour as a great success and his mother being quite animated though quite sick. She died a few days afterwards.[1]

This raised the question of what to do with Oaklea. Initially they planned to let it. In regard to the woods around Oaklea, Bill was in charge, and he had a plan:

> As to the Woods, I am the one most involved in an application for planning permission to put up a largish concrete shed in what we call 'Hackett's Land'. There was a shack there once (possibly you remember it) in which lived a retired Canadian Mounty, alcoholic and looked after by Mother through many years. This new shed will ostensibly serve as a storage building, workshop and office for living in a week or so at a time if and when Oaklea is to be let to tenants. It really is going to be difficult to visit and do anything with the land we own as a whole if we don't prepare somewhere to be our base. I forget if I told how by coincidence we were just finalizing the purchase of what we call The Woods and

> The Field at exactly the time when Mother had her fall, and how we became full owners about two days after she died.[2]

He explains that the idea behind the purchase had originally been to protect his mother from any possible surrounding development, and they decided to stick with the decision.

Bill and his siblings had initially thought of letting the house until someone 'got together enough of both money and will power to buy and manage it: inevitably that person turned out, much faster than planned, to be Mary'.[3] One year later Bill reports on 'Mary having now almost bought Mother's nine acres and the house from the rest of us'. Mary had earlier planned to go and live there together with their mother, but took early retirement instead and moved to live there by herself.[4] These kinds of developments got Hamilton (now 61) thinking about his own retirement (the mandatory retirement age in England at this time was 65), but it seems he did not start seriously planning for that until 1998.

As usual he was involved in many things. There were the trips in 1997 to Valdez and Patagonia and the Mato Grosso expedition. In Brazil Marcio Ayres had been funded to expand the reserve areas in the Amazon, and Bill was helping to study the Amana Lake area. The plan was to propose Amana as a protected area, and in 1998 the Amana Sustainable Reserve was created, the largest in the Brazilian Amazon area (2.35 million hectares).[5] In other words, Hamilton had a great interest in the ecological future of the Amazon. The Bill Hamilton Itinerant Center for Environmental and Scientific Education contributed to the ecological education of the Amazonian population in preservation and defence of their area. ('Defence' against such things as invading fishing boats—in fact, a system had been devised for spotting intruders and reporting them to the authorities).[6]

Back in England again from his Amazon trips, Bill would enjoy himself classifying the exotic plants he had collected. (Peter Henderson reports him spreading these muddy plants all over the boat deck, alarming the neat crew. Henderson found he had to become a diplomat as there were recurring complaints to him: 'Now the Professor has again...').[7] But Hamilton was ecstatically happy collecting and examining these largely unknown plants:

> I am also working on my plant identification album which has been neglected a lot in the last three years. For [this] I am working on odd days at Kew...These days are like days in heaven: I like looking at plants so much, and there I seem to be walking via the folders and their stiff, old sheets, scrawled with copperplate

> dates and names and locations and notes (the last often in Latin), as in the footsteps of so many brilliant naturalists of the overwhelming river (von Martius, Bentham, Humboldt, Bates, Wallace, Spruce...) and perhaps above all, in a niche between mountains of folders behind a cast iron spiral staircase, no mail or telephone calls can reach me.[8]

There was also Bill's ongoing research on various issues, especially his exciting collaborations with Tim Lenton about the Gaia hypothesis, and with Sam Brown on autumn colours. Moreover, he was thrilled that his student Bernie Crespi had been able to find a social thrips in Australia, and called it after him: *Kladothrips hamiltonii.* (This was, in fact, the second organism called after Hamilton.) Hamilton had for a long time been searching for truly social thrips. In fact, his inclusive fitness paper predicted that eusociality in thrips should exist, but because no data existed 'he was forced to go searching himself, not the actions of a typical theoretical biologist'. And looking for them, he had found various social thrips and colonial thrips.[9]

But everything did not go smoothly. There was a rather strange unexpected resistance from the Brazilian authorities when it came to Bill's by now routine request to go to the Amazon to Mamirauá for 1998. He wanted to collect more data to support the Hamilton-Zuk hypothesis, now concentrating on a demonstrable connection between bare skin patches and a history of blood parasites. His research was presumably something along the following lines (this passage refers to one study that Bill wanted to do at this time):

> [A] study of the blood diseases of birds of the varzea...There is a very good Lithuanian bird hematozoa expert who I am trying arrange for Marcio Ayres (with his new large inflow of cash to the project) to invite. Late August might be a good time for him to be working. He says that nesting season is always best since it is the most stressful for birds and that during it parasite titres in the blood always rise, causing one to miss fewer of the latent and sparse cases. Also the land will be coming up—all gurgling, dripping, even stinking—out of the flood water making it easier to erect the mist nets. I am very keen on this project because I believe I may be able to show three things to be connected: the abundance of biting vectors in whitewater varzea habitats, the high incidences of blood parasites, and the presence of wattles and related bare-skin patches on an unusual proportion of the birds. The last two, of course, bear on my theory that most of the exaggerations of sexual selection are honest advertisements of innate freedom from and/or ability to control, chronic disease.[10]

Later it seemed that Bill's research proposal for 1998 had for some reason gone to inappropriate referees—regular parasitologists, who, not surprisingly, did not understand his research.[11] But at the time the difficulties Bill encountered looked like an obstruction by the Brazilian authorities. Bill's student Servio Ribeiro believed that it had something to do with the government's negative attitude to international scientists, and found it to be ironic in the case of Bill, since Bill knew so much about the Brazilian landscape and nature and was so willing to share his knowledge.[12] One year later, however, Bill received a medal from the Brazilian Ministry of Science and Technology, in gratitude for his service to science and contribution to the knowledge of the nature and ecosystem of Brazil. The award ceremony took place in the Brazilian Embassy, and present were his sister Mary and Luisa. Bill reportedly gave an acceptance speech in Portuguese and seemed generally pleased.[13]

But back to Bill's retirement's plans. In 1998 he had been seriously considering his various options. There was the looming question of where to retire. One plan involved the University of Michigan, Ann Arbor, where he believed that he could probably get an appointment put together with the help of the HBACH group and the biology department. The problem was that Luisa was not willing to leave her family, it was too far away. Or he could stay in England, say, in Southampton where Peter Henderson lived and work with him on interesting projects, while travelling as a type of research overseer for the Max Planck institute (he had tentatively discussed some of this with its representatives). Meanwhile his sister Mary had invited him and Luisa to come to live at Oaklea but that did not seem a good idea. Hamilton himself was proud of the concrete shed he had started building and thought that it could serve as a guest house for him and Luisa when they visited Oaklea. Luisa was rather taken aback. She had hopes for a different solution.[14]

In the spring of 1998 at a small dinner party in Oxford, Hamilton suddenly asked me: 'Ullica, how does one get a green card?' I tried to answer but was rather surprised. Luisa seemed somewhat nervous. (As I discovered later, the United States was not really an option for her.) After that Bill and I went on to discuss his impending visit to the Vatican and the meaning of the fact that individuals hostile to sociobiology had also been invited. Later I learnt that in the summer of 1998 he had visited Italy and that there had been a suggestion for a retirement home there too. Bill's last student David Hughes, a wasp specialist, who was visiting his Florence colleague Laura Beani, tells the following story:

> Luisa described an old abandoned house on a hill in Tuscany set among rolling fields of olive trees. Such a property appealed to her as a retirement home for them. However, Bill was increasingly nervous and uninterested in the idea. But his interest was greatly revived when he spotted that wasps had made their nest and examining close he found not only nests but also a very unusual aggregation on the blackberry bushes (*Rubus* spp.) that snaked through a broken window. What was unusual was not only the inactivity of these aggregating wasps but also that most harboured a particular parasite. Correctly identifying it as an insect-Strepsiptera, and supposing its presence accounted for such unusual behaviour, he set about collecting the aggregations. Discussions of domesticity gave way to science and talk of retirement to parasite manipulation of host behaviour. Luisa could but follow! She remarked that many times a walk and conversation in the country with Bill would dramatically be broken off whilst he would hold aloft and remark on some example of life, only eventually returning in a convoluted way to the original conversation.[15]

Bill had brought Mary with him on this trip to Tuscany, maybe as a kind of adviser. In any case, nothing came of this retirement idea. It seems to me that Bill felt that he needed to be close to people who shared his interest in computer modelling, and he needed access to libraries. In other words, he was never really planning to retire from his research. In Tuscany he would have been connected closely to the resources of the University of Florence (including its wasp specialists), and would, as a famous British evolutionist, have attained something of a guru position. But Hamilton saw himself less as a guru than as a working scientist, and with a lot of work still to do. Part of that work was to continue to oversee Oaklea and its surrounding areas. The retirement question was put on hold. But, as Peter Henderson reports, at the end of 1999 Bill had settled on a house in Southampton near Peter's home (and not too far from his younger sister Janet).[16]

What was Bill working on in the late 1990s? He was conducting various collaborative research projects (especially the Gaia project with Tim Lenton and the autumn colour project with Sam Brown). A post doc of his, Jeyaraney (Jeya) Kathiritamby, was following up on those 'lazy' *Strepsiptera* in Tuscany, especially after a similar group had been found under the roof of Laura Beani's house (the meaning of this was pondered in a lengthy email exchange). And of course he was working on the introductory chapters of the second volume of *Narrow Roads* (especially in the late summer of 1998), and his Vatican presentation (a large part of which doubled as chapter 12 of that volume). A sign that he was thinking of a place for retirement even when writing was a flight of fancy in which he saw himself moving to the Juan Fernandez Islands to study its unusual hummingbirds to gather further evidence for the Hamilton-Zuk hypothesis.[17] He was also preparing his paper for a conference in the Bahamas to which he had been invited by the Templeton Foundation (which supports scholarship on the connection between science and religion). That paper was probably the most outrageous Bill had written so far, elaborating a favourite theme in volume 2: the parallel between the Church and the coming superorganism. Bill was probably looking forward to the reaction to that paper.[18]

What Bill himself may have considered the biggest honour afforded him in the late 1990s was the invitation to write the back page endorsement of the new 'variorum' edition of RA Fisher's 1930 book *The Genetical Theory of Natural Selection*. He took this as an almost personal request, as he made the following connection between Fisher and himself:

> This is a book which, as a student, I weighed of equal importance to the entire rest of my undergraduate Cambridge BA course and, through the time I spent on it, I think it notched down my degree. Most chapters took me weeks, some months.
>
> ... And little modified even by molecular genetics, Fisher's logic and ideas still underpin most of the ever broadening paths by which Darwinism continues its invasion of human thought.
>
> Unlike in 1958, natural selection has become part of the syllabus of our intellectual life and the topic is certainly included in every decent course in biology.
>
> For a book that I rate only second in importance in evolutionary theory to Darwin's *Origin* (this as joined with its supplement *Of Man*), and also rate as undoubtedly one of the greatest books of the [twentieth] century the appearance of a variorum edition is a major event...
>
> By the time of my ultimate graduation, will I have understood all that is true in this book and will I get a First? I doubt it. In some ways some of us have overtaken Fisher; in many, however, this brilliant, daring man is still far in front.[19]

One research project that absorbed a lot of Bill's emotional and physical energy in the late 1990s was his continuing pursuit of the polio vaccine theory of the origin of AIDS, encouraged by journalist Edward Hooper, who was working on a detailed reconstruction of the way in which this might have happened and the history of the epidemic. His massive book, *The River* appeared in 1999.

As we have seen, acting by himself Hamilton had not been successful. He had first tried in vain to advocate for the need for testing and then turned to the editors of both *Science* and *Nature*, arguing for having at least a scientific discussion. But his submitted letters had been turned down. No one seemed to want to discuss it—he had met with a wall of silence. Only a medical doctor, Robin Weiss, seemed to take this whole thing seriously. Not even his mother, usually so supportive, had agreed with Bill when she learnt about this. She told him:

> You are going to be very unpopular if you pursue that one—polio of all things, that one is sacred! Anyway, if it's true, it's all happened and what could you do?[20]

Meanwhile, Hamilton had come to find the theory increasingly plausible on theoretical grounds. He was also getting more interested in virulence and in plant and animal vaccines.[21] After being approached by Hooper and learning about his ambitious data-gathering project, Hamilton gave him a personal 'grant' from his Kyoto Prize and began collaboration with him.[22] This resulted in a jointly published chapter in the *Lancet* in 1996 on a famous early case of this epidemic. Hamilton also accepted the invitation to write the Foreword to *The River* (which also contained a chapter on his own involvement with the theory). 'Even if the OPV theory is eventually rejected . . . he has done us a great service', Hamilton observed about Hooper's work in his Foreword. That was also the place for Bill to expound on his own views of science and truth. For Bill it was all déjà vu—he was used to being alone in his views among misunderstanding colleagues. What he may not have considered carefully enough was that this time the issue involved more than just bringing out the scientific truth. Bigger interests were at stake.

The theory needed testing, Hamilton insisted! And if nobody wanted to do this job, he was going to test the theory himself. As such, there was nothing unusual about this—if a theory predicted the existence of some organism, he set out to look for it (for example those colonial thrips that he found, and many instances in the Amazon flora and fauna). The unusual aspect was that the field research to be conducted was to happen in Africa in an area of civil war—not an obvious place for sample collection at the time. People tried to discourage him, but he persisted. He ended up organizing two expeditions to the Congo to try to find support for the theory, one in June–July 1999 together with Ed Hooper and one in January 2000 with Michael Worobey and Jeff Joy.[23]

This is the way Hamilton presented the rationale for his involvement to his close friend Yura Ulehla and wife Blanka in his Christmas letter that year:

12/24/99 Turin, Italy

The reason for being even slower than usual in getting to my Christmas communications is not just the unfinished half dozen projects connected with my work but added to these this year a coming project of going to the Congo (former Zaire) just after New Year to collect chimpanzee shit to examine for virus. Probably *you* will feel *you* are dreaming when you hear this since it is so unlike anything you have ever heard of WDH doing before but the fact is I have already been doing it this year (was in former Zaire in June–July) and now, having turned up nothing last time, must go there again and try to get to the wild adult animals in the forest.

It is all connected with the 'OPV' (Oral Polio Vaccine) theory of AIDS origin which I have come increasingly to feel an unjustly derided and actually very plausible theory of AIDS origin and, since nobody more appropriate seemed willing to take the most obvious step to test it, I decided to do so myself. The idea is that, if it is indeed true that Hilary Koprowski's chimpanzee-bonobo testing camp for his live oral polio vaccines at Kisangani (then Stanleyville) in the (then) Belgian area of Africa was the source of an accidental contamination of the vaccines which were fed to a million Africans in the late 50s, it should be possible to find an appropriate wild SIV (simian immunodeficiency virus) in chimps of the area where Koprowski's assistants collected the animals for his testing camp... It is to these precise areas that I now intend to go together with two young and athletic Canadians and two litres of special RNA buffer solution...

My preliminary trip in June (together with Ed Hooper...) succeeded in turning up 22 specimens of faeces from pet baby chimps in the town, but my expert back in Europe could find no virus in it at all. Either I had collected too little, or I had stored it in a bad way... or there simply was no infection—the latter probably being most likely, since these very young animals could only have caught it in utero or by breast feeding and that is known to be no more than a half chance, even though somewhat higher, seemingly, in chimps than in humans.

'SIV-cpz' is believed never to be common in chimps since only a very few of zoo chimps have ever shown the virus, but since virtually no samples have been taken from wild adult animals, so that the same questions of mother child transmission apply (the zoo animals having very likely been caught as infants), my own feeling is that we perhaps have little indication how common it might be. So far no SIV virus at all has been found in bonobos ('pygmy chimpanzees') so if I can find any it will be a significant event whether or not it turns out to be at all close to the human HIV-1 that causes AIDS. The well known extremely promiscuous sexual activity of the bonobos suggests to me that they might well have SIV as a consequence of their behaviour, or even conceivably as part of its cause (arguing on lines of how the rabies virus 'causes' special transmission behaviour in its hosts, namely that unusual aggressive biting even toward species it doesn't normally attack).

Anyway, basically bonobos only are on the left bank of the Congo river at Kisangani and Chimps only (Pan troglodytes schweinfurthii—again a hardly ever sampled subspecies) are on the right bank. One needs a minimum of a day's walk into the forest...to find them...and, thanks again to that last trip, I am already in touch with interpreters and hunters willing to lead me to wild groups. The hunters, of course, would normally try to shoot the animals and bring their smoked and dried flesh to K market, together with, for sale as a pet, the unfortunate baby of the victim mother if she had one, but of course I will allow no shooting primates of any kind on my trip and I will actually try to persuade the hunters that they will ultimately get more profit from the animals by protecting them and showing them off to tourists, though this may be, unfortunately not an easy thing to persuade realistically with the DRC (Democratic Republic of Congo) in its present war-torn state, there being no tourists in the country whatsoever. Shit and urine, possibly hairs and hair follicles, will be our only objectives but I hope our guides will know how to lead us to those too.

The state of the country is of course a source of concern to us too, but at present there is supposed to be a cease fire between the rebels and the government...Meanwhile the various factions among the rebels are also at an uneasy standoff with each other...However, the factions of the rebels, for want of government troops to fire at..., do occasionally fire at each other in a kind of 'friendly game' spirit while waiting for the big match. So,...we might conceivably get caught up.... Our line this time is going to be that we are on a mission that should be close to the hearts of all Africans of whichever faction—a better understanding of the awful epidemic that has struck them and a step towards a possible vaccine or a cure. This is an idea which I really believe in and which I hope I can persuade them that I believe in...Anyway, once we get beyond the (awful) roads to the back villages and still more to the jungle I think we will be among peacable farmers and hunters and have no reason to fear for safety providing we are careful about local customs and manners and are seen as a source of income through our purchase of food and labour. As I see it they are indeed much the sort of people I am already familiar with in the backwoods of Brazil (indeed largely they may be of the very *same* culture as transported hardly changed across the Atlantic by the slave trade) and with those people I have always got along well. Strong similarity certainly was my impression as I drove around the poor earthen backstreets of Kisangani searching for baby pet chimps during the summer trip.

Well, enough of those past and coming doings. I can tell you from the last trip, it will be exciting enough. This town is the very place about which Joseph Conrad wrote 'The Heart of Darkness' and even to some extent it's back to even more primitive than then when there were at least the big Belgian-run steam boats. Now almost the only functioning kind of boat is the age-old hollowed gigantic tree trunk propelled with paddles that look like huge broad-bladed spears... On the river we went down to the site of the Lindi camp where the Koprowski chimps had been held and the forest had indeed repossessed it and we must search hard amongst overgrown oil palms and through old manioc fields to find rusting fragments of the girders of the two airplane hangars that had housed the 400 chimps that passed through for the 'tests'.

Enough, enough. Other than this new 'African' line of mine the year has been fairly uneventful for me—no trip to Brazil for example—and mainly spent in writing and only a few brevissimo trips elsewhere, for example by invitations to a Max Planck Inst in the Lubeck area of N. Germany, to Wuerzburg, to here (Italy, twice) and to Lausanne, Switzerland. After the last I tried some alps walking and found how unfit (or perhaps more accurately, age-weakened) I have become. Still, I did reach the top of my lesser target mountain and—just—reached the mountain hut that I aimed for before nightfall, so on the whole I am not discouraged about the idea of walking a day into the Congo forests, especially with strong companions and local men beside me, although I hope not to be a laggard...

Your letter this year was a delight, even more full than usual of sardonic Czech humour...

The girls have all gone to their mother in Scotland for Christmas. Helen is just back from a holiday trip to Chile, Ruthie is in mid year of her DPhil on morbilloviruses in Cambridge U...and Rowie is studying art in Leeds and doing quite well I like to think... Today is Christmas Eve and I must ring them all...

Well, a bit late to write Happy Christmas but I am thinking it and I can at least wish you a happy entry to our next THOUSAND YEARS!

From Bill[24]

The tone of Bill's letter to Yura is calm, but we learn that he is planning to enter an area of civil war. We also learn some reasons why he thinks the trip may be fruitful: there is the possibility of finding something significant because so little is known; the mere finding of an SIV virus (even one unrelated to HIV) would constitute a scientific contribution and could be useful towards developing a vaccine or cure. Bill's optimism may have had something to do with an extension of the model system that he had developed earlier with Dieter Ebert, or he may have had tentative theories or hunches of his own that he is not disclosing in this letter. But first he will need the results from the laboratories with which he has pre-arranged to test the findings. One 'model' that appears over-optimistic and under-researched, though, is the suggestion of a similarity between the people he might encounter in this part of Africa and the ones he knows so well from Brazil.[25]

There were dangerous moments during his first trip in the summer of 1999. He got in fairly serious trouble for taking pictures in places where it was forbidden. Here is an excerpt of the email that Bill sent during that trip through a friend of Luisa's, Alessandro Villa, with a Uganda email address.

> Dear Luisa
> The back of the tee shirt that I saw as I walked here said 'Justice for all', well, I don't know how it is for most Congolese, no better than for this lonely musungo I suspect and that [is] no justice at all. On account of a photo of sunset on the Congo river yesterday I have already paid out $250 and expect to pay another $100 in bribes before I am clear and have wasted the entire day chasing people for the next step, my taxi moreover being taken over by the corrupt official as he professed to try to find out where my camera had got to.[26]

Bill goes on to say that the situation has gone from being 'fun' to suddenly becoming very threatening, and that he wants to get out of the country. Ed has left, after he and Ed have 'fallen out a good deal the last few days'. Finally he mentions the name of the man harassing him, an immigration official known at the airport, 'should you wish to direct forces to try to find what he has done with my camera and/or me'. Bill also suggests alerting the Royal Society.

But the next day there is an email from Alessandro telling Luisa that things are OK:

> Dear Luisa, everything OK. I found both [Ed and Bill] healthy and safe! They will come back this evening by plane to Kampala! And they will have a privilege that has never occurred before: bringing with them the chimp faeces they have collected with diplomatic luggage! They will be recorded in history![27]

Much has been made of Bill's last expedition to the Congo to collect evidence for the OPV theory of AIDS. Indeed, this trip had all kinds of dramatic ingredients and potential risks—but so had most of his earlier expeditions. Bill was, by nature, a risk-taker, both intellectually and physically, and did not worry much about potential dangers. He used to tell Luisa when she tried to discourage him from going out to gather data himself: 'Don't worry, I have nine lives!'. At the time of the Congo expedition, he may already have used them up, Luisa sadly noted. His expeditions were usually quite successful (Luisa herself had participated in a couple of them), but this time everything went wrong, she said.[28]

One of Hamilton's young companions became very ill, stung by a poisonous thorn, and Bill himself fell ill soon after with a bad strain of malaria. He was immediately given effective antidotes and he quickly returned to London for further treatment. But as it turned out, malaria was not really the problem. In February 2000 I learnt from EO Wilson that Bill was lying in a coma in a London hospital. He later died on 7 March. It turns out he suffered an unexpected, massive internal hemorrhage shortly upon being admitted to the hospital in late January, but was kept alive with blood transfusions and operations until he suffered what was called 'massive system failure'.[29]

Hamilton's death, after the Congo expedition, has given unwarranted importance to his interest in the OPV theory of AIDS. There is no doubt that he *was* interested in this particular theory, and had started finding it increasingly plausible. But it must also be said that he would *also* have been willing to accept other theories of the origin of AIDS as true, especially some combination of the 'bush meat' theory of the original transfer of chimpanzee SIV virus to humans and its subsequent mutating into HIV-1, and the 'dirty needle' theory of the later epidemiological-seeming transmission and early spread of the HIV-1 virus in Africa. Not only do we know this from what he told close family and friends, but also, this was the way he typically reasoned as a scientist. In fact Bill never committed himself fully to any theory—he kept an open mind in case the theory did not work out. Hamilton was operating with plausibilities, preponderance of evidence, and the like, not absolutes. But at the same time, he was the first to try to gather enough supporting evidence for his favourite theories—in nature, in libraries, and by talking to people. He was his own best critic when it came to creating a convincing case.[30]

So it would be wrong to say that he died 'for' the OPV theory, although he considered it extremely important to find out about the origin of AIDS. For Bill the transfer and transformation of SIV to HIV-1 represented a model case of the dangers of viruses crossing species barriers. He was concerned to warn about this for the future, in relation to such things as xenotransplants, much discussed at the time. What irritated him tremendously was that no one seemed to take the OPV theory seriously. This was one of the cases on which he thought he had plenty of information—he had been following the development of the evidence since the early 1990s when he was originally approached by some individuals who wanted his support.[31]

I asked Ed Hooper what could possibly be the point of pursuing such an old case. His answer was: a class action law suit. Hooper may have had a class action suit in mind, but it is hard to imagine Bill taking the same type of interest in this matter. Although he probably would have enjoyed being proven right about the OPV theory, his main concern was for the truth of the matter to be known.[32]

The second Congo expedition itself might not have been necessary if only the samples collected in Hamilton's first expedition half a year earlier had turned out to be successful. The aim was to collect faeces samples of chimpanzees close to the place of the original vaccination campaign (blood tests were discouraged by primatologists). Unfortunately, though, the samples turned out to be collected from a limited number of individuals and so to be unrepresentative. The premium paid had tempted the villagers to multiply the benefit by rotating pet chimps! Wanting to strike while his visa was still valid, a January expedition appeared to him the best solution. And the Africa trip probably looked especially appealing as an alternative to Brazil after he was prevented from doing his research on parasites there in 1998.[33]

It was this second expedition that put an extra stress on his body, making a pre-existing condition worse. Because there are indications that he had internal bleeding already before, in the form of an ulcer, or diverticulitis (thinning of the duodenal wall). But this is something that, in characteristic Bill Hamilton style, he did not pay serious attention to. He did catch malaria, but was given effective antidotes, with the result that no malaria could be detected when he arrived in London in the Tropical Diseases wing of the hospital. But even if this was the case, malaria takes a toll on the body and had probably made his pre-existing condition worse. There are hints that Bill felt somewhat weak even setting off on the second expedition. In fact, he may have barely recovered from his first expedition, from which he came back with a fever. Bill thought—or hoped?—he was sufficiently recovered, and proudly reported in a letter to his friend Yura that he had been able to climb a mountain in the Alps as a test case, and therefore felt he 'would not be too much of a burden to my young companions'.[34]

At the end of his second expedition, Hamilton may have been sick but he was upbeat. Here is an excerpt from what was very possibly the last email Bill Hamilton sent in his life:

> Dear Luisa
> I am back in Entebbe after a very successful trip but a bit sick again (tail end of malaria?). We fly to Nairobi tomorrow and to London tomorrow night by Kenya Airways.
>
> Could you let Mary know I will be arriving early Saturday morning and am a bit weak so if she could meet me at Heathrow I would be most grateful (Will also try to phone her tomorrow)
>
> About 50 shit samples of different P. t. schweinf: some of P. paniscus may be being collected for us right now.[35]

Hamilton was admitted to the University College and Middlesex Hospital the day afer he returned. It was when he was waiting in the Tropical Medicine wing of the hospital that he asked to lie down, suffered a sudden huge haemorrhage and lost consciousness. One of his doctor sisters said that any place other than the Tropical Medicine wing would have been a better place to identify what was wrong with him.

For the next several weeks Hamilton lay in coma. For some time there seemed to be a 50–50 chance of recovery, although he was not expected to recover fully in any case. His family gathered around and kept watch over him, trying to do something that would wake him up. His daughter Helen sang some favourite songs, and his sister Janet read some of his favourite poems. Christine visited from Scotland. But Hamilton did not show any sign of recognition. Luisa believed, though, that he did show some signs of awareness. (Incidentally, this attempt to stimulate Bill was based on some information from another family member who had been in an unconscious state in a hospital, but later, when he woke up, reported hearing his visitors talking to him—Mary's son James.)[36] But the attempts were unsuccessful as were those of the medical personnel. One after another his organs shut down, until he died of massive organ failure on 7 March.

The funeral was a very private affair, with only close relatives and friends present. There is a poetic point relating to Bill's funeral. Bill had written an intriguing fantasy for a Japanese entomological journal about how he wished to be buried by his favourite beetles in the Amazon jungle, entitled 'My intended burial and why'. How that might have been arranged in practice is anyone's guess. In fact, this was probably one of many fantasies (he used to entertain himself by engaging in various kinds of death fantasies, often involving self-sacrifice of some kind).[37] Instead, he was buried near Oxford close to his beloved Wytham woods. His grave was initially decorated with flowers from the garden of Oaklea, with a large fossil as the headstone. This was the way the people closest to him knew him best: a curious naturalist, who loved and theorized about life in all its forms, be it plants, fossils, insects, birds—or even parasites and pathogens. He had a quest for merging with Nature, being totally united with her. In that famous poetic passage from 'My intended burial' he extends this to being ultimately consumed by Nature:

> I will leave a sum in my last will for my body to be carried to Brazil and to these forests. It will be laid out in a manner secure against the possums and the vultures, just as we make our chickens secure; and this great Coprophaneus beetle will bury me. They will enter, will bury, will live on my flesh; and in the shape their children and mine I will escape. No worm for me nor sordid fly, I will buzz in the dusk like a huge bumble bee. I will be many, buzz even as motorbikes, be borne, body by flying body out in the great Brazilian wilderness beneath the stars, lofted under those beautiful and un-fused elytra which we all hold over our backs. So, finally, I too will shine like a violet ground beetle under a stone.[38]

It is interesting that many obituaries and media reports at the time reported that Hamilton had died of malaria, or complications from malaria. But that was emphatically *not* the case. I received various pieces of information from people who had attended the inquest. He did have a pre-existing condition which had been getting worse. But it also seems that Bill had taken a pill of some sort which caused the bleeding. In order to get the details absolutely right, I will quote from the letter from his sister Janet, who attended the inquest and also got hold of the Coroner's report:

> The Coroner, a lawyer in this case, held an inquest at Westminster Coroners Court. I will also attach a copy of the death cert which gives dates etc.
>
> The Coroner went through the likely causes of death and the possible conditions that might have contributed to the death. This is what you say Michael Rodgers and others remember him doing; I was there too. He concluded, after all the evidence, that Bill died from 'multi organ failure due to upper gastrointestinal haemorrhage due to a duodenal diverticulum and arterial bleed through a mucosal ulcer.' There was nothing by way of contributing factors to put in the second box on the certificate.

> As far as I know the pathologist's report is not available now but some comments from the pathologist who did the autopsy have been passed on to me in a letter from the BBC. They say Prof Sebastian Lucas was the pathologist and remembered the case. His final comment in that letter is that Bill died from a congenital diverticulum which happened to ulcerate, perhaps spontaneously, perhaps because of an external irritant.
>
> The BBC's letter to me also says they asked Michael Worobey about Bill's malaria in Uganda. In essence MW's reply was that he was with Bill when the doctors diagnosed malaria by blood test and gave him injections in the clinic which cured him as confirmed by the blood tests he had in Hosp in London. In my opinion it is greatly to MW's credit that he sorted out Bill's malaria with such exemplary care. I spoke to Michael when Bill was back and he mentioned that Bill had bleeding episodes on the trip; we know these had happened in UK too over the years.
>
> Bill could have had a massive bleed any time anywhere...
>
> There is no way of telling whether Bill took any pills in the day before he collapsed in UCH. He might have, as we all do from time to time for aches, pains etc but if he took something it was not for malaria as he did not have that disease; obviously one cannot die from a disease or complications of a disease that one has not got.[39]

Meanwhile there is admittedly a poetic appeal in the idea that Hamilton, who studied parasite-host interaction, would himself become the target of a parasite, something that was not lost on his modern poetry-writing New Zealand cousin on his father's side, Jeannette Stace. Here is a poem she wrote shortly after his death which neatly capsules her cousin's life:

> kin
>
> my famous cousin insect man Bill Hamilton
> enthralled by them all thrips lice aphids parasites
> and all the social insects bees wasps termites ants
> everywhere he finds his treasures with their secrets
> under rotting bark in childhood Kent in Brazil
> longs for a weta to be sent from New Zealand
> offers a twist to Darwin not just the fittest
> what about altruism some insects may die
> let close kin reproduce their DNA survive
> the questions come quicker than the time for answers

late recognition for the unpretentious man
sometimes forgets to give his once-a-year lecture
in Africa to search out source of HIV
is dead in a few weeks from a parasite[40]

A memorial service was organized by Richard Dawkins on 1 July 2000 in the Chapel of New College, Oxford. It was non-religious except for the inclusion of two anthems, by J Frederick Bridge and Handel, respectively, which had been sung at Darwin's funeral in Westminster Abbey. Bill's family and friends read letters and poems, and told stories. A fund was established to make it possible to archive Hamilton's notes and materials in the British Library, an enormous enterprise valiantly executed by Bill's student Jeremy John under the leadership of Anne Summers.[41]

Before his last expedition Hamilton had been successful in convincing the Royal Society to call a special urgent discussion meeting on the origin of AIDS. The meeting was first postponed. When it was finally convened, the vaccine theory was declared incorrect based on the then available evidence. Had Hamilton been there, he would most likely have pointed out that the important point was the ease with which the virus passed species boundaries—from SIV to HIV, whatever the mode of transmission—and the warning this implied for mankind in regard to such things as xeno-transplantation and future pandemics. He would also most likely have criticized modern medicine for selling out to the pharmaceutical industry.[42]

Another later conference at the Italian Accademia dei Lincei treated the case of AIDS and the problems of a potential pandemic in more detail. At this conference speakers included not only doctor Robin Weiss, Hamilton's original collaborator, but also Ed Hooper, his sister Mary, and Luisa.[43]

Later Michael Worobey, now working with Beatrice Kahn, went back to the Congo to collect and test more samples. His conclusion was that there was no evidence for the OPV hypothesis. This was well publicized in *Nature* and other scientific journals. Ed Hooper did not accept this as an answer. Later Worobey returned and more samples were tested, OPV negative. Hooper continued fighting back.[44]

A number of memorial symposia and special journal editions were later arranged by his students and a number of journals arranged for special issues celebrating Bill Hamilton. A Hamilton memorial website was created. Bernie Crespi considered a volume with Bill Hamilton stories gathered from his students who knew him well. Instead of that volume, we have volume 3 of *Narrow Roads of Gene Land,* containing articles with Prefaces by Hamilton's various student collaborators as well as original articles by Hamilton.[45]

For a long time, those who knew Bill Hamilton had an acute sense of loss. Bill had always been there with help and advice and generous access to the marvellous comparative data in his mind and his card collection. As his friend Peter Grant wrote in his obituary: 'Hamilton's presence is now an absence'. But traces of Hamilton kept reminding people that his ideas were alive. On the signboard in the corridor of the Zoology Department at Oxford I saw a caricature of a mischievous Hamilton grinning and holding up Hamilton's Rule.[46]

But people also remember Bill Hamilton as someone who had a suitable poem for any occasion. A poem that Bill himself might have chosen was the melancholic 'A Shropshire Lad' by his favourite poet, AE Housman. We can be fairly certain that this was an absolute favourite, since he felt inspired to include that poem in his book review of Richard Dawkins' *The Selfish Gene,* a book that he truly appreciated.[47] His daughter Ruth read it at the memorial service.

Bill would certainly also have liked the second poem he chose to quote in that same book review, Wordsworth's poem about Newton, with its evocative final lines:

> ... a mind forever
> Vogaging through strange seas of thought, alone.[48]

He would have felt very honored hearing his friend Richard reading those lines at the Memorial Service, now applying them to him, Bill Hamilton, instead.

But Bill, being Bill, might really have felt most comfortable with his own modest choice, a haiku by Matsuo Basho from his *The Narrow Road to the Deep North,* the book that inspired the title for his own *Narrow Roads of Gene Land.*

In a letter to a close friend, Bill Hamilton revealed that [s]ince first reading it, I have thought how I could find no better banner under which to die than Basho flies for himself in his last haiku, this written a few days before his death:

> On a journey, ill,
> But over withered fields of Autumn,
> Dreams go wandering still.[49]

24

The Edge of Creativity

Scientists as they mature typically assume roles that allow them to multiply their overall scientific effort through advising and managing younger scientists rather than sustaining a dynamic research agenda themselves. Bill Hamilton, who many see as the Darwin of the 20th century, is one of the exceptions. He continued the intense scientific quest he embarked on as a teenager until his accidental death in 2000. By then he had garnered a great number of international honours and prizes. Because of his interdisciplinary orientation many fields claimed him as theirs, including genetics, evolutionary biology, ethology, entomology, parasitology, and evolutionary psychology.

Rather than building on his previous successes and creating an academic empire, as many scientists do, Hamilton operated as an individual. He followed his own research programme, largely self-taught and acting as his own adviser and critic. He set very high standards for his papers, but they often remained obscure to his naturalist colleagues. He was always looking for universal principles across observations from Nature and relishing in evolutionary oddities and paradoxes. For Hamilton, Nature was a big puzzle that needed solving. A physical and intellectual risk taker, he loved challenges, never happier than when faced with an unexpected problem. He retained a dislike of authority throughout his life, had little patience with formalities, and escaped people for the natural world when he could.

Hamilton was a deliberate challenger of prevailing paradigms. First there was the question of altruism. Here Hamilton supplanted the post-war 'good for the species' group selection idea with his kinship theory (or the idea of kin selection). Later, in regard to the explanation of the origin of sexual reproduction, he offered an alternative to the existing mutation elimination explanation in the form of his Parasite Red Queen hypothesis, according to which the main goal of sexual reproduction is to create continuing genetic diversity and in this way fool parasites. At the same time he also gave a new interpretation of sexual selection: the traits selected for are not arbitrary—females want healthy males.

Throughout his academic life, Hamilton preferred to be a path breaker, doing what he did best—introducing new ways of thinking and providing the necessary theories and tools for others to follow. He was happy when others provided more corroborating empirical evidence. Meanwhile he kept moving from field to field, which meant that as his colleagues finally started catching up and engaging with his ideas, he had already moved on to something else—a new unexplored area where he was again the pioneer! Indeed, there is something of the Red Queen about Hamilton's whole scientific approach.

Working as an individual, creative scientist, always on the edge, he suffered the concomitant agony and ecstasy in regard to acceptance of his ideas. Still, he had the good fortune of being left alone to pursue what he wished. For a scientist, he had an almost ideal existence in a difficult world of scientific achievement. His parents and siblings supported him morally. His wife and children provided a background of domesticity, but left him alone when he needed to work. He had many adoring students. He made a career move to America but was lured back to England with a Royal Society Research Professorship. After moving to Oxford he had money to undertake all kinds of research expeditions.

Later, after his wife took up a job as a dentist in the Orkney Islands and Bill, not used to being alone, found another partner, his travel pattern came to include working vacations in Italy.

Even if we can sympathize with Hamilton, his intellectual loneliness, and the lack of understanding he met among advisers and journal referees, we should also note that he was continuously supported in his quest by scholarships and grants of one type or another in his early career. There always seemed to be some superior somewhere who believed in him. And this cannot be attributed to mere 'luck'. He would not have been supported had he not had something to offer, such as naturalist knowledge and theoretical ideas. (I have in mind especially OW Richard's and Richard Southwood's appreciation of him.) Bill was really a walking encyclopedia and thinking machine all in one. From the very beginning he was operating at the level of 'grown-ups', or professional biologists—he was never really anybody's obedient graduate student but mostly his own adviser. Although not without struggle, in the main he was left to do what he wanted, but he kept his part of the contract: he delivered results.

From early on Hamilton followed an internal radar in regard to science and life, self-taught in regard to important scientific matters, and somewhat contemptuous of 'fashion'. It was he who found the problems that he wanted to solve, and he who drew the implications for further research. In this way he constructed an evolving research programme of his own. Over time Hamilton's life increasingly split itself up into what I have called 'Kafka' and 'Bates' modes: deep theorizing on one hand, and intense travel and exploration on the other. Without external disciplining factors such as lectures or administrative duties he could largely do what he wanted, which made the oscillations of this model more intense, especially after he lost his mother and the old Oaklea setting as an anchoring point.

Hamilton was extremely ambitious. He was interested in important unsolved problems, with potential for great satisfaction and glory for the problem solver. Behind his mild mannered appearance, he was very competitive. But he preferred using unusual methods and strategies, sometimes finding the solution by exploring what seemed like sideways and byways. He always knew what he wanted: an uncompromisingly gene-centered answer. This deliberate self-limitation was partly a rational conviction, partly a method, and partly a decision to stand up for biological explanations in a world of political correctness.

Moreover, perhaps surprisingly for someone who routinely worked with approximations, Bill was a kind of perfectionist. He always wanted the best possible model under the circumstances. Sometimes he had to be told by the computer experts assisting him that this was not realistic. But he kept pushing the limits of the possible. His computer expert and collaboratory Brian Sumida described Bill's ambition as follows: 'His objective in any project was to create the best in the world.'[1]

But at the same time, in some respects Bill showed amazing capabilities, beyond what his assistants themselves could achieve. He could handle extremely complex situations in an intuitive way, something akin to a musical maestro, operating at the edge of possibility. And this had to do with his computer program debugging style, of all things. Sumida had never seen anything like it:

> Many debugging strategies exist, mostly centred around following a single logic thread through the program. Bill's method, however, was more macroscopic. He would look at the entire data output to work out what sort of logical error would produce the results, then back reference to the code to find the sector that violated the logic. Most programmers (including myself) work in the reverse fashion, debugging by examining the code, tracking the values of a single variable or a data structure. *Bill's method was like taking a picture of a freeway system from a helicopter, and by examining the patterns, deduce the position and timing of the stop lights from first principles* [italics added].[2]

At one point Bill asked Brian Sumida to develop a simple 'limit-cycle algorithm' that he could explore (he called it 'Cycesh'). It seems that working with it, Bill again used a kind of intuitive multifactor approach, this time similar to playing an instrument to its physical limits:

> When simulating limit cycles, the algorithm can easily blow up (go to infinity), which often happened to me when I ran Cycesh. There is a delicate balance of settings that result in stability. Bill's adroitness in selecting parameter values resembled, to me, the kind of skill displayed by a musician who, while playing an electric guitar on the verge of total feedback is still able to coax beautiful music from the instrument.[3]

In other words, Bill's very strong, almost unrealistic ambitions were in fact offset by the rare set of talents he had at his disposal, some of which often helped him to reach his goal.

His writing style was not particularly user friendly. It was often as rich in detail and subclauses as the variety in the Amazonian rainforest, and just as one organism there could 'hide' among others, so might an insight of Hamilton's remain hidden within one of his complex paragraphs. This is why it took popularizers of Hamilton's ideas, such as Richard Dawkins and Matt Ridley for those ideas to be more widely appreciated. One reason for this was probably Bill's typical ambition to optimize. He optimized in many ways (for instance always combining his conference travels with various side trips). When writing papers, too, he often wanted to combine his literary and scientific ambitions by finding a language that was aesthetically pleasing. This could sometimes get in the way of maximal clarity. Another type of perhaps unwitting 'optimization' was Bill's tendency sometimes to introduce in the same paper just too many unknowns at the same time—new theory, new method, and new organisms—casting some referees into the roles of mountain climbers desperately looking for footholds.[4]

Hamilton treated his students with great respect as fellow explorers. He seldom delegated things, because he was so intensely curious about facts and details of all sorts. One of his last students tells us that Bill, rather than scrutinizing his thesis in detail, asked him to tell him about interesting phenomena that he had observed—and without offering any explanation or theory at the same time. Bill wanted 'just the facts', to put these into his enormous comparative data base, either in his head, or on his 'external grey matter'—the card collection with index cards scribbled full of minuscule notes. In his relentless fact-gathering zeal, Hamilton was very similar to Darwin.[5]

Opening up new fields of inquiry is a demanding course of action for a scientist. It takes a constant flow of new ideas, and at least some of these will have to be workable in practice. As noted earlier, Bill was good on this front, he was a font of ideas, but in the end, he mostly had himself, rather than a regular cadre of students to send out to do work for him and test ideas. The limiting factor, therefore, was his own creativity.

In turn, this creativity required certain conditions to be right. The trips to the Amazon and nature expeditions were a combination of idea and fact gathering missions—he used both to get stimulation for new studies and to find examples that would illustrate some of his existing models. Above all, progress in his research was dependent on absolute concentration and freedom from disturbance. This was true also for the very special position he ended up holding, the Royal Society Professorship. It was renewable every five years, based on accomplishments. As Bill himself noted in 1996, with seeming relief (in a rather Billish fashion): 'My professorship has been through two of its five year renewals'.[6]

Much of Hamilton's later work was done in front of the computer, watching various simulation programmes. And this is where he was at his most creative. The curves generated by the computer directly spoke to him; he used to watch these for hours, whispering to himself.[7] This was his visual imagination operating at its best, unravelling secrets and making discoveries. But was Bill merely watching and thinking? I believe that something was going on here similar to the experience of many artists, craftsmen and designers, a 'seamless and unconscious collaboration of the eye, hand, and mind', described by Juhani Pallasmaa.[8]

> As the performance is gradually perfected, perception, action of the hand and thought lose their independence and turn into a singular and subliminally

> coordinated system of reaction and response. Finally, it is the maker's sense of self that seems to be performing the task as if his/her existential sense exuded the work, or performance. The maker's identification with the work is complete. At its best, the mental and material flow between the maker and the work is so tantalizing that the work seems to be producing itself. This is actually the essence of the ecstatic experience or the creative outburst; artists repeatedly report that they feel that they are merely recording what is revealed to them and what emerges involuntarily beyond their conscious intellectual control. 'The landscape thinks itself in me, and I am its consciousness,' Paul Cezanne confesses.[9]

Exactly this artistic sense of ecstasy is probably how Bill felt as he was having his early 'epiphany' exploring the parasite avoidance theory of sex and discovering the principle for sexual selection. In the process of intense watching, he merged with his own creation.

Over time, however, this pressure to be constantly creative created a dilemma. Bill would have liked to spend more time with his family, but he became increasingly absorbed by his work. In later years he felt more driven than before and had a sense that his creativity was diminishing. At the same time, he felt an urge to do something significant for mankind, his ultimate scientific contribution.[10]

There was a cost to Bill Hamilton's scientific lifestyle. As a scientist, he was constantly operating at the edge of his own creativity, while searching for conditions that would allow it to flourish. Not having established an academic empire, he increasingly put himself at the mercy of his individual capability, with its growing absence of stabilizing and anchoring factors.

At the same time, with his intellectual and physical risk-taking, Bill Hamilton brought back the image of the scientist as hero. And the science he made exciting was evolutionary biology, at a time when its (and Hamilton's) 'competitor field', molecular biology, was again making great inroads. By his example he showed that a scientist could think, say, and do things that on the face of it seemed almost impossible. He was the Indiana Jones of evolutionary science, going off into the jungles with his machete, hacking his way to find hidden treasures and opening up the possibility for truth while conquering great dangers along the way. At the same time he strongly brought back the idea of the social responsibility of the scientist.

But he was surprisingly nonchalant when it came to his own safety. We know about the last expedition, but there are many earlier episodes that were quite dangerous too. What if Bill Hamilton had been lost to the scientific world at a much earlier point? What if he had never been born? What would have happened to the theory of inclusive fitness? Or the parasite theory of sex and sexual selection?

Hamilton himself provided an answer to these questions, with which I believe many scientists would agree. Science is not dependent on the individual scientist. Important scientific truths are going to be discovered sooner or later; they are there to be found, and who discovers them and when is partly a matter of accident. Multiple discoveries are characteristic of science. One person who might as well have been credited with the theory of inclusive fitness, according to Bill, was Ilan Eshel, his good friend, who felt and reasoned about many things in a similar way.[11] A more surprising candidate found by Hamilton—this time for credit for the parasites and sex theory—is the naturalist Edward A Wilson. Wilson was on board a ship on his way with the Scott expedition aimed for Antarctica and the South Pole when he wrote and dispatched from Cape Town a manuscript on the behaviour of the red grouse to the *Proceedings of the Zoological Society of London*. Wilson attributed the birds' behaviour and look (particularly of the legs) to infestation by parasites. Accompanying the manuscript were some unexplained beautiful water colours of the birds. In these Bill perceived some deliberate attention to typically parasitized areas, and he conjectured that Wilson would have continued his discussion in later papers—if he had lived. Unfortunately, he tragically died with Robert Scott on the way to the South Pole. But there was someone else, the zoologist HB Fantham, who also wrote about parasites in that same journal in 1910.[12]

Bill even brings up Charles Darwin as someone who might as well have got lost on one of his expeditions, citing his 'bravado' in the Andes, for example.[13] Bill doesn't tell us what he thinks would have happened in that case—who would have substituted Darwin? But he mentions the great tradition of intrepid British explorers of that time, some of whom perished. In other words: science goes on. Each scientist knows that there will be others following in his or her footsteps, and others before who may have attempted the same thing but the circumstances were not right. Accidents happen to some, but others will follow. As Hamilton puts it himself: 'such historical cases incline me...to allow quite a substantial role for accidents in the evolution of evolution history'.[14]

Science was the best thing that Bill could imagine. It was to science that he devoted his life. He wanted his daughters to become scientists, and two of them did.[15] His students and close colleagues were treated as family members. He lived his life as part of a great brotherhood of science—or perhaps rather a 'siblinghood' which included women scientists as equals.[16] Taking a final look at that somewhat mysterious category of altruism which doesn't involve kin—is it possible that Hamilton here has the scientific community in mind? When he suggests that non-kin altruists may just find themselves in the same habitat because of similar preferences—could this habitat be a university? Is the 'green beard' phenotype perhaps identifiable by the simple fact that someone is a scientist? Because, for Bill, being a scientist is automatically connected to certain values and behaviours, including altruism. And it is important that altruists find one another in order to stand up for truth against oppressive authorities. Bill Hamilton is here touching on a romantic Ur-model of the scientist, and exemplifying it by his own behaviour.

What is the situation today? Are there young Bill Hamiltons around? How can they be recognized? How should they be treated? I believe that it was crucially important for Bill as a young prodigy to have a steady and supportive family to come home to, as well as an organized education at Tonbridge and Cambridge. He did not always like the school, but that was not the point—what was important was for there to be an educational structure against which young Bill could rebel, and which could prevent him from being entirely self-directed. His scientific self was forged exactly in the tension between formal educational requirements and his creative and impulsive mind.

We may also ask: if Bill Hamilton had been a young researcher today, how would he have fared? What kinds of provisions exist today to identify and cherish unusual talent and scientific creativity? The criteria used today may well screen students like Hamilton out. Bill was shy and unforthcoming, although with a strong inner conviction. One strategy in science is to market oneself and to cultivate good social skills. Bill chose a strategy that suited the intense focusing the goes with scientific creative work, and which let him work largely alone, which he wanted. Moreover, he was quite willing to become a teacher or carpenter, in order to be able to continue with his theorizing. What if he had chosen such a route instead of an academic career? Would he have been able to publish, and would we know about him today?

* * *

How has Bill Hamilton's work held up over time? Very well, thank you, would be a fair response. He laid the groundwork for a number of theoretical considerations which have later been taken further by others, often through ever expanding empirical studies. Whole industries have developed around the Parasite Red Queen and the Hamilton-Zuk hypothesis, and these show no sign of abating. Clever biologists are trying to imagine ever new ways in which parasites can be avoided and put these ideas to test in well-designed empirical studies. In 2007 Bill Hamilton made the cover of the journal *Evolution*, which unexpectedly featured a tribute to his early theory of senescence. Other more controversial theories, such as the theory of autumn leaf colours and the cloud hypothesis continue inspiring more daring souls.[17]

But what about Hamilton's 'eldest child', his inclusive fitness theory? There was a kind of indirect assessment in 1988 when *Current Contents* declared Hamilton's 1964 paper the most cited paper ever in the *Journal of Theoretical Biology*. But how did it fare later on? At the time it was launched, it (or rather 'kin selection') became the answer to the post-war paradigm of 'group selection', causing a veritable stampede. But later there were attempts to restore group selection (or group selection of sorts).[18]

Inclusive fitness theory has become a cornerstone of Neo-Darwinian evolutionary explanation, whether called by its real name or under the name of 'kin selection'. And indeed, considering that Hamilton always intended the theory to be an extension (or even 'correction') of Fisher's General Theorem, this seems natural. Also the mathematically hard-nosed population geneticists who had earlier registered dissatisfaction with Hamilton's calculation of inclusive fitness and tried to suggest alternatives finally came to realize that Hamilton had been right within the stated limitations, and that his formulae were in fact even more broadly applicable. Yet his very first paper presenting a shorter version of this idea had been rejected by *Nature*!

That rejection stayed with Bill all his life. For the rest of his life that journal appeared to him as a candle to a moth. He returned to *Nature* again and again, sometimes making it, but always suffering anew when he was rejected. But what was it about *Nature*? Was it just Hamilton's vanity to want to be published in the most famous scientific journal in the world? Yes, of course Hamilton wanted recognition for what he knew to be important contributions, having devoted so much time and energy to arriving at his results. But that was not the central point. Because what is the actual function of a scientific journal? It is to disseminate information to other scientists. In fact, publishing something in a journal is not the final step, it is an invitation or challenge to other scientists to engage with a scientific claim and see if it holds up. And obviously a journal like *Nature*, appearing once a week and reaching out to millions of scientists across the world, would be the ideal place to publish a theory with a claim to universal validity. It would then be up to these scientists to challenge or extend the theory further.

In other words, I believe that what bothered Hamilton most about the rejections was that he was stopped from getting his scientific product out there for the scientific community to see and assess. He had prepared his papers well, 'incubating' them rather than trying to get them published quickly. Still, he was prepared for resistance. Scientists are not known for being easily persuaded. But for a fruitful interchange to take place the ideas in his scientific paper would first need to be made public and reach those inquisitive and sceptical scientific minds. And it was this crucial second step that the editors and referees of *Nature* (especially, but also other journals) had sometimes made impossible. They had, in his view, censored his contributions, prevented them from being assessed by the people for whom they were originally designed. Hamilton disliked censorship in principle, and particularly becoming its target himself.

What, then, would be the ideal situation, from a Hamiltonian point of view? I believe Bill would have liked to arrange as wide an access to his paper as possible for scientists from different relevant fields, acting as his would-be referees. Hamilton was inherently interdisciplinary, so it was not always clear from which field formal referees should be chosen. For instance, as we saw, his parasite theory of sex or the Hamilton-Zuk hypothesis would not easily be accepted by parasitologists. Another problem was that the fate of one of his papers could be decided by only two referees—and that could be a paper that he had worked on for several years. A fairer assessment of the validity of his ideas would be in the court of his scientific peers in behavioural ecology, sociobiology, population genetics, entomology, and other fields. But that was not the way the scientific refereeing system was set up. Journal editors acted like gate-keepers. And sometimes they were mistaken (as Hamilton himself had experienced).[19]

Now what about having not two, but instead over 100 referees assessing the validity of a scientist's work? This doesn't sound possible, or at least not practicable. But this was exactly what happened about a decade after Bill Hamilton's death in a curious little episode that forms a coda to the assessment of his legacy.

It occurred in the form of a response to an article in *Nature* written by three Harvard scientists presenting an unexpectedly strong attack on the theory of kin selection (that is, inclusive fitness). Kin selection was said to have been a ruling paradigm for 40 years but really to be unnecessary and substitutable with regular natural selection theory. The paradigm was said not to have generated any useful empirical results. Inclusive fitness was said to be difficult to calculate. The authors contrasted kin selection with their preferred paradigm, group selection, or rather multi-level selection. Finally, they presented the very validity of kin selection as dependent on the unusual genetic relationships in the Hymenoptera, and Hamilton's view of the origin of eusociality as connected to the haplodiploidy hypothesis.[20]

Mobilizing a variety of counterexamples and counterarguments to the views attributed to the followers of the kin selection paradigm, the authors then brought forth a mathematical model where eusociality could be shown to arise from the cooperation of unrelated insects in a particular ecological situation (involving competition between colonies, a stable food source, and the need for common defence). Relatedness was not needed, what was needed was simply the rise of an allele for eusociality![21]

Although attacking kin selection, the article's real focus was on the origin of eusociality. This was said to come about not by kin selection but by a process of group selection—that is, by inter-colony competition. The discussion was further strongly connected to the older, recently revived, entomological idea of a superorganism.[22]

In fact, Hamilton thought much along similar lines himself. He would have loved the opportunity to join the discussion about the origin of eusociality—one of his standing interests ever since he had been reading Charles Michener. Hamilton did not insist on the role of haplodiploidy for eusociality—as discussed earlier, that was only his early hunch, and he had already backed away from this idea in 1972 in the face of empirical evidence. What he still insisted on, however—but even here he was starting to back away a little too—was that genetic relationship played at least played *some* role in getting eusociality started (see chapter 11). He was convinced about this based on his naturalist's observation and intuition—and the fact that he had been stung by some 1,000 insects, including his own namesake wasp Stelophybia hamiltonii OW Richards! Meanwhile, as a naturalist and with a naturalist's love and understanding for his subjects, he had been developing his own 'underground' theory of the origin of social life.[23]

The 'mass response' from the over 100 internationally known biologists consisted of a handful of letters to *Nature*, the longest signed by over 103 colleagues. The letters basically sorted out (what they saw as) the misconceptions of the original authors. Kin selection was not an alternative to natural selection, it *was* natural selection (as explained by Dawkins 'First Misunderstanding of Kin Selection').

Kin selection *had* generated a mass of empirical research in a number of different subfields (this was listed, with references). Issues pertaining to kin selection and group or multi-level selection could in fact typically be translated mathematically into one another (because of Hamilton's re-derivation of his inclusive fitness theory with the help of the Price Equation in the 1970s). The authors had not shown that relatedness was not important just by leaving it out of their models. Also, the calculation of benefit and cost (represented by b and c) in Hamilton's Rule might well be taking into account ecological factors. And there was more, as mathematical modellers also got in on the discussion. The result? The original authors did not budge.[24]

Some commentators suspected that the original article had not undergone stringent enough refereeing. It had been presented as an 'analysis' rather than an article, which may have been allowing for that.[25] Was this Analysis article *Nature*'s experiment with a new form of contributions that would generate broad discussion among its readers across scientific subfields? If so, the success was stunning. The result was a number of painstakingly argued and politely formulated statements about what appeared to be mistaken beliefs on behalf of the Analysis authors, followed by overviews of the actual state of affairs in respective subfields, and furnished with abundant current references.[26]

For young scientists one can hardly imagine a better source of relevant information than this kind of controversy. This would be the basis of many seminars and dissertations in the near future. Once again the field of evolutionary biology was exciting, and it may have seemed to many of the younger scientists that they had a contribution to make—perhaps already an ambition to sort out the differences between the major opposing parties? At the same time, this controversy became a valuable source of information both for readers and the participating biologists themselves, who may not have been aware of the relevant literature and studies in neighbouring fields. Clearly, too, the controversy solidified the bonds between scientists on both sides, spurring them perhaps to further joint projects. And the mathematical arguments involved may have stimulated some modellers to try to find commonalities between the opponents' views in the language of mathematics.[27]

I cannot but think that Hamilton would have been pleased. He would have seen this mass response as a wish for scientists to pursue the truth. He would have congratulated the editors of *Nature* on opening up this new possibility for scientists to weather different opinions (and even emotions?) instead of just 'refereeing away' contributions (in this case the Analysis). What Hamilton wanted most was for scientific issues to be seriously discussed, and he thought that journals had a special responsibility in this respect. (He did not achieve this himself with his attempt to have an open discussion about the OPV hypothesis of the origin of AIDS—but then again, it was not a purely scientific issue—as he knew quite well.)

What was the verdict of this collective group of over 100 referees, on the correctness and significance of Hamilton's inclusive fitness theory? Hamilton had tried to scout out how natural selection worked, he had come up with the theory of inclusive fitness, and he had been right. And this they now told the world in unison, and in *Nature* at that! There could have been no greater accolade for Bill Hamilton from the people who counted to him most in the world—his scientific peers. Together they had given him a First.

NOTES

Introduction

1 He (Hamilton, 1970; Hamilton, 1975) later reformulated the results in his 1964 paper in what he called 'a vastly more economical and appropriate way' (for details about this reformulation and its implications, see chapter 10 of this book). This assessment of Hamilton's is from the first volume of his collected works, *Narrow Roads of Gene Land I*, p 175. The first two volumes of the collected works, with the subheadings 'Evolution of Social Behaviour' and 'Evolution of Sex', are edited by Hamilton and have introductions to his papers written by Hamilton himself. A third volume, 'Last Words', edited by Mark Ridley, contains co-authored papers and shorter pieces by Hamilton and introductions written by his students and collaborators. In this book, I will be abbreviating references to these volumes as *Narrow Roads*, vol 1, 2, or 3.
2 Barber (1961); his famous article in *Science* got over 500 reprint requests and he also received numerous personal letters from scientists who felt resisted or misunderstood (Barber, personal communication).
3 Dawkins (1976); Ridley (1993).
4 Hamilton (1996), p ix.
5 Barbara Koenig (Department of Biology, Zurich), personal communication.
6 See chapters 21 and 24 of this book.
7 See chapter 10 of this book.
8 *Narrow Roads*, vol 1, p 323.

Chapter 1

1 *Narrow Roads*, vol 3, pp 205–19, at p 208.
2 *Narrow Roads*, vol 3, pp 206, 209.
3 *Narrow Roads*, vol 3, p 206; Mary Bliss, personal communication; Gwen Owen personal communication.
4 *Narrow Roads*, vol 3, pp 222, 208; Mary Bliss, personal communication.
5 Mary Bliss, personal communication.
6 Mary Bliss, personal communication. There was one more child, 'Jimmy', born after Bill, who had a congenital defect.

7 AM Hamilton (1937). On a mandate from the League of Nations after the First World War, the United Kingdom governed Mesopotamia (Iraq) in 1920–32. Civil engineer Archibald Hamilton was sent there in 1927 to complete the Arbil-Rowandruz road in a difficult mountainous area in Kurdistan and ended up spending four years there. A particular feat of his, requiring a lot of engineering ingenuity, was the bridging of the Rowandruz River Gorge (almost inaccessible even by animal caravan), which made the completion of the road possible (see Editor's note at <http://www.amazon.com>).
8 Janet Hamilton and Mary Bliss, personal communication.
9 *Narrow Roads*, vol 2, p 376.
10 Mary Bliss, personal communication; Margaret Ritchie, personal communication; Gwen Owen, personal communication.
11 Mary Bliss, personal communication and visit to Oaklea outside Waimate, New Zealand. Oaklea in Kent is located on a hill 200 metres above the level of the not-too-distant Thames Estuary (*Narrow Roads*, vol 3, p 208).
12 Mary Bliss, personal communication; Marian Luke, personal communication.
13 Gwen Owen; Mary Bliss, personal communication.
14 This may also have meant that the active family members had little time or interest for small talk.
15 Veronica Medana of Eureka Radio, NZ, alerted me to this saying and its meaning.
16 *Narrow Roads*, vol 3, p 209; Mary Bliss, personal communication; Marian Luke, personal communication.
17 See n 12.
18 Mary Bliss, personal communication.
19 Mary Bliss, personal communication. See also Mary Bliss' article 'In Memory of Bill Hamilton' available on the University of Basel website at <http://www.evolution.unibas.ch/hamilton/index.htm> maintained by Dieter Ebert (<http://www.dieter.ebert-at-unibas.ch>; Narrow Roads, vol 2, p 47.
20 Mary Bliss, personal communication. A Nissen hut is a hangar-like shed of stainless steel.
21 Mary Bliss, personal communication and *Narrow Roads*, vol 3, p 222.
22 Mary Bliss, personal communication.
23 Margaret Ritchie, personal communication Bettina also 'sociabalized' landowners.
24 Mary Bliss, personal communication; Janet Hamilton, personal communication.
25 Mary Bliss, personal communication. The challenges that Christian encounters on his journey have names that have later become well-known: House Beautiful, Vanity Fair, The Point of No Return, the Slough of Despair, Valley of the Shadow of Death.
26 Mary Bliss, personal communication.
27 Mary Bliss, personal communication.
28 Mary Bliss, personal communication and her cousin Betty Redmond in Auckland, New Zealand, personal communication.

Chapter 2

1 Gwen Owen, personal communication; *Narrow Roads*, vol 3, pp 205–8.
2 *Narrow Roads*, vol 3, pp 207–8.
3 *Narrow Roads*, vol 3, p 207.

4 See n 3 above.
5 Mary Bliss, personal communication.
6 See n 5 above.
7 Janet Hamilton, personal communication.
8 *Narrow Roads*, vol 3, pp 209, 212.
9 *Narrow Roads*, vol 3, pp 77–8.
10 Mary Bliss, personal communication; Margaret Ritchie, personal communication.
11 *Narrow Roads*, vol 3, pp 212–13.
12 Mary Bliss, personal communication.
13 Mary Bliss, 'In Memory of Bill Hamilton' (see chapter 1 of this book, n 19).
14 See n 13 above.
15 See n 13 above.
16 See n 13 above, and AM Hamilton (1937).
17 *Narrow Roads*, vol 2, pp 126, 209.
18 *Narrow Roads*, vol 2, p 126.
19 *Narrow Roads*, vol 2, p 125.
20 See n 19 above.
21 *Narrow Roads*, vol 2, p 127.
22 Mary Bliss, personal communication.
23 *Narrow Roads*, vol 1, p 12.
24 *Narrow Roads*, vol 2, p 125

Chapter 3

1 Martin Jacoby, who attended Tonbridge at the same time as Bill Hamilton, in his long article 'Bill Hamilton at School: 1949–1954' includes both descriptions of life at the school during that time and comments from students who knew Bill, and also some by Bill's sister Mary and brother Robert. Jacoby, who in fact did not know Bill personally, took upon himself the job of tracking down a number of Bill's schoolmates and asking them for their reminiscences. The schoolmates' often frank comments give us interesting glimpses of Bill as a schoolboy from age 13 to 18. The current chapter could not have been written without being able to lean on Jacoby's valuable research and informative article (retrievable as a pdf file from the Fribourg University website which is devoted to the memory of Bill Hamilton, available at <http://evolution.unibas.ch/hamilton/index.htm>). I was also able to visit Tonbridge and see the various buildings and places mentioned.
2 Jacoby, p 2; letter from Archibald Hamilton to EEA Whitworth (the headmaster at Tonbridge) 29 May 1949; Mary Bliss, personal communication.
3 Jacoby, p 9 (bus), p 12 (Smythe House); Mary Bliss, personal communication.
4 Jacoby, p 9; Mary Bliss, personal communication.
5 Jacoby, p 9 (social test), p 2 (nicknames), p 8 (ordained).
6 Jacoby, p 21.
7 Jacoby, p 19.
8 Jacoby, p 20.
9 Jacoby, p 9.

10 *Narrow Roads*, vol 2, p xxxvii.
11 Helen Hamilton, personal communication.
12 Jacoby, p 3. As it turned out, Bill passed all subjects except French. This is why he later spent some time in France to learn French.
13 Jacoby, p 4.
14 Jacoby, pp 3–4. Mary Bliss, personal communication.
15 Jacoby, p 15 (type of biology at Tonbridge), p 4 (shooting a rabbit).
16 Mary Bliss, personal communication; Janet Hamilton, personal communication; Bill's school reports. Jacoby, pp 11 and pp 3–4. Note, however, that one classmate recalled that 'He was a regular prize-winner for being top of his form' (p 4).
17 Jacoby, pp 11–12.
18 Mary Bliss in Jacoby, p 4.
19 Jacoby, pp 14–15. This particular exam writing style would stick with him for the rest of his studies. According to Simon Hocombe, one of Bill's closest friends at Tonbridge, 'there was so much he felt the need to say that he had usually completed no more than three of the customary five questions of a typical paper before time was up. Three segments of excellence followed by two blanks must have been a nightmare for his masters and examiners to mark.' Probably so. My own belief is that Bill had developed a deliberate risky strategy of dazzle-and-leave-blank largely because he had typically not studied all aspects of the subject in question, using his time to familiarize himself with things *he* found interesting.
20 Jacoby, p 3 (on Oxford and Cambridge). I base my views on my general impression of Tonbridge as an educational institution, and the handwritten, careful notes on the progress of each student, as well as their strengths and weaknesses, written by their teachers in each subject each academic year.
21 'The Red Death' (a short essay) is reproduced in Jacoby, p 5 (from the *Tonbridgian* January 1951, p 295).
22 See n 21 above.
23 See n 21 above.
24 See also Mary Bliss, Jacoby, p 6.
25 Mary Bliss in Jacoby, p 6.
26 This uncompromising attitude later made itself known in the particular persona that Bill as a writer was sometimes to assume. It was Bill's headmaster, the Reverend LH Waddy, nicknamed 'The Arch' (Jacoby, p 2) who called him 'one of the most able, interesting and unfathomable boys' he had ever known, in the headmaster's report for Smythe House, Michaelmas (autumn) term, 1954, Bill's final term at the school. (The report also gives Bill's age: eighteen and a half.) I thank Janet Hamilton for providing this report.
27 Jacoby, p 11.
28 Jacoby, p 16. Bill's sister Mary suggested that cross-country running 'entitled him to go "out of bounds"'.
29 Jacoby, pp 16–18, the *Tonbridgian*, January 1955.
30 Jacoby, p 18.
31 Mary Bliss, personal communication.
32 Jacoby, p 11. Because of his somewhat unusual appearance, one nickname that Bill was given was 'Apeman Hamilton', something that he didn't seem to mind, according to his schoolmates (Jacoby, p 11).

33 Jacoby, p 17.
34 Jacoby, p 14.
35 Jacoby, p 12 (inquisitive), p 13 (Exploders), p 23 (Cadet Corps exercise). Bill did not belong to the Exploders; it was a different Hamilton—JG Hamilton—who 'changed the Fourth Termers' Walk forever' (Jacoby, p 23.)
36 Jacoby, p 22. (It was not sulphur but red phosphorus, according to Bill himself). Bill's father describes the accident in a letter to the headmaster of Tonbridge, EEA Whitworth on 29 May 1949, hoping that Bill would be accepted as a day boy there after passing the entrance exam. He shows a tinge of pride about his son: 'Though he hasn't yet had science as a subject he has a fair chemical knowledge gained from me and from books and perhaps in trying to copy some of my war-time activities in this field he made up an explosive mixture entirely of his own invention and of comparatively simple materials but of quite unusual power and sensitivity. It would have been hard to make anything more powerful and was comparable with the strongest war explosives.' He goes on to say that he has been lenient with Bill, because most boys have done dangerous things and he doesn't want Bill to get discouraged with science or enterprise; in fact he hopes that Bill will eventually be able to carry on as if the accident never happened. However, because the injury was to the right hand's thumb and first two fingers, Bill's writing is at least temporarily impaired, so Archie Hamilton is asking for special consideration in regard to the exam. Meanwhile he assures the headmaster that Bill 'has always been at or about the head of his class in all subjects' and that he 'may still carry this into University life'.
37 *Narrow Roads*, vol 2, p 203. According to Robert Hamilton, it was Dr Coopson (a friend of Bettina's) and Farnbougher hospital. Denmark Hill came later.
38 Jacoby, pp 23–4. See also p 9 on the maybug. As a result of his detective work, Jacoby succeeds in finally clearing up the confusion about the actual involvements and accomplishments of the three contemporary Hamiltons at Tonbridge. This is even systematized in a small comparative table on p 24.
39 Jacoby, pp 18, 15–16; Mary Bliss, personal communication.
40 Jacoby, p 3.
41 Jacoby pp 6–7. In the house report from Bill's last term the new housemaster, RM Williams (who succeeded Waddy) congratulates Bill on 'his success in School Rugger and his leading of the House XV into the semi-finals'. I thank Janet Hamilton for providing this report. On the Athena Society, see Jacoby, p 19.
42 Mary Bliss, personal communication.
43 Jacoby, p 7. The full poem is published there and also in the *Tonbridgian* May 1954, p 24, as well as in *Tonbridge Poets 1953–54*, edited by JM McNeil.
44 On 'The Woodsman', see Hamilton 1996, p 192, n 14. Simon Hocombe reports that '[a]t Tonbridge he/we read poets who were to become his lifelong favourites: Housman, Blake, Masefield, D. H. Lawrence' (Jacoby, p 15).
45 Hamilton later returned to this dark paradigm especially in *Narrow Roads*, vol 2.
46 Mary Bliss, personal communication.
47 Mary Bliss, personal communication.
48 Mary Bliss, personal communication.
49 Mary Bliss, personal communication.
50 Margaret Ritchie, personal communication.

51 Grafen (2004).
52 Mary Bliss, personal communication.
53 What seems to have especially riled Bill was that he was expected to learn the proper way of serving the officers, and that he was expected to address them like a good waiter. Here is a snippet from a letter he wrote to Mary from the Officers' Mess: 'It is all the Yes, sir & No, sir & Right, sir and will you have soup, sir that gets me down (actually, when I have to, I say: "Do you want soup, sir?" because it sounds more boorish)—pandering to all their whims & fussiness. I'm fed up I feel like the chef in "An Anonymous Story"—so much so that I ought to end up hitting one of the officers to complete the likeness—I often feel like it. The sad difference is that I am not a free man & I am not here of my own accord.' (Letter to Mary, 29 June 1955.)

Things did not improve much over time. In a letter to Mary on 14 November 1955 Bill writes: 'It will be an amazing thing if I do survive the next four weeks & get my commission. If I do it will only be by means of a piece of enormous bluff on my part. The trouble is that the few things that I either cannot or will not alter to fool the authorities, they notice at once. Thus I was told that I look bored (the old trouble I used to have with Tom*; nothing in it in fact) & absent-minded (true enough, but I have no intent of trying sacrificing my powers of abstraction) & that I am too slow in everything. This last is the most important thing as far as I am concerned, since it is quite true, but I am afraid there is nothing I can do about [it], much as I would like to.'

[* 'Tom' refers to Bill's housemaster Tom Staveley whom he had for all but his last term, Jacoby, p 2.]
54 More on Waterloo station in *Narrow Roads*, vol 1, p 25.
55 These notebooks are available as part of the Hamilton Archives at the British Library.
56 *Narrow Roads*, vol 3, p 79.
57 See n 56 above.
58 See n 56 above.
59 *Narrow* Roads, vol 3, p 225.
60 Postcard to Mary, 15 April 1958; letter to Charles, 5 October 1958.

Chapter 4

1 The Natural Sciences Tripos undergraduate programme at Cambridge proceeds from giving the students a broader base of options across the sciences the first year (three experimental subjects plus mathematics) towards increased specialization the second year (three subjects from a narrower set of options) and to focusing on a single subject in the third year (see Natural Sciences Tripos, available at <http://www.cam.ac.uk/cambuniv/natscitripos/>).

Bill Hamilton studied botany, zoology, physiology, and mathematics in 1957–59 for Part I of Tripos and genetics for Part II (Bill Hamilton, curriculum vitae, 13 May 1998).

Students usually have final exams at the end of each year, for Part I A, Part I B, and Part II. The highest grade a student can obtain is a 'First' (followed by an 'Upper Second'). The grade obtained in Part II (in Bill's case, genetics) is usually also the grade for the BA degree.

2 Excerpt from Sir Vincent Wigglesworth's *The Life of Insects*, published in 1964, quoted in Hamilton (1996, p 22).
3 Postcard from Bill Hamilton to Mary Bliss, undated, 1958.
4 Hamilton (1999). He refers here to the fact that he didn't get a First but rather an Upper Second in his BA degree at Cambridge.
5 Edwards (2000), p 1423.
6 Box (1985).
7 Anthony Edwards, personal communication.
8 Postcard from Bill Hamilton to Mary Bliss, undated, 1958.
9 Postcard from Bill Hamilton to Mary Bliss, undated, 1958.
10 Postcard from Bill Hamilton to Mary Bliss, undated, 1959.
11 Students are expected to select one subject for their Tripos II. Hamilton selected genetics. The grade in that subject typically determines the final degree for the BA.
12 Anthony Edwards, personal communication.
13 Hugh Ingram, personal communication.
14 Colin Hudson personal communication.
15 Bill's clever cooking style was described to me by Colin Hudson (personal communication). Bill cleverly avoided staying in dorms by simple delay. As he told his godfather 'Uncle' Charles: '[T]oday I moved into new digs (I should have been in College this year, but didn't apply for rooms till too late)' (letter to Charles Brasch, 5 October 1958). The mention of Bill's baby-sitting is in a letter to Charles Brasch, 10 September, 1959.
16 Hamilton (1999).
17 Postcard and letters that were made available to me at Oaklea by Mary Bliss.
18 A collection of postcards made available to me at Oaklea by Mary Bliss. Darwin also used cross-writing in his letters (Alan Grafen, personal communication).
19 Postcard to Mary Bliss.
20 Letter to Charles Brasch, 9 October 1959.
21 See n 20 above.
22 See n 20 above.
23 Robert Hamilton, *A Trip to Anatolia* (typewritten manuscript).
24 See n 23 above.
25 Mary Bliss and Janet Hamilton. Robert wrote up the trouser story.
26 Robert Hamilton, *A Trip to Anatolia.* (Incidentally, it was while they waited a week for the car to be repaired that they stayed at what they mistakenly believed to be the 'student hostel'.)
27 See n 26 above.
28 Letter to Charles Brasch, 9 October 1959.
29 Edwards, personal communication.
30 Edwards, personal communication.
31 *Narrow Roads*, vol 1, p 34. The argument is in Hamilton (1963).
32 Fisher 1930 (1958), p 177, cited in *Narrow Roads*, vol 1, p 8.
33 *Narrow Roads*, vol 1, p 23.
34 Letter to Mary Bliss, 11 November 1959.
35 *Narrow Roads*, vol 1, p 24.

36 Fred Cooke, personal communication. He told me about a tea he attended where Hamilton conversed with Fisher.

Chapter 5

1 *Narrow Roads*, vol 1, p 24.
2 Darwin had regarded altruism as 'one special difficulty, which at first appeared to me insuperable and actually fatal to my whole theory'. There are, however, different views about exactly what he meant by this statement, which he made in relation to worker insects. By 'difficulty', did he mean the phenomenon of worker altruism itself as a type of evolutionary puzzle (since natural selection typically favours greater, rather than lesser, reproduction)? Many biologists have interpreted him this way. Others have argued that Darwin was not primarily worried about the evolution of worker sterility itself. According to them, this represented only a 'minor' difficulty, while his 'major' difficulty had to do with the question of *adaptation*. How was it possible for sterile individuals to continue evolving into several worker castes (such as was the case, eg, in ants)? In 1993 George Williams attempted to clarify the issue: '[t]he modern literature is full of statements to the effect that Darwin saw a special difficulty in the altruism of workers. He did not. His worry was about how the workers could develop adaptations that none of their ancestors had'. (See the excellent article investigating Darwin's special difficulty by Ratnieks, Foster, and Wenseleers, 2010).
3 *Narrow Roads*, vol 1, p 24.
4 *Narrow Roads*, vol 1, pp 14, 4.
5 Letter to Charles Brasch, 19 December 1960.
6 *Narrow Roads*, vol 1, p 5.
7 Letter to Charles Brasch, 19 December 1960.
8 See n 7 above.
9 *Narrow Roads*, vol 1, p 5; letter to Charles Brasch, 19 December 1960.
10 Letter to Charles Brasch, 19 December 1960.
11 See n 10 above.
12 See n 10 above.
13 *Narrow Roads*, vol 1, p 11.
14 Maynard Smith, personal communication.
15 *Narrow Roads*, vol 1, p 25.
16 *Narrow Roads*, vol 1, pp 25–6.
17 See n 16 above.
18 'My kin selection chair' is an expression that he started using only later. He didn't use the term 'kin selection' until the mid-1970s to refer to his own theory; he preferred 'kinship theory' or 'kinship altruism'.
19 He had reasoned along these lines already in Cambridge, hoping to collect ideas from different sources of information (such as anthropological data) and then 'average out of it something relevant' (*Narrow Roads*, vol 1, p 23). This was the type of method Hamilton was to use also later at the early stages of theory-building.
20 See table in *Narrow Roads*, vol 1, p 44. This represents a different version: here we have selfish, altruistic and 'counterselected' behaviour. The 'counterselected' behaviour was later identified with spite (Hamilton, 1970); see chapter 10.
21 His 1964 paper discusses plants (seed sizes especially); see also his comment in *Narrow Roads*, vol 1, p 20.

22 Alan Grafen, personal communication.

23 *Narrow Roads*, vol 1, pp 25, 27. Reading Hamilton's own description of what he did in his brief autobiographical notes or studying the 1964 paper itself, it is not clear exactly how he went about his model building. In fact, his own story may give misleading impressions about what he was actually doing. Therefore I turned to Alan Grafen, who actually discussed this issue with Hamilton and got a sense of the model building going into his work on inclusive fitnes:

> He first worked out from a model the condition under which pairs of relatives behaved altruistically... Each of Bill's trial models was of this kind, so far as I can tell from talking to him and from [*Narrow Roads*]. One model was for altruism to sibs, one for first cousins etc. But each one was onerous, and as a class they were completely novel....
>
> Bill noticed two odd things about each model. First, that gene frequency didn't matter. Second, that all that did matter was the coefficient of relatedness. But these models are large and ungainly and would be very error-prone especially for a non-mathematician. So I'm not surprised they took a long time.

Finally Hamilton was able to formulate a model that represented all of social behaviour simultaneously. Here is Grafen's continuing comment:

> But what Bill did in formalising his final model as published was quite extraordinary. Not content with the separate models, one for each type of ancestral link, he formulates another radically innovative type of model to be able to represent the whole of social behaviour at once (well, subject to additive fitness interactions). One device is to extend Fisher's average excess/effect from the fundamental theorem; another is allowing arbitrary additive fitness interactions but then organising them by relatedness to produce a manageable structure. *And* he gets out a fitness maximisation result (for one locus, but permitting multiple alleles), for inclusive fitness, not simple offspring number. This final part is key to the generality of the result, and flows from Bill's deliberate aping of Fisher's fundamental theorem (though he uses a then-more-modern mathematical result due to John Kingman—a now famous British mathematician—to do more neatly the mechanics that Fisher had gone through).
>
> One key point of the model is that it permits selection on different social interactions at the same time. We may have sibling interactions, double-first-cousin interactions, and so on, all with different costs and benefits, all happening at the same time, caused by the same or different alleles at the one locus; and the result holds that the allele that produces the highest inclusive fitness is favoured. (Nearly all later models by mathematicians are restricted to one type of relative, one b and one c, at a time.)
>
> One of the key points of this general model is that relatedness plays a key role, and one can see where in the argument it plays its part, so we can understand why relatedness should be important, and also see how the rest of the complications cancel out leaving just relatedness as the central relevant concept.

Another important point that Grafen wanted to make had to do with Hamilton's 'discovering' Sewall Wright's correlation coefficient. It may not be obvious from Hamilton's own account that for him the correlation coefficient was actually a real discovery in the sense of an epiphany. As Grafen explains: 'It may seem as if Bill's work was calculating coefficients of relationships, and then he found Sewall Wright's r. In fact the work was the complex and innovative models described above, from which simple fractions emerged. What Bill noticed was that these simple fractions WERE Sewall Wright's r for the relationship in question.'

So, in brief, here is Hamilton's approach, according to Grafen:

'(i) he first worked out from a model the condition under which pairs of relatives behaved altruistically
(ii) then he looked at the conditions, and noticed with an Aha! moment that the condition could be represented in terms of Wright's coefficient of relationship.'

24 For the exact definition of inclusive fitness, see *Narrow Roads*, vol 1, p 38. This concept will be discussed in more detail in chapter 6, where the definition is included in the text.
25 Letter to parents, 5 November 1962.
26 Hamilton's Rule formulates under what conditions altruistic behaviour is likely to evolve. It postulates a donor and a recipient of some kind of altruistic act. A hypothetical gene for altruism increases in frequency if the benefit to a recipient (b) is greater than the cost to the donor (c), taking into account the coefficient of their genetic relationship (r), expressed as the formula $rb - c > 0$. There are many ways of writing Hamilton's Rule.
27 *Narrow Roads*, vol 1, p 3.
28 Letter to parents, 5 November 1962.
29 See n 27 above.
30 Letter to parents, 5 November 1962.
31 Letter to Colin Hudson, 18 January 1963.
32 See n 31 above.
33 Letter to Colin Hudson, 22 January 1963 (continuation of 18 January letter).
34 See n 33 above.
35 Blest (1963).
36 GC and DC Williams (1957).
37 Letter to Williams, 14 May 1963.
38 Williams (2000).
39 *Narrow Roads*, vol 1, p 229.
40 Provine (1972); Barkan (1990).
41 Cf Lynn (1996), pp 11–12. The Geneticists' Manifesto (Muller, 1939) was published in *Nature* on September 16, 1939. It is reproduced eg in Kristol and Cohen (2002), pp 22–5.
42 Lynn (1996), pp 9–10; Muller (1935).
43 Fisher (1930); Lynn (1996), pp 9–12.
44 Haldane (1930); Huxley (1932).
45 Provine (1972); Barkan (1990); Lynn (1996), pp 14–15.
46 Lynn (1996), pp 14–15; Wolstenholme (1963).
47 Penrose was also rising internationally, becoming the President of the International Society of Human Genetics.
48 *Narrow Roads*, vol 1, pp 4, 13–19.

49 Bill's official adviser, Cedric Smith, was in a sensitive position as well. Having started as a lecturer, he was on the way to becoming a Weldon Professor of Biometrics (a title he attained in 1964). It was clear that he had to look good, and having Hamilton around as a student talking about altruism was obviously awkward. Moreover, Smith, too, just like Penrose, was a Quaker. In other words, Bill Hamilton unknowingly presented a living threat to academic careers.
50 *Narrow Roads*, vol 1, pp 13–19 (on Bill feeling disliked in the Galton lab).
51 *Narrow Roads*, vol 1, p 25.

Chapter 6

1 I visited Barbados and Colin Hudson in November 2004.
2 See *Narrow Roads*, vol 1, p 25, and chapter 5 of this book.
3 Letter to Colin Hudson, 8 October 1962, and *Narrow Roads*, vol 1, p 27 ('emotional pendulum', '3 years of alienation', 'extreme idleness').
4 See, eg, *Narrow Roads*, vol 1, p 25.
5 Kevles (1983), p 5.
6 For instance, parts of p 190 of *Narrow Roads*, vol 1.
7 Letters to Colin Hudson, 7 March and 9 June 1963. See also *Narrow Roads*, vol 1, p 297.
8 *Narrow Roads*, vol 3, p 80.
9 See n 8 above.
10 See n 8 above.
11 See n 8 above; and letter to Colin Hudson, 7 March 1963; see also *Narrow Roads*, vol 1, p 297.
12 As published in *Narrow Roads*, vol 1, pp 31–82.
13 *Narrow Roads*, vol 1, p 38.
14 Grafen (1982).
15 *Narrow Roads*, vol 1, p 38.
16 *Narrow Roads*, vol 1, p 49.
17 *Narrow Roads*, vol 1, p 48.
18 See n 17 above.
19 See n 16 above.
20 *Narrow Roads*, vol 1, p 45.
21 See n 17 above.
22 *Narrow Roads*, vol 1, p 27.
23 I thank Charles Uth at IIT for discussions about the thinking of engineers in regard to approximations. Mathematicians may of course make various types of approximations too, but I have in mind some basic differences in attitudes.
24 *Narrow Roads*, vol 1, pp 21–2.
25 Bill generally used anthropomorphism as a research tool.
26 *Narrow Roads*, vol 1, pp 52–3.
27 *Narrow Roads*, vol 1, p 20. See further discussion in chapter 11 of this book.
28 *Narrow Roads*, vol 1, p 55.
29 Jimmy died in fact during an operation for pyloric stenosis, according to Janet Hamilton, correcting Bill's information in *Narrow Roads*, vol 2, pp 477–8. For the Burghers of Calais, see eg *Narrow Roads*, vol 1, p 360.
30 Rowena Hamilton, personal communication; Grant (2002). Note that Bill did not see himself as an atheist but as an agnostic (letters to Colin, 28 August 1962, 13 July 1971).

31 See chapter 10 of this book.
32 *Narrow Roads*, vol 1, p 2.

Chapter 7

1 Bill was stationed in Kerr's lab in Rio Claro, from which he made various field expeditions. One of his ambitions during this trip was to learn Portuguese.
2 *Narrow Roads*, vol 3, pp 80–1.
3 *Narrow Roads*, vol 3, p 81.
4 See n 3 above.
5 *Narrow Roads*, vol 1, p 20. As for travels, we know for instance that in June 1963 he was in the state of Goias (Hughes, 2002).
6 *Narrow Roads*, vol 3, p 80.
7 Hughes (2002).
8 See n 7 above.
9 See n 7 above.
10 Kerr, personal communication.
11 See n 10 above.
12 Bates, *Naturalist on the River Amazons* (1863), revised edition 1962.
13 See Henderson, Hamilton, and Crampton (1998) and chapter 19 for Hamilton's further explorations of this topic.
14 *Narrow Roads*, vol 2, p 683.
15 *Narrow Roads*, vol 2, p 679.
16 Rinderer, Oldroyd and Sheppard (1993).
17 See n 16 above.
18 *Narrow Roads*, vol 3, p 82.
19 See n 18 above.
20 See n 18 above.
21 See n 18 above. Hamilton later realized that inbreeding was even more common in other types of enclosed spaces such as galls, enclosed flowers, figs, and reused former pupal cases.
22 Bill appears to have different reminiscences about when he got the editor's letter—was he still in England or had he already left for Brazil? Compare *Narrow Roads*, vol 1, p 29, with *Narrow Roads*, vol 3, p 80.
23 *Narrow Roads*, vol 1, pp 29–30 ; Hughes (2002).
24 Colin used to come over to England regularly to see his family there. Over the years he and Bill had made plans to travel somewhere together. Colin had earlier suggested in a letter that he and Bill travel together in Eastern Europe, but after two forays there (with bike and car, respectively) Bill had declined.
25 See *Narrow Roads*, vol 3, p 311, for a related insect catching incident.
26 Here is the parents' thinking behind this, as seen in a letter from Bettina, 2 May 1964. Bill has told her in an earlier letter about his plans, and she comments:

> What a stupendous trip that is going to be. Fairly simple you say as far as Brasilia... And after Brazilia [sic] the jungle, at least it seems you intend to go & look for it... So I'd advise you to use that £200 Dad sent out after all. I thought I'd persuaded him to wait till you asked for it but he said no, it must be there if he wants it. He's worth an awful lot more to us than that so he must have a 100%

safe motor car. So please see that it is & that you have all the spares you can possibly need on the road. (*Narrow Roads*, vol 3, p 421)

27 *Narrow Roads*, vol 1, p 78; *Narrow Roads*, vol 2, p 81. Many of these specimens were later to be exhibited in the Natural History Museum in London and written up as part of Richards' major taxonomic work on South American wasps.
28 *Narrow Roads*, vol 1, pp 29–30.
29 Marian Luke, personal communication; Mary Bliss, personal communication; Bill, letter to Colin, 4 March 1964. For a reference to this particular scenario from *Out of the Night*, see Maynard Smith (1972), p 67.
30 Colin Hudson, personal communication.
31 See n 30 above.
32 Wynne-Edwards (1962).
33 Especially 'nasty' examples. Bill and Colin were in correspondence about competition and sibling rivalry. Contra Colin, who played such things down, Bill insisted that examples of these behaviours existed. Now he had found some usable nasty ones in Wynne-Edwards' book. Colin dryly noted how admirable it was 'to see the detached naturalist's stance'.
34 John Maynard Smith told me in an interview that he had totally forgotten the meeting. See more about the meeting in Segerstrale (2000), pp 61–2.
35 Maynard Smith (1964), Letter to the Editor in response to Wynne-Edwards (1963).
36 See n 35 above.
37 Wilson (1994), p 320.
38 Wilson (1994), pp 320–1.

Chapter 8

1 See chapters 10 and 13 of this book for a discussion of the earlier papers and chapter 11 for Hamilton's theory of social evolution in insects.
2 *Narrow Roads*, vol 3, p 81.
3 *Narrow Roads*, vol 3, p 86.
4 Mary Bliss, personal communication.
5 Kitching (2000).
6 See n 5 above.
7 *Narrow Roads*, vol 1, p 86.
8 *Narrow Roads*, vol 1, p 87.
9 *Narrow Roads*, vol 1, p 88.
10 See n 9 above.
11 *Narrow Roads*, vol 1, p 89.
12 *Narrow Roads*, vol 1, pp 89–90.
13 See n 9 above.
14 See chapter 5 of this book. The Taiwanese study came in handy as the empirical basis for Part 8 of Hamilton's senescence paper in 1966.
15 *Narrow Roads*, vol 3, p 82.
16 *Narrow Roads*, vol 3, p 162.
17 *Narrow Roads*, vol 3, p 83; for a photograph with text, see *Narrow Roads*, vol 1, p 130.
18 Hamilton says that he believes his opening passage in his 1967 paper presents Fisher's idea more clearly than Fisher himself did. In particular, if a population is grouped,

then Fisher's assumption of a 1:1 ratio of a panmictic population with unrestricted competition for mates does not hold (*Narrow Roads*, vol 1, p 131).

19 See chapter 11 of this book.

20 *Narrow Roads*, vol 3, p 417. See chapter 11 of this book for a discussion of haplodiploidy and eusociality.

21 *Narrow Roads*, vol 1, p 140.

22 He had started research on this even before his first trip to Brazil, but continued his work there. His model organism in this case was a small parasitoid, *Melittobia acasta*. As he put it: 'For a long time…I thought of sex ratio under subdivision as my "Melittobia problem"' (*Narrow Roads*, vol 1, p 139).

23 *Narrow Roads*, vol 1, p 135.

24 *Narrow Roads*, vol 1, p 137.

25 *Narrow Roads*, vol 1, pp 136–7.

26 *Narrow Roads*, vol 1, p 132.

27 *Narrow Roads*, vol 3, p 82.

28 See n 27 above.

29 See n 27 above.

30 *Narrow Roads*, vol 1, p 133. On the development of evolutionary game theory, see Segerstrale (2000), pp 102–3.

31 Haig (2002); Burt and Trivers (2008).

32 *Narrow Roads*, vol 1, p 134.

33 *Narrow Roads*, vol 1, p 141.

34 Williams (2000).

35 Segerstrale (2000), pp 89–90.

36 Hugh Ingram, personal communication. Hugh Ingram had together with Bill Hamilton and Colin Hudson spent a week together at the Semerwater expedition in 1958 at Cambridge (see chapter 4 of this book).

37 The fellow student slipped on a bird's nest, and Leco had used an old piton, which did not hold. Letter to Yura, 26 October 1967; Mary Bliss, personal communication.

Chapter 9

1 *Narrow Roads*, vol 1, p xxxviii. For a detailed description of the expedition and of Mato Grosso, see A Smith (1972).

2 Colin Hudson, personal communication.

3 Maynard Smith (1989), p 205 (originally Maynard Smith, 1976).

4 Alan Grafen, personal communication (about the *viva*) and Richard Southwood, personal communication (about the DSc).

5 Cedric Smith, letter to Mary Bliss, 21 April 2000.

6 Letter to Yura Ulehla, 6 July 1968.

7 A Smith (1972), p 143; Richard Southwood, personal communication.

8 *Narrow Roads*, vol 1, p 254. For more description of Hamilton's insect work in Mato Grosso, see A Smith (1972), pp 142–3.

9 *Narrow Roads*, vol 2, p 366.

10 A Smith (2000).

11 See n 10 above.

12 Mary Bliss, Christine Hamilton, personal communication.
13 Letter to Yura Ulehla, 16 February 1970.
14 Letters to Yura Ulehla, 16 February 1970.
15 Letter to Yura Ulehla, 12 December 1971.
16 Christine Hamilton, personal communication; Mary Bliss, personal communication.
17 *Narrow Roads*, vol 1, p 187.
18 See n 17 above.
19 *Narrow Roads*, vol 1, pp 186–7.
20 See n 17 above.
21 See n 20 above.
22 Ripley (1969) quoted in Tiger and Robinson (1991).
23 See n 17 above.
24 For the Prisoner's Dilemma, see p 243 in this book. When extended to many individuals this kind of situation leads to such things as overuse of resources, coined as 'The Tragedy of The Commons' (Hardin, 1968).
25 *Narrow Roads*, vol 1, p 218.
26 *Narrow Roads*, vol 1, pp 218–19.
27 *Narrow Roads*, vol 1, p 189.
28 Se n 27 above.
29 See n 27 above.
30 *Narrow Roads*, vol 1, p 191.
31 *Narrow Roads*, vol 1, p 188.
32 See n 31 above.
33 *Narrow Roads*, vol 1, p 222.
34 See n 33 above. Here Hamilton refers to Haldane (1932).
35 *Narrow Roads*, vol 1, p 223.
36 Postcard to Mary, 3 November 1988.
37 Trivers (2002), p 10.
38 *Narrow Roads*, vol 1, p 263.
39 Trivers (2002), p 11.
40 See n 39 above.
41 Hamilton indicates that he went to Harvard before the Washington conference, Trivers assumes it was after ((2002), p 10).
42 Trivers (2002), p 10.

Chapter 10

1 Richard Southwood, personal communication.
2 See n 1 above.
3 Letter to Yura Ulehla, 16 February 1970.
4 In the republication of his 1964 two-part paper in Williams' book (1971) Bill had been able to correct a couple of minor errors.
5 This he introduced in a short paper in *Nature*, Hamilton (1970).
6 *Narrow Roads*, vol 1, pp 272–3.
7 *Narrow Roads*, vol 1, pp 300–1. He simply declared 'Reciprocal altruism as defined by Trivers has no need of relatedness' (p 301), while resisting the temptation to strongly

voice his disagreement with Trivers' incorrect-seeming use of 'altruism'. (For Hamilton, as we have seen, for an act to be truly altruistic it should be possible to extrapolate it all the way to suicide).

8 I will return to the *Annual Review* paper in chapter 11 of this book.

9 *Narrow Roads*, vol 1, p 233. It had become a 'Citation Classic'.

10 *Narrow Roads*, vol 1, pp 230–1; Williams (1964).

11 Frank (1995). As a result of the injury Price was dependent on regular doses of thyroxine.

12 *Narrow Roads*, vol 1, p 320, 173–5.

13 *Narrow Roads*, vol 1, p 320; Frank (1995).

14 *Narrow Roads*, vol 1, pp 172–3. Also, Price said he had never understood or used statistics. For him, the whole thing was a miracle (*Narrow Roads*, vol 1, p 322). For a more detailed discussion of the Price equation, see Frank (1995), pp 375–9.

15 *Narrow Roads*, vol 1, pp 172, 332; Frank (1995), p 379.

16 *Narrow Roads*, vol 1, pp 173–5, 179; Frank (1995), p 375. In 1996, in his autobiographical notes, Hamilton says about his 1970 paper in *Nature*, inspired by Price: '[H]is much earlier, rather patronizing remarks about kin selection and spite had induced me to work out the account that is in this paper. I had been delighted with the results no so much because of spite, which I even now regard as practically a non-starter for important evolutionary effects . . . but because I could reformulate all the results that I had had before in my 1964 paper in a vastly more economical and appropriate way.' (*Narrow Roads*, vol 1, p 175).

But how could spite ever be favoured by evolution? This has to do with the idea of 'negative relatedness'—meaning that an actor and a recipient are less related to each other than a random member of the population. Negative relatedness of this sort can be expected to lead to selection for spiteful behaviour. As Frank (1995), p 375, succinctly puts it: 'Spite can be favored because the product of negative relatedness and a negative benefit to a recipient (harm) is positive, thus benefit multiplied by relatedness can outweigh the cost.' He mentions a couple of examples of spite, including the *Medea* allele that uses kin recognition to destroy individuals that are negatively related to itself. Later research on spite research based on Hamilton has been extended (eg Blackman, 2004; Gardner and West, 2004). Gardner and West (2004) suggest that the conditions under which spite can be expected to evolve are '1) when actors can inflict damage to others at little cost to themselves (many examples come from nonreproductives among eusocial insects), 2) when kin recognition is possible (either genealogical kin or by means of "greenbeard" phenotypic markers), 3) when competition is mostly local, or the population is viscous' (p 1201).

17 *Narrow Roads*, vol 1, p 175.

18 Se n 17 above.

19 See n 17 above.

20 *Narrow Roads*, vol 1, p 176.

21 Maynard Smith (1976/1989), p 205.

22 *Narrow Roads*, vol 1, pp 320–1, 325.

23 *Narrow Roads*, vol 1, p 322.

24 *Narrow Roads*, vol 1, p 323.

25 *Narrow Roads*, vol 1, p 322.
26 Maynard Smith, letter to Price, 24 October 1972. Albert Somit, personal communication; *Narrow Roads*, vol 1, p 324.
27 Price, letter to Maynard Smith, 19 October 1972; *Narrow Roads*, vol 1, pp 324–5; Albert Somit, personal communication.
28 *Narrow Roads*, vol 1, p 325. Letter to George's brother Edison Price , 15 February 1975. On George Price's suicide, see *Narrow Roads*, vol 1, p 174; and Segerstrale (2000), p 68.
29 Albert Somit, personal communication.
30 Letter to Edison Price, 15 February 1975.
31 Price, letter to Maynard Smith, 19 October 1972; *Narrow Roads*, vol 1, pp 321, 320.
32 See n 30 above.
33 See n 30 above.
34 Letter from Price to Maynard Smith, 19 October 1972.
35 Letter for Maynard Smith to Price, 24 October 1972. A look at the Maynard Smith and Price, 'The Logic of Animal Conflict' paper in *Nature* 1973 shows that the proper corrections were made (eg Levins was correctly cited for group selection) and Hamilton (1967) was indeed recognized (Maynard Smith and Price, 1973).
36 *Narrow Roads*, vol 1, p 318.
37 Hamilton (1975); *Narrow Roads*, vol 1, p 316.
38 *Narrow Roads*, vol 1, pp 316, 329 (vignette quotation of Washburn).
39 See more on Washburn in Segerstrale (2000), pp 147–50.
40 Washburn (1978).
41 *Narrow Roads*, vol 1, pp 330–1. Hamilton suggests in his 'Innate Social Aptitudes of Man' paper (Hamilton, 1975) that Darwin left an 'open problem' when he discussed the evolution of courage and self-sacrifice in man. Such features would be counter-selected within groups, but selected for in the struggle between groups. Hamilton saw this open problem as the starting point for his own argument. See also *Narrow Roads*, vol 1, pp 332–6.
42 *Narrow Roads*, vol 1, pp 334–5.
43 *Narrow Roads*, vol 1, p 336.
44 *Narrow Roads*, vol 1, pp 336–7.
45 *Narrow Roads*, vol 1, p 337.
46 See n 45 above.
47 See n 45 above.
48 *Narrow Roads*, vol 1, p 324.
49 There are different views of the efforts to bring back group selection. According to some, this had not been done it in the right way (eg Maynard Smith (1998), Trivers (1998) in their comments to *Unto Others* 1998; Grafen, personal communication). Hamilton's own view on the attempt to bring back group selection, voiced in 1996, was that group selection is a weak force which 'doesn't exorcise harsher facets of natural selection'. He warned that those with 'fervent belief' in selection at the group level are started on a course straight toward Fascist ideology (in fact referring to Sober and Wilson (1998)) (*Narrow Roads*, vol 1, p 192).
50 David Sloan Wilson did some statistics on the relationship between the citations to Hamilton's two papers, 1964 and 1975 respectively. Consulting the *ISI Web of Science* for the period 1985–2007, he found that Hamilton's 1975 paper was much less cited

than his 1964 paper, the ratios of the 1964/1975 papers varying between 11.6 (1990, 2000) and 25.4 (1995). No clear trend appears—if anything the result for 2007 (15.8) is lower than for 1985 (21.6). (DS Wilson, email, 31 December 2007 to a group of theoretical biology colleagues engaged in a sustained email exchange about group selection and kin selection.) This was in a response to Steven Frank's email saying that 'the theory of IF [inclusive fitness] has grown and developed into a broader theory of causal analysis that nicely handles anything one can throw at it' (Frank, email, 31 December 2007). According to DS Wilson, while this may be correct for theoretical biology, it is not true for evolutionary biology as a whole (which is why he mobilized the Web of Science to prove his point).

Chapter 11

1 *Narrow Roads*, vol 1, pp 424, 418. Bill tells his mother how the girls are speaking Portuguese to each other at the same time as they are learning to speak English (Letter to Bettina, 28 February 1976).
2 The Hamiltons were in Brazil from June 1975–February 1976 (*Narrow Roads*, vol 1, p 435). After that they travelled to Chile by jeep (to collect seeds for the Forestry Commission), followed by a visit to the Eberhards in Colombia and Colin Hudson and his wife in Barbados. After that the plan was for Bill to give a paper at the Bicentennial Celebration of the Academy of Sciences in Philadelphia (Letter to Bettina, see n 1 above). This schedule is a typical Bill plan. Bill is writing from Chile.
3 *Narrow Roads*, vol 1, pp 465, 450–1.
4 *Narrow Roads*, vol 1, p 441.
5 *Narrow Roads*, vol 1, pp 464–8.
6 *Narrow Roads*, vol 1, pp 429–30.
7 *Narrow Roads*, vol 1, p 426.
8 *Narrow Roads*, vol 1, p 427.
9 *Narrow Roads*, vol 1, p 424, Warwick Kerr, personal communication.
10 *Narrow Roads*, vol 1, p 425.
11 *Narrow Roads*, vol 1, p 389.
12 *Narrow Roads*, vol 1, p 393; *Narrow Roads*, vol 3, p 83; Hughes (2002), p 87. Victoria Taylor (PhD, 1975) also wrote a large number of articles on *Ptinella* (referred to in many of Hamilton's papers). However, why did she leave to go off to study elephants in Africa, Hamilton later wondered? Well, one reason was certainly the smallness of this organism, specifically his suggestion that she go smaller still by attempting to find an egg of *Ptinella*! This did not look very practical as a project (Victoria Taylor, personal communication). (Hamilton's suggested projects were indeed sometimes impractical, but if followed, usually brought interesting results.)
13 *Narrow Roads*, vol 1, p 388. Hamilton also thought *Ptinella* was suggestive of the beginning of a termite like social development (*Narrow Roads*, vol 1, p 407).
14 Hamilton (1978), published as part of chapter 12 in *Narrow Roads*, vol 1.
15 *Narrow Roads*, vol 1, pp 390–1, 394.
16 *Narrow Roads*, vol 1, pp 395–6.
17 *Narrow Roads*, vol 1, pp 396–7.
18 See n 17 above.

19 *Narrow Roads*, vol 1, p 386.
20 *Narrow Roads*, vol 1, pp 388–9.
21 For a picture and discussion of this huge beetle, see *Narrow Roads*, vol 1, p 386. (I was shown Bill Hamilton's original by Christine Hamilton.) The story about being buried by a carrion beetle is in Hamilton (1991) (in Japanese), later published as Hamilton (2000) in English, and to be found in *Narrow Roads*, vol 3, as chapter 3 (pp 73–87). For more on the carrion beetle, see chapter 19 of this book. I thank Alan Grafen for suggesting adding the expression 'typical "earnestly-meant"' for Bill's style.
22 *Narrow Roads*, vol 1, pp 388–9.
23 *Narrow Roads*, vol 1, p 58.
24 See n 23 above.
25 *Narrow Roads*, vol 1, p 70.
26 *Narrow Roads*, vol 1, p 71.
27 See n 26 above.
28 *Narrow Roads*, vol 1, pp 282–3.
29 *Narrow Roads*, vol 1, p 284. Later he found both aphid soldiers and colonial thrips.
30 *Narrow Roads*, vol 1, p 274.
31 *Narrow Roads*, vol 1, p 275.
32 *Narrow Roads*, vol 1, p 276.
33 *Narrow Roads*, vol 1, pp 276–7. This was thought to be the case when Hamilton wrote the paper, but is challenged by recent work.
34 *Narrow Roads*, vol 1, p 281.
35 *Narrow Roads*, vol 1, p 282.
36 See n 35 above.
37 *Narrow Roads*, vol 1, pp 290–1.
38 *Narrow Roads*, vol 1, p 291.
39 *Narrow Roads*, vol 1, p 292.
40 *Narrow Roads*, vol 1, p 300.
41 *Narrow Roads*, vol 1, p 301.
42 See n 41 above.
43 *Narrow Roads*, vol 1, p 302.
44 *Narrow Roads*, vol 1, pp 265–6.
45 *Narrow Roads*, vol 1, p 389.
46 Wilson was very supportive of Hamilton in 1965, see chapter 5 of this book.
47 Gould (1978); eg Maynard Smith (1975, 1966).
48 As Hamilton himself describes the situation in his autobiographical notes, there were not many good examples of altruism available when he wrote his 1964 paper, so social insects and warning cries of animals became his main examples. He muses:

> Perhaps because of the space I devoted to social insects, for a long time the combined paper was cited largely by people concerned with my arguments about them, whereas actually these arguments had mostly been afterthoughts hardly even started until the whole of what became Part I of the paper had been written. (*Narrow Roads*, vol 1, p 20)

49 *Narrow Roads*, vol 1, p 265.
50 See chapter 24 of this book.

Chapter 12

1 Wilson (1975); Maynard Smith (1975).
2 Maynard Smith (1975).
3 Hamilton (1976a).
4 See n 3 above.
5 Maynard Smith (1964); Maynard Smith (1965).
6 Strangely, Maynard Smith refers to 'Hamilton's papers in 1963' although the two-part paper was published in 1964. He may mean Hamilton's 1963 *and* 1964 papers but it doesn't solve the mystery.
7 Maynard Smith (1976); Dow (1976); Haldane (1955).
8 Maynard Smith (1958). We now learn that he *refereed* Hamilton's papers in 1963.
9 Haldane (1955).
10 Hamilton (1976b).
11 Mitchison (1976).
12 As we saw, Maynard Smith acknowledged that he had been the referee.
13 Maynard Smith (1975).
14 See chapter 7 of this book.
15 Maynard Smith (1958); Maynard Smith (1966).
16 Maynard Smith (1966), p 27. I am indebted to Steven Frank for alerting me to this passage, especially the 'entertaining twist'. Incidentally, the Blest moth example is one of Hamilton's own cases of altruism in Part II of his 1964 paper (cf *Narrow Roads*, vol 1, p 51).
17 I was told this by, among others, Steven Stearns (Department of Biology, Yale) and David Penney (Department of Biology, Massey University, New Zealand). Both had taken long car trips with Bill.
18 See chapter 2 of this book.
19 Trivers (1985), p 46.
20 Trivers (1985), pp 46–7.
21 Wright (1951).
22 Haldane (1955).
23 Haldane (1932). That book contains an Appendix where Haldane attempts to derive altruism from group selection, an exercise that Hamilton, as an undergraduate, had declared to have failed.
24 Haldane (1955).
25 Haldane (1955), p 131.
26 Hamilton (1963); Kropotkin (1930). For a description of Kropotkin's reception in London, see Mayr (1992). Incidentally, Kropotkin was a member of the Royal Society.
27 Edwards (1966).

Chapter 13

1 *Narrow Roads*, vol 1, p 486.
2 Also, he had been an invited speaker at international conferences and colloquia.
3 Students complained about the irrelevance and incomprehensibility of his lectures and asked that the results from his course be disregarded. *Narrow Roads*, vol 1, p 425, and Alan Grafen, personal communication.

4 *Narrow Roads*, vol 1, p 485.
5 Richard Southwood, personal communication.
6 Incidentally, Bill knew this very well himself, or soon learnt it, judging by his later habit of calling invited conference papers and book chapters his 'wild oats'.
7 Southwood was indeed making a career as an administrator, moving up to become a Dean at Imperial College and in 1979 becoming the Chair of the Zoology department at Oxford University.
8 *Narrow Roads*, vol 1, p 485. As we saw (chapter 11 of this book), he had just discovered that he actually enjoyed teaching the right kind of students.
9 The sociobiology controversy was reported on in such places as *Science* and *Nature* and *The Scientist*—journals that Hamilton regularly perused, for example, Lewin (1976), Wade (1976).
10 Wilson (1994) devotes a chapter to what he calls 'the molecular wars'.
11 For an account and analysis of the sociobiology controversy, see Segerstrale (2000).
12 Segerstrale (2000), chapter 3.
13 Hamilton (1976a).
14 See, eg, Segerstrale (2000), p 211. For an example, see the cover article in *Time Magazine*, 1 August 1977.
15 This was also true for some of the political critics' writings about sociobiology.
16 As noted, Hamilton and Trivers had met already in 1969 at Harvard.
17 Trivers himself joked about this (personal communication). But it was of course not only Trivers' personal presence at Harvard but the fact that his ideas, directly related to Hamilton's, were gaining fame through his publications. With his early papers (Trivers 1971, 1972, 1974) Trivers may have been a sort of early and lucid academic 'explicator' of Hamilton even before the eruption of the sociobiology controversy in 1975. At Harvard, it was actually the anthropology department led by Irven De Vore, that had early on embraced the new Hamiltonian paradigm.
18 Trivers (2002), p 167. Hamilton had not paid enough attention to the males. When they were factored in, the mathematics worked out alright. If the queen got to decide the sex ratio, it would be 1:1. With full worker control, it would be 1:3. Trivers and Hare's empirical result showed values between 1 and 3, suggesting a struggle between the queen and laying workers (fitting with the idea of 'parental conflict').
19 Trivers and Hare (1976). The quote is from that paper reprinted in Trivers (2002), p 196. The term 'kinship theory' was being launched in this paper (Trivers, 2002, p 166).
20 Trivers (2002), pp 155–6.
21 See n 20 above. Hamilton published the corrections in an Addendum to his chapter in Williams (1971), republished in *Narrow Roads*, vol 1, pp 80–1.
22 For the *Nature* article reporting on Trivers and Hare, see May and Krebs (1976).
23 Dawkins (1976). For Neo-Darwinism, see Segerstrale (2002).
24 Moreover, 'the gene's eye view' became a unifying conceptual tool (Segerstrale, 2006). This idea was also supported by George Williams.
25 Dawkins (1989), p 90. For a discussion of what could be called 'Hamilton's lag', see Segerstrale (2000), pp 89–90; Dawkins (1989), pp 325–9.
26 Tinbergen (1963). For the ethological response, see Segerstrale (2000), pp 93–4.
27 *Narrow Roads*, vol 1, p 330.
28 Grafen (2000).

29 Grafen (2006), Edwards (2000). Eshel and Feldman (2001).
30 *Narrow Roads*, vol 1, p 485.
31 *Narrow Roads*, vol 1, pp 485–6.
32 *Narrow Roads*, vol 1, p 369.
33 *Narrow Roads*, vol 1, p 375.
34 *Narrow Roads*, vol 1, p 370.
35 *Narrow Roads*, vol 1, pp 370–1.
36 *Narrow Roads*, vol 1, pp 379–80.
37 *Narrow Roads*, vol 1, pp 372–3.
38 *Narrow Roads*, vol 1, pp 373–4.
39 *Narrow Roads*, vol 1, p 377.
40 *Narrow Roads*, vol 1, pp 379, 385.
41 *Narrow Roads*, vol 1, p 384.

Chapter 14

1 Incidentally, it seems that Ann Arbor expected Harvard to be making Hamilton an offer of a permanent position, and was eager to snatch him first. This was the impression I got from Richard Alexander. I checked with EO Wilson: no such offer had actually been planned at the time (Wilson, personal communication).
2 *Narrow Roads*, vol 2, p 17.
3 Trivers (2002), p 10.
4 MCZ News, Spring 1978. That same issue also featured a report about the publication of EO Wilson's *On Human Nature*.
5 Cover story in *Time* magazine, 1977, pp 54–63.
6 Fox (1975).
7 According to EO Wilson (1994), p 331, Margaret Mead stood up, defending the right of existence of sociobiology and condemning the spirit of book burning.
8 Hamilton (1975), pp 149–50, or *Narrow Roads*, vol 1, pp 345–6. This passage was also quoted in Washburn (1977).
9 Washburn (1977).
10 Robin Fox, letter to Bill Hamilton 30 March 1977 (dated as 30 March 1976). To this was appended copies of relevant pages of the Newsletter, as well as his own letter to the editor sent in response to Washburn's letter. Fox is also suggesting that Bill respond himself.
11 Reynolds (1980), p 39, italics added. Interestingly, although these anthropologists are concern with racism, Hamilton reports that Trivers called this Hamilton's 'fascist' paper (presumably indirectly reporting on the discussion among anthropologists) *Narrow Roads*, vol 1, p 316.
12 Washburn 1978/80, p 276.
13 Hamilton, personal communication.
14 I recall that Trivers referred to Hamilton's 'racist' paper. Perhaps both versions are correct.
15 *Narrow Roads*, vol 1, pp 330–1.
16 As mentioned before, paradoxically, Washburn was someone that Hamilton admired, because of his biologically oriented views of early human history (*Narrow Roads*, vol 1, p 317).

17 See the pastoralist quote.
18 See n 17 above. Hamilton also refers to Eshel (1972).
19 In 1996 Hamilton says that he has hardly changed his views in regard to the offending quote, except that he could have mentioned that the effect could be partly cultural (*Narrow Roads*, vol 1, pp 317–18).
20 Trivers (2002), p 11. Hamilton suggested the term 'return effect' altruism for Trivers' nonhuman examples, but Trivers didn't budge.
21 See the pastoralist quote.
22 See chapter 11 of this book.
23 Hamilton suggests that Washburn's attack on him might actually have been caused by his felt need to strongly distance himself from 'wrong' views, as a result of Bill's quotation of him. He had become more 'incorrect' looking than he could tolerate (*Narrow Roads*, vol 1, p 317).
24 Lewontin (1977).
25 Hamilton (1977a).
26 See n 25 above.
27 Hamilton (1976c). For a comparison of scientific styles, see Segerstrale (2000), chapter 13.
28 Letter to Bettina, 17 May 1977.
29 Note that Hamilton presents the sociobiology controversy as having originated around his work, rather than around EO Wilson and his book. An interesting analysis of an 'ultimate' rather than 'proximate' kind? A simple explanation may have been that he felt the need to brag to his mother, his close supporter.
30 In 1977 the Ann Arbor Collective for the Science for the People had, among other activities, just published a book called *Science as a Biological Weapon*. See Segerstrale (2000), chapter 2, for an account of the controversy around Harvard and the role played by the Boston Science for the People.
31 This former graduate student was quite upset recalling the hostile atmosphere. In his department, Richard Alexander was leading the expansion of what he called 'Darwinian anthropology' (applying Darwinian principles directly to anthropological phenomena). Alexander may in fact have been more interested in various aspects of human sociobiology than EO Wilson (see Alexander, 1980, 1987). Interestingly, Alexander is said to have predicted the characteristics of a possible altruistic mammal—and voilà, in 1974 the naked mole rat was found.
32 Dawkins (1979). Hamilton himself referred to Sahlins' Marxism as 'religious' and did not feel the need to comment (*Narrow Roads*, vol 2, p 107).
33 For the reasoning of the critics of sociobiology about 'bad science' see Segerstrale (2000) pp 41–2; chapters 10 and 11 of this book.
34 For more on this incident and the AAAS meeting, see Segerstrale (2000), pp 22–4.
35 Gould and Lewontin (1989).
36 Maynard Smith, letter to Hamilton, October 1977.
37 Letter to Maynard Smith, 19 October 1977.
38 Maynard Smith, letter to Hamilton, 27 October 1977. (Incidentally, Maynard Smith and Price did quote Hamilton's 1967 paper, though they did not quote his idea of 'unbeatable strategy'.) For Maynard Smith's own explanation of his behaviour as a referee, see Segerstrale (2000), p 64.

39 *Time*, 1 August 1977, p 56.
40 Hamilton (1977b).
41 Hamilton (1977b), p 975–6.
42 Hamilton (1977b), pp 976.
43 Hamilton (1977b), p 976. See the similarity between Hamilton's discussion here and that in his 1975 paper in regard to terminology for various types of selection.
44 Hamilton (1977b), p 977. Hamilton is here trying to reiterate his insight from using the Price equation to rederive his inclusive fitness formula.
45 Hamilton (1977b), p 977. Hamilton here refers to Wilson's *Sociobiology*, p 120.
46 Hamilton (1977b), p 976. Compare with Hamilton's discussion of Darwin in his 1975 paper.
47 Hamilton (1977b), p 976.
48 Hamilton (1977b), p 976.
49 Trivers (2002), p 164. Also, as Hamilton might have seen, the haplodiploid hypothesis and Trivers' paper were in fact celebrated by Wilson in chapter 20 of *Sociobiology*.

Chapter 15

1 *Narrow Roads*, vol 2, pp 4–5, 7. Hamilton draws a parallel between the cardinal and himself and a thrush that sang for Thomas Hardy.
2 *Narrow Roads*, vol 2, p 5. Peter and Rosemary Grant, Christine Hamilton, Mary Bliss, personal communication.
3 *Narrow Roads*, vol 2, pp 18, 487.
4 Grant (2002); Queller (2001, 1985);Alexander personal communication. See also Frank (1995).
5 *Narrow Roads*, vol 2, p 58. For the origin of sexual selection, see *The Ant and the Peacock* (Cronin, 1991).
6 *Narrow Roads*, vol 2, pp 358, 54.
7 *Narrow Roads*, vol 1, p 358.
8 *Narrow Roads*, vol 2, pp 417, 429.
9 *Narrow Roads*, vol 1, p 362. This was also Williams' view. For Hamilton, see n 15.
10 *Narrow Roads*, vol 1, pp 353–4.
11 *Narrow Roads*, vol 1, pp 358–9.
12 *Narrow Roads*, vol 1, p 363.
13 *Narrow Roads*, vol 1, p 364.
14 Hamilton was invited to review Bell's book for *BioScience*, Hamilton (1982).
15 *Narrow Roads*, vol 2, pp 11, 428–9.
16 *Narrow Roads*, vol 2, p 12.
17 Hamilton, Henderson and Moran (1981).
18 *Narrow Roads*, vol 2, pp 57–8.
19 *Narrow Roads*, vol 2, p 52.
20 See n 19 above.
21 *Narrow Roads*, vol 2, pp 53–4.
22 *Narrow Roads*, vol 2, pp 52–7, quote at p 57.
23 *Narrow Roads*, vol 2, quotes at pp 56–7, 61.
24 *Narrow Roads*, vol 2, pp 63–5.
25 *Narrow Roads*, vol 2, p 65.
26 See n 25 above.

27 *Narrow Roads*, vol 2, pp 65–6.
28 *Narrow Roads*, vol 2, pp 68–70, quote on pp 69–70.
29 *Narrow Roads*, vol 2, p 70.
30 *Narrow Roads*, vol 2, p 71.
31 See n 30 above.
32 *Narrow Roads*, vol 2, p 72.
33 See n 32 above.
34 See n 32 above.
35 Hamilton and Zuk (1982).
36 Letter to Staffan Ulfstrand, 14 June 1979.
37 See n 36 above.
38 *Narrow Roads*, vol 2, p 66. The helpful editor was Pehr Enckell who had been planning a special issue of *Oikos* with Staffan Ulfstrand. It certainly helped the readability of the article that it ended with a summarizing section called 'Conclusions'.
39 Ulfstrand, personal communication, and *Narrow Roads*, vol 2, p 66.
40 *Narrow Roads*, vol 2, pp 95–6, 101–2, 108; Grafen (2006).
41 *Narrow Roads*, vol 2, pp 95–7, 108; Rick Michod, personal communication; Michod and Hamilton (1980), Michod (1982).
42 Letter to Maynard Smith, 23 October 1980 in response to a letter from Maynard Smith inquiring if Hamilton could serve as a Ph.D. examiner for Mike Orlove.
43 See n 42 above.
44 Letter from Maynard Smith to Hamilton, 14 November 1980 (handwritten).
45 Again, this is similar to what Maynard Smith told me (and no doubt lots of others) in interview.
46 *Narrow Roads*, vol 2, p 327; Zuk (2000).
47 *Narrow Roads*, vol 2, p 170.
48 *Narrow Roads*, vol 2, pp 174–6. Sickle cell anemia on p 175.
49 *Narrow Roads*, vol 2, p 232.
50 Hamilton obtained the term 'sosigonic' from a classically trained colleague at New College, JB Hainsworth (*Narrow Roads*, vol 2, p 719).
51 *Narrow Roads*, vol 2, p 233.
52 Eshel and Hamilton (1984); *Narrow Roads*, vol 2, pp 238–52; periodic reversal on p 249.
53 Eshel and Hamilton (1984); *Narrow Roads*, vol 2, p 250.
54 *Narrow Roads*, vol 2, p 171.
55 *Narrow Roads*, vol 2, pp 264–6.
56 *Narrow Roads*, vol 2, p 266.
57 *Narrow Roads*, vol 2, pp 266–7.

Chapter 16

1 Letter to Colin Hudson, 8 April 1981.
2 *Narrow Roads*, vol 2, pp 127–8.
3 *Narrow Roads*, vol 2, p 127. Hamilton thought of the group as 'groping for a new discipline, perhaps a bit like that which now sometimes takes the name "complexity theory"' (*Narrow Roads*, vol 2, p 129).

4 Axelrod (2011), p 9.
5 *Narrow Roads*, vol 2, p 120.
6 See n 5 above.
7 Axelrod (2005), p 8.
8 See n 7 above.
9 *Narrow Roads*, vol 2, p 124.
10 *Narrow Roads*, vol 2, p 144.
11 Axelrod and Hamilton (1981).
12 *Narrow Roads*, vol 2, pp 146–7.
13 *Narrow Roads*, vol 2, pp 150–1.
14 *Narrow Roads*, vol 2, p 152.
15 *Narrow Roads*, vol 2, p 123.
16 *Narrow Roads*, vol 2, p 126.
17 *Narrow Roads*, vol 2, pp 152–3.
18 *Narrow Roads*, vol 2, pp 153–4.
19 See n 14 above.
20 See n 14 above.
21 *Narrow Roads*, vol 2, pp 133–4.
22 Axelrod (2005), p 8.
23 *Narrow Roads*, vol 2, p 123.
24 *Narrow Roads*, vol 2, p 126.
25 Axelrod (2005), p 8.
26 Axelrod (2005), p 9.
27 Axelrod (2005), p 10.
28 *Narrow Roads*, vol 2, p 122.
29 Axelrod (2005), p 10.
30 See n 26.
31 Trivers (2002), p 52.
32 Trivers (2002), p 53.
33 See n 32 above; see also pp 24–5 for Bill's game theoretical interpretation.
34 *Narrow Roads*, vol 2, p 131.
35 Hamilton liked the romantic sound of Axelrod's 'the shadow of the future' (*Narrow Roads*, vol 2, p 121). For a later definition of this concept, see, for example, Axelrod (2000), p 18.
36 See n 32 above.
37 Trivers (2002), p 55.
38 Trivers (2002), p 54.
39 *Narrow Roads*, vol 2, p 156.
40 See n 39 above and Trivers (2002), p 54.
41 Hamilton discusses this in *Narrow Roads*, vol 2, pp 132–3. See also Axelrod (2000) on advances in cooperation theory; Trivers (2002), p 55, for a useful overview of new models that relax the rules of the Prisoner's Dilemma; and Axelrod (2011) for the discussion of some more recent models and especially the development of 'generous' tit-for-tat and 'contrite' tit-for-tat to take care of the problem of 'noise' (that is, misunderstandings, and misimplementations of intentions) in the iterated Prisoner's Dilemma game.
42 *Narrow Roads*, vol 2, p 142.

Chapter 17

1 Bill's (H)BACHS colleague Carl Simon, personal communication. Simon also provided me with a map to help me find my way around the place.
2 Hamilton and Zuk (1982).
3 Southwood, personal information.
4 *Narrow Roads*, vol 2, p 303.
5 Bill went over to Oxford to discuss this with Southwood.
6 Dawkins (2001), p xv, and Christine Hamilton, personal communication.
7 *Narrow Roads*, vol 2, pp 303–4, 307–9. The student for whom he wrote that recommendation letter was Nancy Moran. According to her, the description of the case is not quite correct, and she asked me to note the following:

> '[I]n his version, it is the grad student (me) who comes to him to complain about this letter. I remember very clearly though a somewhat more complicated story, as follows: I had never heard of the contents of the letter, when one day Bill came down to my office in the Museum and told me that Bev Rathcke (a professor in the same department but in another building) had come to him to tell him that she had gotten wind of his letter for me, and that she found it inappropriate and sexist (I don't know her exact words of course). He was adamant that he found her reaction to be silly, a denial of the (to him) obvious fact that women were worse at math, and that she should recognize that it was in fact a positive statement to point out that I was better than average for my sex. Later, I became friends with Bev, partly because I stayed at U Michigan for six months as a lecturer after getting my degree, and was on somewhat more equal terms with faculty during that time. So I heard her side of the story, which was that she had been on a field course in Costa Rica and that she heard about the letter from a grad student there, someone who had been the student member of a search committee at the University of Washington...
>
> In NRG II, this whole episode was included as one element in what Bill saw as a rising tide of political correctness in the US, part of what made him move. So either story, with me or with Bev as the complainer, would have served the purpose... But I am certain he was wrong, I remember it all very clearly and I also know that I was never really offended at what he wrote, though I disagreed with his premise about women and math. In fact, I know that he did mean it as a compliment. I think the one thing I said to him, when he told me about what Bev said, was that clearly this version of the letter seemed counterproductive, since it brought attention to him rather than to me, so maybe he could just cut that sentence. I really needed a job!' (Nancy Moran, email, 15 October 2011).

8 Haig, Pierce, and Wilson (2000). Robert told me the novel was inspired by him.
9 This was suggested to me by Mark Ridley, personal communication and Alan Grafen, personal communication. But was it Christminster or Wessex? Hamilton, when mentioning Hardy, discusses how '[o]f all Oxford colleges, New College would perhaps be called the most "Wessex-oriented" in its traditional intake' and goes on to discuss how a scene from *Jude the Obscure* (one of his favorite books) might have been set in a particular still existing Oxford tavern (*Narrow Roads*, vol 2, p 315).
10 *Narrow Roads*, vol 2, p 309.

11 See chapter 15 of this book for 'sosigonic'.
12 *Narrow Roads*, vol 2, p 311.
13 See n 12 above.
14 Dawkins (2001).
15 *Narrow Roads*, vol 3, pp 7–8.
16 Janet Hamilton, personal communication.
17 John Maynard Smith also attended the 1986 Dahlem conference. I have a report on Hamilton's behaviour at that conference from Uli Schmetzer, Department of Biology, University of Zurich. My informant's impression was that at this conference Hamilton went around looking gloomy, while Maynard Smith was a very charming participant and later ended up inviting people to his hotel room for drinks from his mini-bar (which he later paid for). Moreover, in November 1987, at the 'Nobel Symposium' at Gustavus Adolphus College, Maynard Smith disagreed with Hamilton who in his presentation predicted a build-up of deleterious genes in the human species. He said he believed that we would be able to solve this problem in the future through genetic engineering (for a transcript, see *Narrow Roads*, vol 2, pp 456–7; Hamilton's paper 'Sex and Disease' is republished as chapter 12 in *Narrow Roads*, vol 2). Elsewhere Hamilton noted that he would not even have accepted the invitation to this symposium, had it not been that he wanted to visit Ann Arbor on the way to talk to Bob Axelrod. (*Narrow Roads*, vol 2, p 450).
18 No doubt Hamilton, as usual, was able to negotiate the content of his contributed papers.
19 *Narrow Roads*, vol 2, pp 261–4, and chapter 8, *Narrow Roads*, vol 2, 'Instability and cycling of two competing hosts with two parasites'.
20 *Narrow Roads*, vol 2, p 275.
21 See *Narrow Roads*, vol 2, pp 370–6 on Imanishi and the Kyoto Symposium.
22 Hidaka, personal communication; Aoki, personal communication.
23 *Narrow Roads*, vol 2, p 393.
24 *Narrow Roads*, vol 2, p 402.
25 *Narrow Roads*, vol 2, p 382.
26 *Narrow Roads*, vol 2, p 384.
27 Hamilton notes that 'the picture has changed rather dramatically . . . there are now known examples of highly developed altruism, even of sterile castes, within separated clones' (*Narrow Roads*, vol 2, p 394). He provides a table of these cases on p 395.
28 *Narrow Roads*, vol 2, p 405.
29 See n 28 above.
30 *Narrow Roads*, vol 2, p 406.
31 See n 30 above.
32 *Narrow Roads*, vol 2, p 407.
33 Steven Frank, Letter to Bill Hamilton, 10 May 1985.
34 Steven Frank, Letter to Bill Hamilton, 4 September 1986.
35 Note entered in handwriting as part of the chronology for the development of the modelling of the HAMAX paper (see more details of this chronology, chapter 18 of this book). (This note presumably refers to his presentations and writings touching on issues of health and disease.)
36 *Narrow Roads*, vol 2, p 322.
37 *Narrow Roads*, vol 2, p 346.

38 See n 37 above.
39 *Narrow Roads*, vol 2, p 350.
40 *Narrow Roads*, vol 2, p 356.
41 *Narrow Roads*, vol 2, p 357.
42 Ulfstrand, personal communication.
43 *Narrow Roads*, vol 2, pp 535–44 (chapter 13).

Chapter 18

1 The paper he presented at this conference was 'Pathogens as Causes of Genetic Diversity in Their Host Populations' (1982). It was later published in the conference volume.
2 *Narrow Roads*, vol 2, pp 65–6, 68, and see Axelrod's comment on the following page in this chapter.
3 For details about the three-dimensional Perspex cubes see Sumida, *Narrow Roads*, vol 3, p 5. The 'spaghetti' comment was from DS Wilson, personal communication.
4 Zuk (2000). For details about the 'fakir' bath model, see *Narrow Roads*, vol 2, pp 182–3. and for the 'floating ball model' see *Narrow Roads*, vol 2, pp 653–4 and 642–3.

Hamilton realized that his models were not appreciated as much as he would have liked them to be. He believe he knew why:

> Hardly any evolutionist seems to like my 'floating ball' metaphor for how recombination aids defence against parasites…But it is intended not so much for evolutionists and mathematicians as for outsiders who (like me) never had a course in population genetics…The image annoys precise minds because the gene-frequency space notion that I use—the weighed cube or hypercube that has been inflated into a ball—is easily seen by any mathematician or population geneticist as hopelessly inadequate to represent the full dynamic of the system. Where I use a cube as the primary idea, there are really needed 8 dimensions for a truthful image, or 16 when the parasites are represented numerically, or 24 if the asexual hosts are represented too…And all this is for just three loci, still far short of the numbers needed for firm support of sex. *For me, however, it is exactly this enormity of the full problem that make* [sic] *the simple shadows in my three-dimensional representation so much better than nothing; these shadows cannot be denied to give clues to the vast mobile hyperballs of the full process.* (p 642, italics added)

He recognized additional problems, too, such as the need to describe 'more peculiar "motions" of the ball in the water than simple rotation'. These, however, could be visualized with special computer simulations (Bill here mentioned especially Akira Sasaki's excellent work, p 643).

5 Axelrod (2005), p 12. See also Holland (1975, 1992).
6 This could be made very visually compelling with the help of computer graphics. Here is the description of one such model of host-parasite coevolution, which Bill sent as a piece of artwork to his good friend Naomi Pierce:

> 'Here's an example of the sort of cycle and sort of picture I produce, but not a nice one. I am always hoping to discover the face of God as I generate these cycles, but here I seem to have discovered a rather nasty supercilious eye that

watches me. The green trajectory is of the parasite population: it swings wider because having several generations to each generation of hosts. The pink is the host population, nicely herded to near the centre by the sheep-dog-like parasite population racing around, and consequently with no danger of any allele going extinct, which of course I like.... (Letter to Naomi Pierce, dated 21 October 1986).

7 'Valley crossing' is 'a hang-gliding' transition ... across a potentially endless mountainscape of attempted optimization. (*Narrow Roads*, vol 2, 128). For the Kyoto conference, see chapter 17 of this book.

8 This is from a sheet of brief notes documenting the chronology of events in the preparation of the HAMAX paper and submitting it for publication. It may have been prepared specially for *Narrow Roads*, vol 2, or it may have been to document in detail the collaboration between Hamilton and Axelrod. (I was given a copy of these notes by Luisa Bozzi, they are probably now part of the Hamilton archive.)

9 Axelrod (2005), p 13. For typical difficulties in checking agent-based models for accuracy, see Axelrod (2005), p 13, n 19.

10 *Narrow Roads*, vol 2, p 576.

11 *Narrow Roads*, vol 2, p 602–3.

12 Axelrod (2005), p 17.

13 *Narrow Roads*, vol 2, p 561.

14 *Narrow Roads*, vol 2, p 605.

15 *Narrow Roads*, vol 2, pp 604–5.

16 *Narrow Roads*, vol 2, p 609.

17 Axelrod (2005), p 17.

18 *Narrow Roads*, vol 2, p 602 (Reiko). Later, after reading similar referees' responses to other authors, Hamilton realized that this reaction was a typical one (*Narrow Roads*, vol 2, pp 614, 637).

19 *Narrow Roads*, vol 2, pp 601–2.

20 Axelrod (2005), p 15.

21 *Narrow Roads*, vol 2, p 607–8.

22 Axelrod (2005), p 17.

23 See n 21 above.

24 Bill had again clearly overestimated the referees' capability to follow his train of thought, forcing them away from their existing knowledge into new scenarios requiring quite new assumptions. Unlike Axelrod, who when presenting a new idea typically strove hard to connect new ideas with their already existing knowledge, Hamilton often neglected this potentially important pedagogical touch and just expected readers to follow him along into an unknown (perhaps even unrecognizable) landscape (we saw Trivers commenting on something akin to this earlier in chapter 16). (Sometimes, though, more adventuresome referees were able to appreciate what Hamilton wanted to do.)

25 Axelrod (2005), p 18.

26 Axelrod (2005), note on p 17.

27 *Narrow Roads*, vol 2, p 609; Axelrod (2005), p 16.

28 *Narrow Roads*, vol 2, p 413.

29 *Narrow Roads*, vol 2, p 422–3.

30 *Narrow Roads*, vol 2, pp 423, 425.

31 *Narrow Roads*, vol 2, p 414.
32 See n 31 above.
33 John Maynard Smith had supported the publication of Kondrashov's paper in *Nature*, earning a note of thanks from Kondrashov in his paper.
34 It seems that Hamilton refers to the same thing in *Narrow Roads*, vol 2, p 637. There he also presents the other referee's brief report in full.
35 See also the references in Seger and Hamilton (1988).
36 Perhaps this contributed to Hamilton's irritation with John Maynard Smith at the Nobel Symposium at Gustavus Adolphus College in November 1987 (see chapter 17 of this book and references there).
37 The relevant references here are Read and Harvey (1989a), Zuk (1989), Hamilton and Zuk (1989), and Read and Harvey (1989b), see also John (1997). Andrew Read, a doctoral student in the department and friendly with both Hamilton and Paul Harvey (his adviser) found himself in a particularly difficult situation. He had earlier written an article applying the Hamilton-Zuk hypothesis to bird song, and found support for it (Read, 1987), after which he went on to write a longer article for *American Naturalist*. This had already been submitted, when his adviser suggested that the results could be an artfact of the method used and suggested that Read try using a different method, which would seem to avoid potential pitfalls of subjective evaluation by employing instead a panel of reviewers who would be using bird books. There was also the potential methodological problem that correlations might have arrived due to phylogeny (see *Narrow Roads*, vol 2, p 794). Read accepted the challenge, did a new study, and found that his earlier correlations did not hold up any longer. He ended up retracting his journal submission and instead publishing a critical methodological commentary in *Nature* together with Harvey. Hamilton understood Read's position and did not take offence (Read, personal communication). Harvey had just co-edited a book on methodology relevant to this issue (Pagel and Harvey, 1989). See more on this at length in *Narrow Roads*, vol 2, pp 793–825, 'Our Paper Then and Now' (Appendix to chapter 6). I learnt about Hamilton's very strong interpretation of the criticism from Luisa Bozzi (personal communication).
38 In the mid-1970s, Robert had married 'a headhunter's daughter' (as he proudly described it) in Borneo. Bettina attended the wedding, held in a longhouse.
39 Peter and Rosemary Grant, (personal communication). See also Grant (2002).
40 Already Hermann Muller and before him Haldane had been calculating this 'mutation load' as the increase in deleterious mutations per person per generation.
41 It requires a mutation rate of at least one mutation per individual per generation (*Narrow Roads*, vol 2, p 619).
42 *Narrow Roads*, vol 2, p 616.
43 *Narrow Roads*, vol 2, p 620.
44 *Narrow Roads*, vol 2, p 619.
45 *Narrow Roads*, vol 2, p 657 (truncating selection), p 617 (mate choice).
46 *Narrow Roads*, vol 2, p 626.
47 *Narrow Roads*, vol 2, pp 627, 658.
48 *Narrow Roads*, vol 2, p 626.
49 *Narrow Roads*, vol 2, p 627.

50 Hurst, *Narrow Roads*, vol 2, vol 3, p 95.
51 *Narrow Roads*, vol 2, pp 640–1 (n 36).
52 *Narrow Roads*, vol 2, p liv.
53 See chapter 22 of this book.

Chapter 19

1 Letter to Yura Ulehla, 9 April 1989.
2 See n 1 above.
3 See n 1 above.
4 Ebert, *Narrow Roads*, vol 3, pp 189–94.
5 See n 4 above.
6 See n 4 above.
7 See n 4 above.
8 Henderson, *Narrow Roads*, vol 3, pp 308–9.
9 Henderson, *Narrow Roads*, vol 3, pp 310–11.
10 *Narrow Roads*, vol 3, pp 336–7.
11 See n 10 above.
12 *Narrow Roads*, vol 3, p 327.
13 *Narrow Roads*, vol 3, p 337.
14 *Narrow Roads*, vol 3, pp 327–8.
15 *Narrow Roads*, vol 3, p 314.
16 Nevo (2001).
17 See n 16 above.
18 The resulting paper was called 'Extraordinary multilocus gene organization in mole crickets, Gryllotalpidae' (Nevo et al, 2000).
19 For instance as he is flying over Arizona on his way to the meeting with Michod (see chapter 14 of this book).
20 See his warm descriptions of various insects encountered in unusual places (for example, his various apartments) in his autobiographical notes.
21 The Babassu discussion is in *Narrow Roads*, vol 2, pp. xxix–xxx, 498, and elsewhere. 'Why shouldn't plants be altruistic, too?' is in *Narrow Roads*, vol 1, p 21.
22 *Narrow Roads*, vol 3, p 206.
23 The suggestion about Bill's tendency to model one woman on another that he knew, or attributing to people sociobiological grounds for behaviour was given to me by Bill's friend Luisa Bozzi.
24 For jungle impressions see, for example, Ribeiro (2000); for an instance of subdued dinner communication, see A Smith (2000).
25 See pictures taken by Marcio Ayres, on the Hamilton website of Basel University, available at <http://evolution.unibas.ch/hamilton/index.htm>.
26 *Narrow Roads*, vol 3, p 205.
27 As suggested, for example, by Cambridge psychologist Simon Baron-Cohen in various writings (such as Baron-Cohen et al, 2001) and also by University of London sociologist Christopher Badcock (for example, Badcock, 2004).
28 Baron-Cohen and colleagues conducted studies of engineers and scientists and identified some typical features of Asperger's Syndrome (for example, Baron-Cohen et al,

2001). But the syndrome exists in various degrees. So, for instance, according to Baron-Cohen, 'Passion, falling in love and standing up for justice are all perfectly compatible with Asperger's Syndrome...What most people with Asperger's Syndrome find difficult is casual chatting—they can't do small talk.' That was for instance the case with Newton. On the other hand, Einstein had a good sense of humour—a feature not found in people with severe Asperger's. This was pointed out by a critical, psychiatrist (Dr Glen Elliott at the University of California at San Francisco), who noted 'One can imagine geniuses who are socially inept and yet not remotely autistic.' (Muir, 2003).

29 See also Badcock and Crespi (2006). In the final paragraph of their article, they actually state that Bill Hamilton, whom they both knew very well, is an example of someone with Asperger's syndrome. Badcock told me that he and Bill often discussed the Asperger syndrome together. Badcock believed that he himself, too, had the syndrome and added that Asperger's Syndrome people like to talk to each other. From our discussion I got the feeling that Bill may have started to think of himself along these lines (Badcock, personal communication). This was the way Bill was presented:

> Indeed, individuals at the highest-functioning end of this spectrum, not least among them Isaac Newton and William D. Hamilton, may have driven the development of science, engineering, and the arts, through mechanistic brilliance coupled with perseverant obsession. (Badcock and Crespi (2006), p 14)

The aim of the Badcock-Crespi paper, however, was much more general. The authors' 'imprinted brain hypothesis' suggested that an imbalanced genomic imprinting in the development of the brain may have provided an evolutionary basis for the aetiology of autism. In this they agreed with Baron-Cohen's view that rather than constituting a disorder or disability, the autism spectrum is merely an expression of human cognitive diversity. (Bernie Crespi kindly sent me the proofs of their joint paper.)

30 For example, Trivers (2002, p 10) (on the microphone), Dawkins (*Narrow Roads*, vol 2, pp xv–xvi) (on scribbling), Trivers (2002), see chapter 9 of this book), and Andrew Bourke (on stampede, personal communication).

31 Many have told me this. There is a written testimony by John Maynard Smith (2000) and by Tim Brown (*Narrow Roads*, vol 3, p 255).

32 For example, on long car trips (see Steven Stearns' essay, 'Three days with Bill Hamilton' on the University of Basel website).

33 See, for example, *Narrow Roads*, vol 1, p 191.

34 For instance, Bill is reported to have jumped over fences around Dunedin, New Zealand, in 1995 and climbed a tree in Impruneta close to Florence in 1998.

35 As seen in chapter 5 of this book, for example.

36 Hamilton in fact chose this Kafka quote as the vignette for chapter 2 (dealing with his big 1964 paper) in *Narrow Roads*, vol 1, p 11.

37 *Narrow Roads*, vol 1, p 12.

38 Letter to Naomi Pierce, 21 October 1986.

39 Ribeiro (2001).

40 See n 39 above.

41 See n 39 above.
42 Ribeiro mentions also getting introduced to other Brazilian scholars visiting Oxford and enjoying the special community that emerged this way.

Chapter 20

1 'Dental Surgery and Scout Bus Wrecked', *Oxford Mail*, 12 June 1991.
2 Christine Hamilton, personal communication.
3 Christine Hamilton, personal communication; letter to Yura Ulehla, 12 December 1992.
4 The hair story was told to me by various people at Oxford University and others who had visited during this time, among others Sebastian Bonhoeffer and Laurence Hurst. Christine Hamilton herself provided the crucial piece of information that she used to cut Bill's hair. (Incidentally, this hair-cutting—which took place outdoors—almost seemed like some kind of 'event'—at least, there are two photos of this procedure.)
5 Many have offered me details about how Bill was coping by himself (not very well) and how his students tried to cheer him up, among others Katrina Mangin, personal communication. Nigella Hillgarth's thesis was on coccids in pheasants; they affect the 'cockerel type' ornaments of the species (*Narrow Roads*, vol 2, pp 672, 696). Bill had prepared ingenious-seeming cages for the pheasants in her study, but they kept escaping.
6 Letter to Yura Ulehla, 12 December 1993. Interestingly, this letter is dated 12 December 1992, while it is obvious that it must be 1993 because of its content—the description of the prizes. This is something that occasionally happens with Bill Hamilton in his letters.
7 See n 6 above.
8 Nancy Moran, personal communication.
9 Letter to Staffan Ulfstrand 1993 (no date, this is written on the back of a postcard).
10 Letter to Yura Ulehla, 12 December 1993.
11 Christine Hamilton, personal communication. The interpretation of Bill's ambivalence about house buying is mine.
12 I thank the Crafoord Foundation for kindly providing some newspaper clippings and other materials for me.
13 Birgitta Tullberg, Department of Biology, University of Stockholm, personal communication.
14 See Hamilton (1963, 1964, 1967, 1972).
15 See chapter 5 of this book.
16 Ulfstrand, personal communication. Christine, meanwhile, conversed with the king.
17 According to Hurst 'Hamilton gave a strikingly novel lecture about sphagnum bogs and Gaia. He concluded that while sphagnum bogs may be a level of selection, Gaia wouldn't work. I had not the slightest inkling that he was thinking about such things' (Hurst, *Narrow Roads*, vol 3, p 94). See more about Hamilton and Gaia in chapter 21 of this book.
18 Ulfstrand, personal communication.

19 The cover letter to *Science*, 17 January 1994, is available online as part of University of Woollongong (Australia) sociologist Brian Martin's website: <http://www.bmartin.cc/dissent/> or retrievable from Ed Hooper's website: <http://www.aids.origins.com>.
20 By 'Letter' here Hamilton refers to the actual article he wanted published, 'Aids Theory vs. Law Suit'. It is available online in the Martin collection (see n 19 above) and in print in a somewhat revised form (see n 22 below).
21 Letter to Daniel Koshland (editor of *Science*), 24 February 1994.
22 This letter did not change the mind of *Science's* editor-in-chief. Part of one of the referee's letters, Review # 2, is available as part of the Martin archive above. On 11 March, Bill wrote to *Nature's* editor John Maddox and he tried to get his Letter published in *Nature*, but was rejected. His unpublished Letter is reprinted in *White Death* by Julian Cribb (1996). (See also Hooper, *Narrow Roads*, vol 3, p 242).
23 At the Evanston meeting (where I was also giving a paper, he likened HBES to a secret society).
24 *HBES Newsletter*, vol 1, no 3, June 1991, p 5.
25 *Narrow Roads*, vol 2, pp 466–7.
26 For 'cassandra' units, see *Narrow Roads*, vol 2, p 468. For truth, see p 484.
27 See his interview with Franz Roes (1996).
28 See n 27 above.

Chapter 21

1 Bill had been briefly interviewed by Luisa at a conference in Florence in 1988 and told her about his theories about the evolution of sex and sexual selection. Later that year she visited Oxford to finish the interview and gather more material for her article in *La Stampa*.
2 Letter to Yura Ulehla, 22 December 1994. Bill here uses the expression 'last year', but he clearly refers to 1994, since it is still around Christmas. Also, Christine's contract was to expire in May 1994, so in early 1994 it was not exactly at the end of her two years yet, although it may have felt so to Bill.
3 Christine had hoped that they would spend Easter there with her parents.
4 The person with the photo was Laura Beani, Luisa's friend and biology book co-author, and the wife of the conference organizer, Francisco Dessi. Both Beani and Dessi were professors at the University of Florence; Beani also collaborated with colleagues at Oxford. Bill was in Castiglioncello together with Christine on the way to the Kyoto prize ceremony.
5 Luisa was amicably separated from her architect husband.
6 Letter to Luisa, 24 February 1994.
7 Letter to Yura Ulehla, 22 December 1994.
8 See n 7 above.
9 See n 7 above.
10 Jeanette Stace (Bill's first cousin, living in Wellington, NZ), personal communication.
11 Christmas letter to Yura, 1997 (exact date not available, first page missing).
12 See n 11 above.
13 Hamilton's Fyssen Prize lecture is 'Born the Slave of the Queen of Life'", reprinted in *Narrow Roads*, vol 3. He discusses the Baldwin effect on p 224. See chapter 19 of this book for more on the Baldwin effect and its role in the Amazon.
14 Christmas letter to Yura, 1997 (see n 11 above).

15 See n 11 above. There are several photos from Bill's election to the Academy of Finland and the dinner afterwards. One of the photos of Hamilton with the Kyoto prize around his neck is reproduced on the cover of the *New Zealand Science Review*, vol 60 no 4, 2003. For Bill's general reaction to prizes, see Zuk (2000) and Mary Bliss, personal communication.

16 Christmas letter to Yura, 1997, and Rauno Alatalo, personal communication. For a description of the Tvarminne skating incident, see Jacobus Boomsma (2001).

17 For details about the Amazon project, see chapter 19 of this book and *Narrow Roads*, vol 3, chapter 20 (Henderson, 2005). For Gaia and the cloud hypothesis, see *Narrow Roads*, vol 3, chapter 15 (Lenton, 2005). For autumn colours, see *Narrow Roads*, vol 3, chapter 17 (Brown, 2005). These projects are described briefly below. A further important collaboration was the one with Akira Sasaki, who together with Hamilton discovered the phenomenon of an 'evolutionary pacemaker', further explaining the workings of the HAMAX model, see *Narrow Roads*, vol 3, chapter 18 (Sasaki, 2005). The work with Sasaki (on meta-populations of co-evolving hosts and parasites) involved particularly interesting decorative patters, which Bill called 'tomato attractors'.

18 *Narrow Roads*, vol 3, p 257. It was a pleasant meeting, however, and James Lovelock was appreciative of the lovely, gluten-free cake that Luisa, visiting at this time, had prepared for tea (Lovelock is gluten-intolerant) (Luisa Bozzi, personal communication). For more on the Hamilton-Lenton-Lovelock collaboration, see Tim Lenton's introduction to *Narrow Roads*, vol 3, chapter 15, and the two papers that follow, Hamilton (1995) and Hamilton and Lenton (1998). See also Lenton's review paper in *Nature* (Lenton, 1998) explaining misunderstandings of Gaia and the way to reconcile it with evolutionary concerns.

19 Lovelock much appreciated Hamilton and was pleased about the collaboration (personal communication). Note, however, that despite what turned out to be a successful collaboration with Lenton and Lovelock, Hamilton was originally looking for a different and more universal principle. He saw Daisyworld as a special case (Lovelock's daisies alter the environment in the same way at the individual and global level). For Hamilton, much stronger evidence was needed, and that would be obtained if Gaia-like properties emerged independently in a model (*Narrow Roads*, vol 3, p 263; see also Hamilton (1996c). In particular, it would be important to know whether a 'Genghis Khan type organism' could in principle arise somewhere and do harm. This was, indeed, what Hamilton later started working on with Peter Henderson in a model called Damworld, reporting success to Lenton in March 1999 (see *Narrow Roads*, vol 3, p 263).

20 See Lenton, *Narrow Roads*, vol 3, pp 258, 260, 262. They decided to submit to *Science* since Lenton was involved in his own discussion with *Nature* about a review article on Gaia and natural selection (eventually published, see n 18 above). *Science* returned their manuscript without a review. Finally it was published in *Ethology, Ecology and Evolution* (*Narrow Roads*, vol 3, pp 261–2). See Hamilton (1998) and Hunt (1998) for articles on the cloud hypothesis.

21 See *Narrow Roads*, vol 3, p 351, for 'cheekiness', and Motluk (2001) for the *New Scientist* report. This paper was submitted to *Nature* 'in autumn of 1998, naturally' said Brown (p 354), but was rejected, '"bonkers" being a not infrequent response' (Brown, p 355).

22 The professor was John McCulloch, an evolutionary biologist at Cambridge. This attempt at a Darwinist analysis of British history was published in *New Scientist* in 2001, and cited

in English language newspapers. I have a clipping from *Sydney Morning Herald*, 22 June 2001, entitled 'Off with Their Heads but Mind the Bloodline' (sent to me by Mary Bliss).

23 Luisa Bozzi, personal communication.
24 Alexander (2000).
25 Angier (2000).
26 Trivers (2002), p 55.
27 See n 26 above.
28 See n 25.
29 Zuk (2000).
30 For Hamilton's 'chosen group' see, for example, *Narrow Roads*, vol 3, p 77; for his view of the need for simulation, see *Narrow Roads*, vol 3, p 268.
31 For some of Bill's experiences, see, for example, *Narrow Roads*, vol 2, pp 95–6.
32 Roes (1996).
33 See n 30 above.
34 *Narrow Roads*, vol 2, p 609. Hamilton admits that he later came to rely more on simulation for guiding him through difficult issues (*Narrow Roads*, vol 1, p 138).
35 Sasaki, *Narrow Roads*, vol 3, p 374. For Bill's amazement at colour polymorphism in aphids in 1986, see Sasaki's description in *Narrow Roads*, vol 2, p 370.
36 *Narrow Roads*, vol 2, p 15.
37 Trivers (2002), pp 155–6.
38 Grant (2002), p 392.
39 Trivers (2002), p 11.
40 See n 29.
41 *Narrow Roads*, vol 2, p 414.
42 See n 41 above.
43 *Narrow Roads*, vol 2, pp 414–15.
44 *Narrow Roads*, vol 2, p 425.
45 *Narrow Roads*, vol 1, pp 259–61 (the quote is compiled from these pages).
46 Mary Bliss, personal communication.
47 Laura Beani, personal communication.
48 For the libraries, see *Narrow Roads*, vol 1, pp 139–40. For the plumber impersonation, see *Narrow Roads*, vol 1, p 431.
49 Bernie Crespi, personal communication. The visit to the Niah caves was in 1991.
50 *Narrow Roads*, vol 1, p 229.
51 Uncle Charlie used to send Bill poetry and other books, and books with collections of paintings. He also invited Bill to write a review on a book on dolphins in his journal, Bill's very first publication (*Narrow Roads*, vol 1, p 1).
52 *Narrow Roads*, vol 2, p 737.
53 *Narrow Roads*, vol 2, p 104.

Chapter 22

1 *Narrow Roads*, vol 1, p vii.
2 See n 1 above, and Sarah Hrdy, personal communication.
3 *Narrow Roads*, vol 3, also contains some important prize acceptance talks and book reviews by Hamilton.
4 Grafen (2000).

5 Peter Grant found a sentence over 150 words long (Grant, 2002). Both Mark Ridley and Alan Grafen encouraged me to read Hamilton's footnotes too, to get a sense of his sometimes style of digression upon digression.
6 For example, Nevo (2002).
7 *Narrow Roads*, vol 1, p viii.
8 These cubes were produced by Brian Sumida (see his explanation in *Narrow Roads*, vol 3, p 5).
9 See *Narrow Roads*, vol 2, chapters 6 and 16, and the long Appendices to each.
10 See chapter 10 this book and chapter 9 in *Narrow Roads*, vol 1.
11 See n 10 above.
12 *Narrow Roads*, vol 2, p xxxi.
13 *Narrow Roads*, vol 2, p 677.
14 *Narrow Roads*, vol 2, pp 6, 55, 347.
15 *Narrow Roads*, vol 2, p 556.
16 For example, *Narrow Roads*, vol 2, pp 347, 556.
17 For the discussion in the 1960s and 1970s, see Segerstrale (2000), chapter 5.
18 See n 17 above.
19 *Narrow Roads*, vol 1, p 196.
20 See *Narrow Roads*, vol 2, chapter 12.
21 *Narrow Roads*, vol 2, p 484.
22 *Narrow Roads*, vol 2, pp xlvii, 465.
23 *Narrow Roads*, vol 2, pp 488–9.
24 *Narrow Roads*, vol 2, p 489.
25 *Narrow Roads*, vol 2, p 484, and his article 'Technosuperlife' (Hamilton, 1999).
26 For example, *Narrow Roads*, vol 2, p 484, and 'Technosuperlife'.
27 *Narrow Roads*, vol 2, p 459.
28 *Narrow Roads*, 'Technosuperlife'.
29 *Narrow Roads*, vol 2, pp 601, 493, 460–1, 471–3, 476; Singer (1994).
30 *Narrow Roads*, vol 2, p 475.
31 Naomi Pierce, personal communication, Haig (2003).
32 *Narrow Roads*, vol 1, p 17.
33 Maynard Smith, personal communication, Mary Bliss, personal communication.
34 *Narrow Roads*, vol 2, pp 275–6, letter to Colin Hudson, 20 October 1961.
35 *Narrow Roads*, vol 2, p xlvi.
36 Haig (2003).
37 Mary Bliss, personal communication.
38 *Narrow Roads*, vol 2, pp xxviii–xliv, and Hamilton, personal communication.
39 For the intent of the paper see *Narrow Roads*, vol 2, p xli. The content of the paper is in fact roughly identical with chapter 12 in *Narrow Roads*, vol 2, as Hamilton says himself, admitting it is 'a rather rambling and long essay', *Narrow Roads*, vol 2, p xli).
40 *Narrow Roads*, vol 2, p xliii. Bill was indeed given much less time than he had planned for the paper, and the time keeping was strict. (*Narrow Roads*, vol 2, p xliii).
41 Cavalli-Sforza's paper was entitled 'The Meaning of Nature'.
42 Levinson (1998).
43 Gereon Wolters, personal communication; and Wolters (2000).

44 Trivers (2002), p 53.
45 Mark Ridley, personal communication; Sarah Bunney, personal communication.

Chapter 23

1 Letter to Yura, 26 December 1996. She was 89. His fathers died in 1973 at age 74.
2 See n 1 above.
3 See n 1 above.
4 Letter to Yura, Christmas 1997 (no exact date, first page missing). Mary Bliss, personal communication.
5 Letter to Yura, 26 December 1996; Moura and Gama (2004).
6 Executive summary for the Mamiraua Project 1995.
7 Henderson, *Narrow Roads*, vol 3, p 311.
8 Letter to Naomi Pierce, 6 April 1996.
9 Email to Crespi, 14 July 1998 and Hughes (2002), p 87.
10 Letter to Yura, 26 December 1996.
11 Luisa Bozzi, personal communication.
12 Ribeiro (2000).
13 Luisa Bozzi, personal communication; Mary Bliss, personal communication.
14 Luisa Bozzi, personal communication.
15 Hughes (2002).
16 Henderson, *Narrow Roads*, vol 3, chapter 16, p 314, on Southampton.
17 See also Kathiritamby, *Narrow Roads*, vol 3, chapter 6, on Strepsiptera.
18 'Technosuperlife: Its Forerunners and Lies', paper prepared for the Evolution, Purpose and Meaning Symposium by the John Templeton Foundation at Lyford Cay, Nassau, the Bahamas (4–6 February 2000), WDH Archive, British Library (Hamilton, 1999).
19 From the back cover of the Variorum edition of Fisher's book.
20 Letter to Yura, 24 December 1999.
21 Letter to Patrick Bateson, 10 October 1999. Nevo (2002) on vaccines.
22 Chapter 37 in *The River* (1999). The quote is from the Foreword to the book (p xxxi). The *Lancet* article is Hooper and Hamilton (1996), reproduced in *Narrow Roads*, vol 3.
23 Hughes (2002) noted this do-it-yourself attitude. See more on the expedition in Hooper, *Narrow Roads*, vol 3, chapter 14. Also Lusia Bozzi, personal communication.
24 Letter to Yura, 24 December 1999.
25 He sent a similar letter to his sister Margaret where he added a sentence about something along the lines of 'the Hamiltonian can-do spirit'. In other words, Bill was confident he would be able to cope.
26 Email to Luisa, 7 July 1999.
27 Email from Alessandro Villa to Luisa, 8 July 1999.
28 It should also be noted that this time Bill did not have the help of Peter Henderson, who sometimes rescued him in difficult situations (Luisa Bozzi, personal communication).
29 Mary Bliss, personal communication. Mary was sitting with her brother in the University College Hospital, waiting for the results of the malaria test, when he said he felt weak and asked for a bed to lie down. That is when he suddenly lost a substantial amount of blood.

30 Luisa Bozzi, personal communication about Bill's willingness to accept alternative theories and the 'window' he kept open.
31 See Hamilton's Foreword to *The River* and Hooper (2005), *Narrow Roads*, vol 3, chapter 14. Hamilton had also been in Uganda in 1995 and seen for himself the families ravaged by AIDS (Foreword, pp xxix–xxx).
32 Ed Hooper, personal communication. and Hooper (2004). As for Bill Hamilton, see his Foreword to *The River*. Colin Hudson told me that when he visited Bill in Oxford in 1999, Bill had shown him the book with a very serious face, given him the Foreword to read and told him, 'What do you say about *that*?'.
33 I have this information from Luisa Bozzi (personal communication). I was also told by her that in regard to the cancelled Brazil expedition, this time Bill's proposal had been, surprisingly, rejected. Apparently this time it had got into the hands of the 'wrong' type of reviewers—that is, 'real' parasitologists—who could not appreciate it and who did not get Bill and his work with the Mamiraua project.
34 I learnt about Bill's bleeding from Michael Worobey and Luisa Bozzi (personal communication) and about the duodenal ulcer from his sister Janet. The statement that malaria takes a toll on the body came from a doctor in Entebbe with whom Luisa was in contact. See more on these matters, later in this chapter.
35 Email to Luisa Bozzi, 27 January 2000.
36 Mary Bliss, personal communication.
37 Rowena (Rowie) Hamilton, personal communication.
38 This was originally published in the Japanese journal *The Insectarium* (1991) and later translated into English and published in *Ethology, Ecology and Evolution* (Hamilton, 2000) and in *Narrow Roads*, vol 3. According to Akira Sasaki, the Japanese article made Bill Hamilton one of the best loved and known scientists in Japan (*Narrow Roads*, vol 3, p 370). A shorter version was published in *TLS*, 11 September 1992.
39 Janet Hamilton, email to author, 2 January 2012. (Prof Lucas' comment has been paraphrased by author.)
40 This poem was given to me by Jeanette Stace herself in 2003. It was later included in a small book of her poems and haikus.
41 For the programme at the Memorial Service, see the University of Basel's Bill Hamilton memorial website. The anthem is reproduced at the beginning of *Narrow Roads*, vol 2. For details about the archives in the British Library, see Jeremy John's chapter in *Narrow Roads*, vol 3, John (2005).
42 For the published Proceedings of this Royal Society conference, see Weiss and Wain-Hobson (2001). For details about the conference and for references to other relevant articles, see Hooper, *Narrow Roads*, vol 3, chapter 14. Many of the articles are available on <http://aids.origins.com> (this is a website with materials about the OPV theory and AIDS. It also includes a number of Hamilton's letters to *Science* in 1994, see chapter 20). Incidentally, I asked John Maynard Smith, who had attended the conference, what he thought about its conclusions. He told me that he found them convincing (Maynard Smith, personal communication).
43 'Origin of HIV and Emerging Persistent Viruses'. Tavola rotunda nell'ambito della Conferenza annuale della Ricerca (Roma, 28–29 settembre 2001). Accademia Nazionale dei Lincei, Roma, 2003. Many of the articles from this meeting are available on <http://aids.origins.com>.

44 Worobey, Hahn et al (2004); Hooper (2004); and *Narrow Roads* , vol 3, chapter 14. See also Hooper at <http://aids.origins.com>.
45 Crespi, personal communication. For symposia, see among others *Behavioral Ecology* Vol 12, No 3 (2001).
46 Grant (2006). Many obituaries are to be found on the University of Basel website. Dawkins' obituary is also reproduced as the Preface in *Narrow Roads*, vol 3.
47 Hamilton (1976c).
48 Wordsworth could actually see Newton's statue from his window at Cambridge.
49 Letter to Naomi Pierce, 26 February 1995. Bill here uses another version of Basho's title. On different possible translations, see Sumida (2005), *Narrow Roads*, vol 3, p 1.

Chapter 24

1 The quote is from Sumida (*Narrow Roads*, vol 3, p 3).
2 *Narrow Roads*, vol 3, pp 4–5.
3 *Narrow Roads*, vol 3, p 7.
4 Dieter Ebert told me that he had learnt how *not* to write a paper by watching Bill.
5 Ribeiro (2000). On Darwin, see Browne (2002), p 11.
6 *Narrow Roads*, vol 1, p lv.
7 For example, as described by Akira Sasaki, *Narrow Roads*, vol 3, p 374.
8 The expression is from the book *The Thinking Hand* by the Finnish architect and philosopher Juhani Pallasmaa (Pallasmaa, 2005, p 82).
9 See n 8 above.
10 In some places in the *Narrow Roads of Gene Land* volumes he talks about his 'diminishing neurons' or the loss of his capacity to identify insects at a distance (for example, *Narrow Roads*, vol 1, p 138.
11 See, for example, *Narrow Roads*, vol 2, pp 232–4.
12 *Narrow Roads*, vol 2, pp 806–10; Fantham on p 808.
13 *Narrow Roads*, vol 2, p 809.
14 See n 13 above.
15 The two oldest, Helen and Ruth Helen is an ecologist, Ruth has a Ph.D. in parasitology from Cambridge, and Rowena has a first class degree in Fine Arts.
16 Lisa Lloyd (personal communication). The point about that women scientists was Lisa Lloyd's (personal communication).
17 *Evolution*, vol 61, no 6, 2007.
18 For instance, Sober and Wilson (1998).
19 Hamilton was later sometimes able to turn around decisions, and he was particularly keen on helping younger colleagues in the struggle with referees and editors. See, for example, *Narrow Roads*, vol 2, pp 612–14, and Stouthamer, *Narrow Roads*, vol 3, chapter 2. Bruce Waldman also expressed to me his gratitude to Hamilton for fighting for him (personal communication).
20 Nowak, Tarnita, and Wilson (2010).
21 See n 20 above. Similar general ideas were also expressed by EO Wilson earlier in a number of articles, for example, Wilson (2005, 2008).
22 See nn 20 and 21 above. For more on the superorganism, see the book of that name by Wilson and Holldobler (2008). Incidentally, Holldobler did not share Wilson's recent position on kin selection (Pennisi, 2009).

23 See chapter 11 of this book.

24 See Abbott et al (2011), Boomsma et al (2011), Strassmann et al (2011), Ferriere and Michod (2011), Herre and Weislo (2011) and Dawkins (1979). My brief summary here is extracted from the lead letter, Abbott et al. The original author's response is Nowak, Tarnita and Wilson (2011).

25 This was much discussed in the 'blogosphere'.

26 And a lot of blogs, citing the opinions of such people as Richard Dawkins and Robert Trivers (for instance on Jerry Coyne's website, 'Why evolution is true').

27 Hughes (2011) attempted such a reconciliation, or at least an overview.

GLOSSARY

adaptation

A product of evolution, for instance a particular anatomical structure, behavior, or trait that is well-adjusted for fulfilling a certain function in a certain environment. Also: the process of adjustment or modification involved in the production of this structure or behavior.

allele

Alternative form of a gene occupying a particular locus on a chromosome. Since only one allele can occupy a locus, different alleles can be said to "compete" with each other.

altruism

Any behavior that promotes another organism's "fitness" (number of offspring) at the expense of its own fitness. (Note that "altruism" is here used in a technical sense to refer strictly to the outcome of an organism's *behavior*; it does not involve motives.) A puzzle for Darwinian theory until Hamilton developed the idea of "inclusive fitness" in 1964.

arrhenotoky

In Hymenoptera, the common mode of reproduction, whereby fertilized (diploid) eggs develop into females and unfertilized (haploid) eggs develop into males (compare with thelotoky).

Baldwin Effect (also called genetic assimilation)

Here I quote Hamilton (*Narrow Roads*, vol 3, p 224):

> The idea carries the prediction that a species able to be plastic in its lifetime adaptation has much more chance of acquiring radical genetically endowed characters of the right kind in its descendants—these through natural selection and eventually replacing the plasticity... If striving in a strange environment allows an organism to survive there at all, it may be giving to natural selection about *its only chance* to find the genes that can eventually complete the conquest of a new way of life. Thus

species in which individuals have plastic, effort-driven responses, whether these are automatic or by trial and error, are able to evolve more rapidly.

Hamilton calls the effect "semi-Lamarckian" because it involves striving.

clones

Unlike the case of relatives, whose sharing of genes is probabilistic, clones (e.g. monozygotic twins) are certain to carry identical genes (unless a mutation has intervened).

coefficient of relationship

A mathematical expression for the relatedness of two individuals. In his (1964) paper Hamilton discovered that Sewall Wright's coefficient of relationship closely approximated the probability that a particular individual's gene would be carried by a relative. In 1970, Hamilton replaced the correlation coefficient with the regression coefficient.

deme

A local population of closely related interbreeding organisms.

ethology

The study of whole patterns of animal behavior in natural environments. Ethologists are typically interested not only in questions about (adaptive) function but also (proximate) causation, development, and evolutionary history. "Functional ethologists" and sociobiologists have typically concentrated on only the first of these questions.

evolutionarily stable strategy (ESS)

An evolutionary stable strategy (ESS) is a pattern of behavior (a "strategy") which is "evolutionarily stable"; that is, when it is the dominant one in the population, it will prevail against any "invading" alternative behavior pattern. Natural selection tends to produce populations of organisms that are evolutionarily stable. ESS is associated with Maynard Smith and Price 1973; it was inspired by Hamilton's "unbeatable strategy," 1967.

eusociality

The highest grade of sociality in social insects. It is characterized especially by the existence of a sterile caste of workers who help raise the offspring of their sister, the queen.

fitness

In biology, "fitness" refers to the number of offspring produced by an organism. (It involves no evaluation of any other type of "fitness.") "Inclusive fitness" was the concept that Hamilton created to adjust RA Fisher's concept of fitness for the effects of relatives on one another's reproduction (fitness).

game theory

Game theory is a tool for systematic analysis of strategic interaction. A "game" is a social situation in which the outcome of an individual's actions depends on the action of other individuals. A game involves a set of "players" that can choose between (clearly specified) alternative strategies (eg, cooperate or don't cooperate). The various (well-defined) combinations of the players' chosen strategies result in (predetermined) "payoffs" (outcomes, rewards) for each player. Traditional game theory involved the study of mathematical models of cooperation and conflict between rational decision makers, and the payoff was utility. In evolutionary game theory the payoff is always fitness (future reproductive success). Game theory is a popular tool for modeling and predicting behavior both of humans and animals.

gene's eye view

One way of understanding the reasoning of Neo-Darwinism is to see it from a "gene's eye perspective" (Dawkins): what ultimately counts is not the survival of the individual organism itself but rather the survival of copies of its genes. Since relatives share genes, copies of a particular gene can (with various degrees of probability) be found in them as well, which is why the idea of "kin altruism" makes sense. (Note however that Hamilton's inclusive fitness theory was extended even to formally unrelated individuals, which possessed "superkinship" traits and Dawkins' "greenbeards.")

genetic algorithm

An optimization technique based on artificial intelligence that generates adaptive responses to changing environments. Genetic algorithms (GAs) are powerful heuristic search algorithms which use biological evolution as a model for their search process. The algorithms emulate Mendelian populations. They are able to produce "fitter" individuals over time by "mating" the fittest individuals from the current generation to produce a new generation. The foundation of the GA is John Holland's *Adaptation in Natural and Artificial Systems* (1975). Hamilton used the GA in his work on HAMAX and later.

genetic relatedness

The probability that two individuals possess copies of the same gene (inherited by a common ancestor). This probability of sharing genes is ½ for parents and offspring or for full siblings. It is 1/8<change to same fraction style as '1/2' in line above> for first cousins (the principle is to go through the whole chain of related individuals and multiply all the probabilities of the individuals that are sharing genes—so for two first cousins, starting with a particular cousin, it is ½ for his or her relationhip to a parent, times ½ for this parent's relationship to his or her sibling, times ½ for this siblings' relationship to his or her own offspring—the second first cousin = 1/8<change to same fraction style as '1/2', above>).

genotype

All or part of the genetic constitution of an individual. A particular genotype may result in various observable traits, the *phenotype*, such as green eyes, short wings, and dark coloring. Population-genetic theory typically regards the alleles (alternative possible genes at the same locus) of a genotype as resulting from a random process of recombination.

gradualism

The evolution of new species by gradual accumulation of small genetic changes over long periods of time. Also a theory of evolution that emphasizes this.

group selection

A process of natural selection applying to groups rather than individuals. An early assumption was that individuals sacrificed themselves "for the good of the group," for which Wynne-Edwards first formulated a possible mechanism in 1962. Most biologists, however, soon declared group selection an unlikely phenomenon and preferred the new paradigm of kin selection (one exception was EO Wilson). While many, following Maynard Smith's early formulation, continued seeing kin selection as opposed to group selection, Hamilton himself (following Price) in 1975 included both of these among the alternative ways in which altruism could evolve by natural selection.

haplodiploidy

A system in which males stem from unfertilized eggs and are haploid (have only one set of chromosomes from their mother), and females arise from fertilized eggs and are diploid (have two sets of chromosomes, from father and mother). Haploid males pass on *all* their genes to their daughters. Almost all social and non-social Hymenoptera are haplodiploid.

haplodiploidy thesis

The suggestion that male haplodiploidy is an important explanation for the tendency towards eusociality in Hymenoptera because of the closeness in genetic kinship that the haplodiploid system introduces.

Hamilton's Rule

Explains how natural selection can favor altruism between genetic relatives. This can happen if the reduction in fitness (number of offspring) of the donor (cost = c) is more than made up for by the increased fitness (number of offspring) of the recipient (benefit = b). An important consideration therefore is the coefficient of genetic relatedness between donor and recipient (r). One formulation of the rule is $br - c > 0$. There are many ways of expressing the rule. Note that b and c can refer to purely ecological factors.

Hamilton-Zuk hypothesis

Suggests that sexual "runaway" selection is not based on arbitrary criteria for female choice (as suggested by Fisher), but rather reflect female preference for healthy males. Male displays therefore advertise health. (In birds, bright feathers and costly ornaments demonstrate freedom from parasites, and showiness of a species tends to correlate with high load of parasites.)

Hymenoptera

The large order of insects that includes ants, bees, wasps, and sawflies.

inclusive fitness

This concept, developed by Hamilton in 1964 and further in 1975 (based on the Price Equation), explains how it is possible for natural selection to favor altruism. This can happen if the benefits of altruism can be made to fall on individuals who are likely to be altruist rather than random members of the population. A typical case is a group of relatives ("kin selection"), but Hamilton intended "inclusive fitness" to be a broader concept than both "kin selection" and "group selection." It can even involve unrelated individuals (see "superkinship trait" or "green beard" effect).

In practice, however, the concept of inclusive fitness has been typically used to formalize the reasoning about natural selection in kin groups. Natural selection can favor altruism between genetic relatives if the reduction in fitness (number of offspring) of the donor is more than made up for by the increased fitness (number of offspring) of the recipient. The reasoning here is that an individual's genes are represented also in relatives, in proportions corresponding to their genetic relatedness. Technically, inclusive fitness of an organism is typically measured as its own fitness plus the effect of its behavior on its relatives' reproduction minus its relatives' effect on its own reproduction, multiplied by its genetic relatedness to each related organism.

kin selection

A process of selection in which individuals are postulated to behave altruistically towards relatives with whom they (probabilistically) have genes in common. The term "kin selection" was launched by Maynard Smith in 1964. Kin selection has often been contrasted with group selection, following Maynard Smith's 1964 formulation, but Hamilton's (1975) reformulation of inclusive fitness included both kin selection and group selection among the ways in which altruism could evolve by natural selection.

kinship theory

Used by Trivers as a term for Hamilton's theoretical contribution in 1964.

linkage disequilibrium

When there is linkage and interaction between genetic loci on a chromosome, they may produce certain stable combinations, "linkage disequilibria," noticed as particular combinations of alleles that appear at a non-random level in a population. Under these conditions the usual assumptions and calculations of population genetics, which is based on free competition between alleles, do not hold.

locus

The position of a gene on a chromosome. There may for instance be a locus for eye color, with alternative genes (alleles) coding for different colors.

macroevolution

Large-scale processes over long time spans, having to do, for example, with species formation or morphological changes in the fossil record. In contrast, microevolution deals with processes of selection, adaptation, development, and the like (in Neo-Darwinism expressed as a change in gene frequencies, using the language of population genetics).

meiosis

Meiosis is a central part of sexual reproduction. It involves a process of cell division which gives rise to daughter cells, gametes, with half as many chromosomes; these then fuse in sexual reproduction to restore the original number of chromosomes.

Modern Synthesis

The reconciliation between Darwinism and Mendelian genetics in the early decades of the 20th century. In this way many branches of evolutionary biology were reformulated in the language of population genetics, which expresses evolution (or rather microevolution) mathematically as a change in gene frequencies in a population.

mutualism

An interaction between members of different species which is mutually beneficial (e.g., cleaner fish and its host).

Neo-Darwinism

The expression of evolutionary power in the mathematical language of population genetics.

panmictic population

Freely interbreeding. This is typically the default, baseline assumption about natural populations, but not always met in nature, where mate competition can be local.

phenotype

The visible properties of an organism that are produced by an interaction between the genotype and the environment.

phenotypic plasticity

An organism's ability to modify its phenotype as a response to changes in the environment. The term has typically been used to refer to plasticity during development, but can in principle refer also to plasticity that occurs later (eg, a behavioral change).

pleiotropy

The term for one gene influencing multiple traits, eg, eye color and leg length, simultaneously.

polymorphism

The coexistence of two or more physical forms of a species. In many cases, natural selection maintains polymorphisms in a stable balance in a species.

population genetics

Field that considers evolution mathematically in terms of populations, or gene pools. Population genetics expresses (micro) evolution as a change in gene frequencies in a population. Selection, for instance, is expressed population-genetically as the increase of one genotype at a greater rate than another in the population. Other processes for altering gene frequencies which may also be mathematically formulated are mutation pressure, meiotic drive, genetic drift, and gene flow.

reciprocal altruism

Form of altruism among unrelated individuals that involves altruistic behavior by one individual to another in the expectation of future reciprocation of altruistic acts by the other individual. The concept is associated with the work of Trivers (1971). Hamilton was not happy with it since for him altruistic behavior had a self-sacrificial aspect. In fact, for Hamilton, it needed to have the potential for suicide as the possible outcome when taken to the extreme.

recombination

The creation of new gametic associations of existing alleles (genes) at different loci through the mechanisms of crossing-over and reassortment of chromosomes. Recombination is a fundamental aspect of sexual reproduction.

regression

Probabilistic "dilution" of identical genes in relatives further and further removed.

Sewall Wright Effect

According to Sewall Wright, a population is best able to make rapid evolutionary advance if it is divided in small quasi-isolated demes.

sociobiology

EO Wilson defined sociobiology in 1975 as "the scientific study of the biological basis of social behavior in all kinds of organisms including man." Many biologists preferred to call the new field developing around the genetics of altruism and the application of game-theoretical principles "behavioral ecology" or "functional ethology" instead. Professional "human sociobiologists" often call themselves Darwinian anthropologists or evolutionary psychologists. (See Segerstrale (2000) for a discussion.)

supergene or superkinship trait (or green beard effect)

The ability of unrelated individuals who possess the same gene, eg "for" altruistic behavior to perceive the presence of like genes in another individual could be due to "something like a supergene affecting 1) some perceptible feature of the organism, 2) the perception of the feature, and 3) the social response consequent upon what was perceived....If some sort of attraction between likes for purposes of co-operation can occur the limits to the evolution of altruism...would be very greatly extended" (*Narrow Roads* vol 1, p 54). Hamilton speculated that this might come about, eg, on the basis on a linked phenotypical trait (later called "green beard" by Dawkins), or a similar habitat preference.

thelytoky

A mode of reproduction whereby females produce daughters from unfertilized eggs.

"twofold cost" of sexual reproduction

For Hamilton this was the fact that a sexually reproducing species loses half of its biomass through the production of males, compared with an asexual species that produces entirely female offspring. (This is a consequence of the sex ratio in species in which only the females rear offspring.) However, George Williams regarded the twofold cost as the cost of *meiosis*, that is, loss of half the genetic material, following Maynard Smith (1971).

REFERENCES

Abbott, P et al, 2011. Inclusive fitness theory and eusociality. *Nature* 471, 24 March, E1–E4.

Alexander, RD, 1980. *Darwinism and Human Affairs.* London, Pittman.

Alexander, RD, 1987. *The Biology of Moral Systems.* New York, de Gruyter.

Alexander, RD, 2000. WD Hamilton remembered. *Natural History* 109, 44–6.

Alexander, RD and DW Tinkle, eds, *Natural Selection and Social Behavior.* New York, Chiron Press, pp 363–81.

Angier, N., 2000. William Hamilton dies: an evolutionary biologist. *New York Times,* 10 March.

Anon., 2000. Professor WD Hamilton. *The London Times,* 9 March. (Attributed to John Maynard Smith.)

Anon., 2001. Off with their heads, but mind the bloodline. *The Sydney Morning Herald,* 22 June.

Axelrod, R, 1984. *The Evolution of Cooperation.* New York, Basic Books.

Axelrod, R, 2005. *Agent-Based Modeling as a Bridge between Disciplines* (online version). Later published in L Tesfatsion and KL Judd, eds, 2006. *Handbook of Computational Economics, vol 2, Agent-Based Computational Economics,* Handbook in Economics Series. Amsterdam, Elsevier.

Axelrod, R, 2012. Launching 'The evolution of cooperation'. *Journal of Theoretical Biology* 299, 21–4.

Axelrod, R and WD Hamilton, 1981. The evolution of cooperation. *Science* 211, 1390–6. (Republished in WD Hamilton, 1996. *Narrow Roads of Gene Land,* vol 1. London, Spektrum Academic Publishers.)

Badcock, CR, 2004. Mentalism and mechanism: the twin modes of human cognition. In C Crawford and C Salmon, eds, *Human Nature and Social Values.* Mahwah, NJ, Lawrence Erlbaum Associates, pp 99–116.

Badcock, C and B Crespi, 2006. Imbalanced genomic imprinting in brain development: an evolutionary basis for the aetiology of autism. *Journal of Evolutionary Biology* 19 (4), 1007–32.

Barber B, 1961. Resistance of scientists to scientific discovery. *Science* 34, 596–602.

Barkan, E, 1992. *The Retreat of Scientific Racism.* Cambridge, Cambridge University Press.

Baron-Cohen, S et al, 2001. The autism-spectrum quotient (AQ): evidence from Asperger syndrome/high-functioning autism, males and females, scientists and mathematicians. *Journal of Autism and Developmental Disorders* 31, 5–17.

Bates, Henry Walter, 1863. *The Naturalist of the River Amazons*. London. (A later edition, amongst others, was published by University of California Press, 1962.)

Bell, G, 1982. *The Masterpiece of Nature—The Evolution of Genetics and Sexuality*. Berkeley, University of California Press.

Blackman, S, 2004. Spite: evolution finally gets nasty. *The Scientist* 18, 14.

Blest, AD, 1963. Longevity, palatability and natural selection in five species of moth. *Nature*, 197, 1183–6.

Bliss, M, 2001. In memory of Bill Hamilton: hazards of modern medicine. Presented at the meeting Origin of HIV and Emerging Persistent Viruses Accademia Nazionale dei Lincei, Rome, September. (Available online at William D Hamilton Memorial Website maintained by Dieter Ebert, <www.unifr.ch/biol/ecology/hamilton/hamilton.html>.

Boomsma, JJ, 2000. Thoughts about 'Hamiltonian inspiration'. <http://www.unifr.ch/biol/ecology/hamilton/hamilton/boomsma.hmtl>.

Boomsma, J et al, 2011. Only full-sibling families evolved eusociality. *Nature* 471, 24 March, E4–E5.

Brown, S, 2005. A view from Mars. In M Ridley, ed, *Narrow Roads of Gene Land*, vol 3. Oxford, Oxford University Press, pp 349–56.

Browne, J, 2002. *Charles Darwin: The Power of Place*. Princeton, Princeton University Press, p 11.

Burt, A and R Trivers, 2008. *Genes in Conflict: the Biology of Selfish Genetic Elements*. Cambridge, Mass, Belknap Press.

Cronin, H, 1992. *The Ant and the Peacock*. Cambridge, Cambridge University Press.

Dawkins, R, 1976. *The Selfish Gene*. Oxford, Oxford University Press.

Dawkins, R, 1979. Twelve misunderstandings of kin selection. *Zeitschrift fur Tierpsychologie* 51, 184–200.

Dawkins, R, 1982. *The Extended Phenotype: The Gene as Unit of Selection*. Oxford and San Francisco, Freeman.

Dawkins, R, 1989. *The Selfish Gene*. Second edition. Oxford, Oxford University Press.

Dawkins, R, 2001. Foreword to William D Hamilton, *Narrow Roads of Gene Land: The Collected Papers of WD Hamilton*, vol 2, xi–xix. Oxford and New York, Oxford University Press.

Dow, M, 1976. Letter to the editor (response to Hamilton). *New Scientist* 71, 22 July, 195.

Ebert, D, 2005. In M Ridley, ed, *Narrow Roads of Gene Land*, vol 3. Oxford, Oxford University Press, pp 189–94.

Edwards, AWF, 1994. The fundamental theorem of natural selection. *Biological Reviews* 69, 443–74.

Edwards, AWF, 2000. The genetical theory of natural selection. *Genetics* 154, 1419–26.

Eshel, I, 1972. On the neighbor effect and the evolution of altruistic traits. *Theoretical Polulation Biology* 3, 258–77.

Eshel, I and M Feldman, 2001. Individual selection and altruistic relationships: the legacy of WD Hamilton. *Theoretical Population Biology* 59 (1), 15–20.

Eshel, I and WD Hamilton, 1984. Parent–offspring correlation in fitness under fluctuating selection, *Proceedings of the Royal Society: B* 222, 1–14.

Ferriere, R and R Michod, 2011. Inclusive fitness in evolution. *Nature* 471, 24 March, E6–E8.

Fisher, RA, 1930. *The Genetical Theory of Natural Selection*. Oxford, Clarendon Press. (Second edition, 1958, Dover; and Variorum edition, 1999, ed JH Bennet, Oxford University Press.)
Flynn, R, 1996. *Dysgenics*. Praeger, Westport, Conn.
Ford, EB, 1945. *Butterflies*. New Naturalist Series. London, Collins.
Fox, R, ed, 1975. *Biosocial Anthropology*, New York, John Wiley & Sons.
Frank, SA, 1995. George Price's contributions to evolutionary genetics. *Journal of Theoretical Biology* 175, 373–88.
Gardner, A and S West, 2004. Spite and the scale of competition. *Journal of Evolutionary Biology* 17, 1195–203.
Gould, SJ, 1977. Caring goups and selfish genes. *Natural History* 86 (12), 20–4.
Gould, SJ, 1978. *Ever Since Darwin*. New York, WW Norton.
Grafen, A, 1982. How not to measure inclusive fitness. *Nature*, 298, 425–6.
Grafen, A, 2000. Obituary of William D Hamilton. *The Guardian*, 9 March.
Grafen, A, 2004. William Donald Hamilton. *Biographical Memoirs of Fellows of the Royal Society* 50, 109–32.
Grafen, A, 2005. William Donald Hamilton. In M Ridley, ed, *Narrow Roads of Gene Land*, vol 3. Oxford, Oxford University Press, pp 423–58.
Grafen, A., 2006. The intellectual contribution of *The Selfish Gene* to evolutionary theory. In A Grafen and M Ridley, eds, *Richard Dawkins: How a Scientist Changed the Way We Think*. Oxford, Oxford University Press, 66–74.
Grant, P, 2002. William D. Hamilton: biographical memoirs. *Proceedings of the American Philosophical Society*, 146 (4), 388–94.
Haig, D. 2002. *Genomic Imprinting and Kinship*. New Jersey, Rutgers University Press.
Haig, D. 2003. The science that dare not speak its name. (Review of *Narrow Roads of Gene Land* II). *Quarterly Review of Biology*, 78, 327–35.
Haig, D, NE Pierce, and EO Wilson, 2000. William Hamilton (1936–2000). *Science* 287, 31 March, 2438.
Haldane, JBS, 1925. *Daedalus or Science and the Future*. London, Dutton.
Haldane, JBS, 1932. *The Causes of Evolution*. London, Longman's Green.
Haldane, JBS, 1955. Population genetics. *Penguin New Biology* 18, 34–51.
Hamilton, AM, 1945. *The Road through Kurdistan: The Narrative of an Engineer in Iraq*. London, Faber and Faber. (First published 1937.)
Hamilton, WD, 1963. The evolution of altruistic behaviour. *American Naturalist* 97, 354–6.
Hamilton, WD, 1964. The genetical evolution of social behaviour. *Journal of Theoretical Biology* 7, 1–52.
Hamilton, WD, 1967. Extraordinary sex ratios. *Science* 156, 477–88.
Hamilton, WD, 1970. Selfish and spiteful behaviour in an evolutionary model. *Nature* 228, 1218–20.
Hamilton, WD, 1971a. Geometry for the selfish herd. *Journal of Theoretical Biology* 31, 295–311.
Hamilton, WD, 1971b. The genetical theory of social behavior, I and II. In G Williams, ed, *Group Selection*. Chicago and New York, Aldine/Atherton, pp 23–87.
Hamilton, WD, 1971c. Addendum. In G Williams, ed, *Group Selection*. Chicago and New York, Aldine/Atherton, pp 87–9.

Hamilton, WD, 1971d. Selection of selfish and altruistic behaviour in some extreme models. In JF Eisenberg and WS Dillon, eds, *Man and Beast: Comparative Social Behavior*. Washington, DC, Smithsonian Institution Press, pp 57–91.

Hamilton, WD, 1975. Innate social aptitudes of man: an approach from evolutionary genetics. In R Fox, ed, *ASA Studies 4: Biosocial Anthropology*. Malaby Press, London, pp 133–53.

Hamilton, WD, 1976a. Letter to the editor. *New Scientist* 71, 1 July, 40.

Hamilton, WD, 1976b. Letter to the editor (response to Dow). *New Scientist* 71, 22 July, 195.

Hamilton, WD, 1977a. The play by nature (Review of *The Selfish Gene* by R Dawkins). *Science* 196, 757–9.

Hamilton, WD, 1977b. Review of EO Wilson, *Sociobiology: The New Synthesis*. *Journal of Animal Ecology* 46, 975–83.

Hamilton, WD, 1977c. The selfish gene. *Nature* 267, 102.

Hamilton, WD, 1980. Sex versus nonsex versus parasites, Oikos 35, 282–90.

Hamilton, WD, 1982. Unraveling the riddle of Nature's masterpiece (Review of Graham Bell's *The Masterpiece of Nature*). *Bioscience* 32 (9), 745–6.

Hamilton, WD, 1986. Instability and cycling of two competing hosts with two parasites. In S Karlin and E Nevo, eds, *Evolutionary Processes and Theory*. New York, Academic Press, pp 645–68.

Hamilton, WD, 1993. Between Shoreham and Downe: seeking the key to natural beauty. Inamori Foundation Kyoto Prize Commemorative Lecture. (Republished in M Ridley, ed, 2005. *Narrow Roads of Gene Land, vol 3, The Collected Papers of W. D. Hamilton*, Oxford, Oxford University Press.)

Hamilton, WD, 1994. AIDS theory vs. lawsuit. Unpublished letter to *Science*, 27 January. (Published in slightly revised for as an appendix in J Cribb, 1996, *The White Death*. Sydney, Angus and Robertson, pp 254–7.

Hamilton, WD, 1996. *Narrow Roads of Gene Land: The Collected Papers of WD Hamilton. Vol 1: Evolution of Social Behaviour*. Oxford and New York, WH Freedman.

Hamilton, WD, 2000. My intended burial and why. *Ethology Ecology and Evolution* 12, 111–22. (Originally published in *The Insectarium* 28, 238–47, in Japanese. Reprinted in vol 3 of *Narrow Roads*, pp 73–88.)

Hamilton, WD, 2001. *Narrow Roads of Gene Land: The Collected Papers of WD Hamilton, vol 2: Evolution of Sex*. Oxford and New York, Oxford University Press.

Hamilton, WD and SP Brown, 2001. Autumn tree colours as a handicap signal. *Proceedings of the Royal Society B* 268, 1489–93.

Hamilton WD and TM Lenton, 1998. Spora and Gaia: how microbes fly with their clouds. *Ethology Ecology and Evolution* 10, 1–16.

Hamilton, WD and M Zuk, 1982. Heritable true fitness and bright birds—a role for parasites. *Science* 218, 384–7.

Hamilton, WD and M Zuk, 1989. Parasites and sexual selection: Hamilton and Zuk reply. *Nature* 341, 289–90.

Hamilton, WD, PA Henderson, and NA Moran, 1981. Fluctuation of environment and coevolved anatagonistic polymorphism as factors in the maintenance of sex. In RD Hardin, G, 1968. The tragedy of the commons. *Science* 162, 1243–8.

Hamilton, WD, R Axelrod, and R Tanese, 1990. Sexual reproduction as an adaptation to resist parasites (a review). *Proceedings of the National Academy of Sciences USA* 87, 3566–73.

Harvey, P and M Pagel, 1991. *The Comparative Method in the Analysis of Comparative Data*. Oxford Series in Ecology and Evolution. Oxford, Oxford University Press.

Henderson, P, 2005. Life, evolution and development in the Amazonian floodplain. In M Ridley, ed, *Narrow Roads of Gene Land*, vol 3. Oxford, Oxford University Press, pp 307–14.

Henderson, P, WD Hamilton, and W Crampton, 1998. Evolution and diversity in Amazonian floodplain communities. In DM Newbery, HH Prins, and ND Brown, eds, *Dynamics of Tropical Communities, The 37th Symposium of the British Ecological Society*. Oxford, Blackwell, pp 385–419. (Reprinted in *Narrow Roads* vol 3, 315–48.)

Herbers, J, 2009. Darwin's 'one special difficulty': celebrating Darwin 200. *Biology Letters* 5, 214–17.

Herre, EA and WT Wcislo, 2011. In defence of inclusive fitness theory. *Nature* 471, 24 March, E8–E9.

Holland, J, 1975. *Adaptation in Natural and Artificial Systems*. Cambridge, Mass, MIT Press. (Second edition published in 1992.)

Hooper, E, 1999. *The River: A Journey Back to the Source of HIV and AIDS*. Penguin, Harmondsworth and Little, Brown, Boston, Mass.

Hooper, E, 2004. Why the Worobey/Hahn 'refutation' of OPV/AIDS theory is wrong. <www.aidsorigins.com>.

Hooper E and WD Hamilton, 1996. 1959 Manchester case of syndrome resembling AIDS. *Lancet* 348, 1363–5.

Hughes, D, 2002. The value of a broad mind: some natural history meanderings of Bill Hamilton. *Ethology, Ecology and Evolution* 14, 83–9.

Hughes, D, 2010. Recent developments in sociobiology and the scientific method. *Trends in Ecology and Evolution* 26, 57–8.

Hunt, L, 1998. Send in the clouds. *New Scientist*, 28–33.

Hurst, L, 2005. Sex, sexes and selfish elements. In M Ridley, ed, *Narrow Roads of Gene Land*, vol 3. Oxford, Oxford University Press, pp 89–97.

Huxley, A, 1932. *Brave New World*. New York, Harper & Row.

John, JL, 1997. Seven comments on the theory of sosigonic selection. *Journal of Theoretical Biology* 187, 333–49.

John, JL, 2005. Because topics often fade. In M Ridley, ed, *Narrow Roads of Gene Land*, vol 3, pp 399–422.

Kathirithamby, J, 2005. Further homage to Santa Rosalia. In M Ridley, ed, *Narrow Roads of Gene Land*, vol 3. Oxford, Oxford University Press, pp 117–27.

Kitching, R, 2000. Spoken comment to Robyn Williams of Australian ABC Radio National's Ockam's Razor, 18 June, <http://home.austarnet.com.au/stear/abc_death_of_greatness.htm>.

Kondratshov, AS, 1988. Deleterious mutations and the evolution of sexual reproduction. *Nature* 336, 435–40.

Krebs, J and RM May, 1976. Social insects and the evolution of altruism. *Nature* 260, 4 March.

Kropotkin, P, 1902. *Mutual Aid: A Factor of Evolution*. London, Heinemann.
Lenton, TM, 1998. Gaia and natural selection. *Nature* 394, 439–47.
Levinson, S, 1998. Report on the Vatican conference 1998, Max Krant internal newsletter.
Lewin, R, 1976. The course of a controversy. *New Scientist*, 13 May.
Lewontin, RC, 1977. Caricature of Darwinism. (Book review of *The Selfish Gene*.) *Nature* 266, 283–4.
Lovelock, JE, 1995. *The Ages of Gaia—A Biography of Our Living Earth*. Second edition. Oxford, Oxford University Press.
Lynn, R, 1996. *Dysgenics*. Westport, Conn, Praeger.
May, RM, 2001. Memorial to Bill Hamilton. *Philosophical Transactions of the Royal Society, London B*, 365, 785–7.
May, R and J Krebs, 1976. Social insects and the evolution of altruism. *Nature* 160, 4 March.
Maynard Smith, J, 1966. *The Theory of Evolution*. Second edition. London, Penguin Books, p 391. (First edition published 1958.)
Maynard Smith, J, 1964. Group selection and kin selection. *Nature* 201, 1145–7.
Maynard Smith, J, 1965. The evolution of alarm calls. *American Naturalist* 99, January-February, 59–63.
Maynard Smith, J, 1972. Eugenics and utopia. *On Evolution*. Edinburgh, Edinburgh University Press, 61–81.
Maynard Smith, J, 1974. The theory of games and the evolution of animal conflicts. *Journal of Theoretical Biology* 47, 209–21.
Maynard Smith, J, 1975. Survival through suicide. *New Scientist* 28, 496–7.
Maynard Smith, J, 1976a. Evolution and the theory of games. *American Scientist* January-February, 41–5. (Reprinted in Maynard Smith, 1989, *Did Darwin Get It Right?* New York: Chapman and Hall, 201–15.)
Maynard Smith, J, 1976b. Letter to the editor (response to Hamilton). *New Scientist* 71, 29 July, 247.
Maynard Smith, J, 1978. *The Evolution of Sex*. Cambridge, Cambridge University Press.
Maynard Smith, J, 1982. *Evolution and the Theory of Games*. Cambridge, Cambridge University Press.
Maynard Smith, J, 1998. The origin of altruism. (Review of E Sober and DS Wilson's *Unto Others*). *Nature* 393, 639–40.
Maynard Smith, J and G Price, 1973. The logic of animal conflict. *Nature* 246, 15–18.
Mayr, E, 1992. Haldane's Causes of Evolution after 60years. *Quarterly Review of Biology* 67, 175–86.
Michod R, 1982. The theory of kin selection. *Annual Review of Ecology and Systematics* 13, 23–55.
Michod RE and WD Hamilton, 1980. Coefficients of relatedness in sociobiology. *Nature* 288, 694–7.
Mitchison, N, 1976. Letter to the editor (response to Hamilton). *New Scientist* 71, 5 August, 300.
Moran, N, N Pierce, and J Seger, 2000. WD Hamilton, 1936–2000. *Nature Medicine* 6, 367.
Motluk, A, 2001. Leaf me alone. *New Scientist* 14 July. (Also accessible at <www.newscientist.com>.)

Muir, H, 2003. Did Einstein and Newton have autism? *New Scientist*, 3 May, p 10.

Muller, HJ et al, 1939. The geneticists' manifesto. *Nature* 16 September. (Reprinted as HJ Muller, 2002. Social biology and population improvement. In W Kristol and E Cohen, eds, *The Future is Now*. Maryland, Rowman and Littlefield, Lanham, pp 22–32.

Muller, HJ, 1935. *Out of the Night*. New York, Vanguard Press.

Nevo, E, 2001. WD Hamilton—evolutionary theorist: life and vision (1936–2000). *Theoretical Population Biology* 59, 21–5.

Nowak, MA, CE Tarnita, and EO Wilson, 2010. The evolution of eusociality. *Nature* 466, 1057–62.

Nowak, MA, CE Tarnita, and EO Wilson, 2011. Nowak et al reply. *Nature* 471, 24 March, E9–E10.

Pallasmaa, J, 2009. *The Thinking Hand*. Chichester, John Wiley & Sons, Ltd.

Pennisi, E, 2000. In search of biological weirdness. *Science* 290, 1077–9.

Pennisi, E, 2009. Agreeing to disagree. *Science* 323, 706–8.

Price, GR, 1970. Selection and covariance. *Nature* 227, 520–1.

Price, GR, 1972. Extension of covariance selection mathematics. *Annals of Human Genetics* 35, 485–90.

Provine, WP, 1973. Geneticists and the biology of race crossing. *Science* 182, 790–6.

Queller, DC, 2001. WD Hamilton and the evolution of sociality: Hamilton Symposium. *Behavioral Ecology* 12 (3), 261–3.

Ratnieks, F, K Foster, and T Wenseleers, 2011. Darwin's special difficulty: the evolution of 'neuter insects' and current theory. *Behavioral Ecology and Sociobiology* 65, 481–92.

Read, AF, 1987. Comparative evidence supports the Hamilton and Zuk hypothesis on parasites and sexual selection. *Nature* 328, 2 July, 68–70.

Read, AF and P Harvey, 1989a. Reassessment of the comparative evidence for the Hamilton and Zuk theory on the evolution of secondary sexual characters. *Nature* 339 (22) June, 618–20.

Read, AF and P Harvey, 1989b. Read and Harvey reply. *Nature* 340, 13 July, 105.

Rennie, J, 1992. Living together. (Trends in parasitology.) *Scientific American* January, 122–33.

Reynolds, V, 1980. Sociobiology and the idea of primordial discrimination. *Ethnic and Racial Studies* 3 (3), 303–15.

Ribeiro, S, 2000. William D. Hamilton, 1936–2000, remembered by his last Brazilian student. *Antenna* 24, 119–21.

Ridley, M, 1993. *The Red Queen*. Harmondsworth, Penguin.

Ridley, M, ed, 2005. *Narrow Roads of Gene Land. Vol III: The Collected Papers of WD Hamilton*. Oxford, Oxford University Press.

Rinderer, TE, BP Oldroyd, WS Sheppard, 1993. Africanized bees in the US. *Scientific American* 269, 84–90.

Roes, F, 1996. If you have a simple idea, state it simply. Interview with Bill Hamilton. Online at <http://www.froes.dds.nl/HAMILTON.htm>. (Also published in 2000 as: In his own words. *Natural History* 109, 46–7.)

Rose, MR et al, 2007. Hamilton's forces of natural selection after forty years. *Evolution* 61, 1265–76.

Sahlins, M, 1976. *The Use and Abuse of Biology*. Ann Arbor, University of Michigan Press.

Sasaki, A, 2005. In M Ridley, ed, *Narrow Roads of Gene Land*, vol 3. Oxford, Oxford University Press, pp 369–75.

Sasaki, A, WD Hamilton, and F Ubeda, 2002. Clone mixtures and a pacemaker: new facets of Red Queen theory and ecology. *Proceedings of the Royal Society B* 269, 761–72.

Seger, J and WD Hamilton, 1988. Parasites and sex. In RE Michod and BR Levin, eds, *The Evolution of Sex: An Examination of Current Ideas*. Sunderland, Mass, Sinauer Associates, pp 176–93.

Segerstrale, U, 2000. *Defenders of the Truth: The Battle for Science in the Sociobiology Debate and Beyond*. Oxford, Oxford University Press.

Segerstrale, U, 2002. Neo-Darwinism. In M Pagel, ed, *Encyclopedia of Evolution*. Oxford, Oxford University Press, 107–10.

Segerstrale, U, 2006. An eye on the core. In A Grafen and M Ridley, eds, *Richard Dawkins: How a Scientist Changed the Way We Think*. Oxford, Oxford University Press, pp 75–91.

Smith, A, 1971. *Mato Grosso: Last Virgin Land. An Account of the Mato Grosso, Based on the Royal Society and Royal Geographical Society Expedition to Central Brasil, 1967–9*. London, Michael Joseph.

Smith, A, 2000. WD 'Bill' Hamilton. *The Guardian*, 22 March.

Stearns, S, 2000. Three days with Bill. Available on the memorial website at Basel University, managed by Dieter Ebert <http://evolution.unibas.ch/hamilton/index.htm>.

Strassmann, JE et al, 2011. Kin selection and eusociality. *Nature* 471, 24 March, E5–E6.

Sumida, B, 2005. Oku no hosomichi. In M Ridley, ed, *Narrow Roads of Gene Land*, vol 3. Oxford, Oxford University Press, pp 1–9.

Tiger, L, and MH Robinson, 1991. Introduction. In MH Robinson and L Tiger, eds, *Man and Best Revisited*. Washington, DC, Smithsonian Institution Press, pp xvii–xxiii.

Tinbergen, N, 1963. On aims and methods of ethology. Zeitschrift fur Tierpsychologie 20, 410–33.

Trivers, RL, 1971. The evolution of reciprocal altruism. *Quarterly Review of Biology* 46, 35–57.

Trivers, RL, 1974. Parent-offspring conflict. *American Zoologist* 14, 249–64.

Trivers, RL, 1976. Foreword to *The Selfish Gene*. Oxford, Oxford University Press, pp v–vii.

Trivers, RL, 1985. *Social Evolution*. Menlo Park, Benjamin Cummings.

Trivers, RL and H Hare, 1976. Haplodiploidy and the evolution of the social insects. *Science* 191, 249–63.

Trivers, RL, 1998. As they would do to you. (Review of E Sober and DS Wilson's *Unto Others*. *The Skeptic* 6 (4), 81–3.

Trivers, RL, 2000. Obituary: William Donald Hamilton (1936–2000). *Nature* 404, 828.

Trivers, RL, 2002. *Natural Selection and Social Theory*. Oxford, Oxford University Press.

Wade, N, 1976. Sociobiology: troubled birth for new discipline. *Science* 191, 1151–8.

Washburn, SL, 1977. Sociobiology. *Anthropology Newsletter* 18 (3), 3.

Washburn, SL, 1978. Human behavior and the behavior of others. *American Psychologist* 33 (5). (Reprinted in A Montagu, ed, 1980. *Sociobiology Examined*, Oxford, Oxford University Press, pp 254–82.

Weiss, RA, 2001. Poliovaccine exonerated. *Nature* 410, 1035–6.

Weiss, RA and S Wain-Hobson, eds, 2001. Origins of HIV and the AIDS epidemic, Philosophical Transactions of the Royal Society London, 256, 771–977.
West, S and A Gardner, 2010. Altruism, spite and greenbeards. *Science* 327, 1341–4.
Williams GC, 1966. *Adaptation and Natural Selection*. Princeton, NJ, Princeton University Press.
Williams GC, ed, 1971. Group Selection. Aldine Atherton, Chicago.
Williams, GC, 1975. *Sex and Evolution*. Princeton, NJ, Princeton University Press.
Williams, GC, 1993. Review of H Cronin's *The Ant and the Peacock. Quarterly Review of Biology*, p 412.
Williams, GC, 2000. Some thoughts on William D Hamilton (1936–2000). *Trends in Ecology and Evolution* 15 (7), 302.
Williams, GC and DC Williams, 1957. Natural selection of individually harmful social adaptations among sibs with special reference to social insects. *Evolution* 11, 32–9.
Wilson, DS, Sober E, 1994. Reintroducing group selection to the human behavioral sciences. *The Behavioral and Brain Sciences* 17 (4), 585–654.
Wilson, EO, 1971. *The Insect Societies*. Cambridge, Mass, Harvard University Press.
Wilson, EO, 1975. *Sociobiology: The New Synthesis*. Cambridge, Mass: Harvard University Press.
Wilson, EO, 1994. *Naturalist*. Washington, DC, Island Press.
Wilson, EO, 2005. Kin selection as the key to altruism: its rise and fall. *Social Research (New York)* 72, 159–68.
Wilson, EO, 2008. Giant leap: how insects achieved altruism and colonial life. *Bioscience* 58, January, 17–25.
Wilson, EO and B Holldobler, 2009. *The Superorganism*. New York, WW Norton.
Wolstenholme, G, ed, 1963. *Man and His Future (Based on the CIBA conference)*. London, Churchill.
Worobey, M, et al, 2004. Contaminated polio vaccine theory refuted. *Nature* 428, 820.
Wynne-Edwards, VC, 1962. *Animal Dispersion in Relation to Social Behavior*. Edinburgh, Oliver & Boyd.
Wynne-Edwards, VC, 1963. Intergroup selection in the evolution of social systems. *Nature* 200 (4907), 623–6.
Zuk, M, 1989. Validity of sexual selection in birds. *Nature* 340, 13 July, 104–5.
Zuk, M, 2000. A career of many colors. *Parasitology Today* 16, 457–8.

INDEX